Environmental Accounting

Environmental Accounting

Edited by:

Quan Cui

Delve
PUBLISHING

www.delvepublishing.com

Environmental Accounting

Quan Cui

Delve Publishing

2010 Winston Park Drive,

2nd Floor

Oakville, ON L6H 5R7

Canada

www.delvepublishing.com

Tel: 001-289-291-7705
001-905-616-2116

Fax: 001-289-291-7601

Email: orders@arclereducation.com

ISBN: 978-1-77361-117-4 (Hardcover)

DECLARATION

Some content or chapters in this book are open access copyright free published research work, which is published under Creative Commons License and are indicated with the citation. We are thankful to the publishers and authors of the content and chapters as without them this book wouldn't have been possible.

ABOUT THE EDITOR

Quan Cui obtained her PhD from Beijing Normal University in 2015. She worked as environmental impact assessment engineer for years. Aiming at linking science to policy, her study interest spans from ecosystem modelling to water resources management. She published peer-reviewed articles about using machine learning models in inflow forecasting, CO_2 management and watershed ecosystem health assessment.

TABLE OF CONTENTS

LIST OF CONTRIBUTORS

Jasper Montana
Department of Geography, University of Cambridge, Downing Place, CB2 3EN, UK

Camino Liquete
European Commission, Joint Research Centre (JRC), Directorate D – Sustainable Resources, Via Enrico Fermi 2749, 21027 Ispra, VA, Italy

Angel Udias
European Commission, Joint Research Centre (JRC), Directorate D – Sustainable Resources, Via Enrico Fermi 2749, 21027 Ispra, VA, Italy

Giulio Conte
European Commission, Joint Research Centre (JRC), Directorate D – Sustainable Resources, Via Enrico Fermi 2749, 21027 Ispra, VA, Italy

Bruna Grizzetti
European Commission, Joint Research Centre (JRC), Directorate D – Sustainable Resources, Via Enrico Fermi 2749, 21027 Ispra, VA, Italy

Fabio Masi
RIDRA Srl, Via La Marmora 51, Florence 50121, Italy

S. Vialet Chabrand
School of Biological Sciences, University of Essex, Colchester, CO4 3SQ, UK

J.S.A. Matthews
School of Biological Sciences, University of Essex, Colchester, CO4 3SQ, UK

O. Brendel
EEF, INRA, Université de Lorraine, F-54280 Champenoux, France

M.R. Blatt
Laboratory of Plant Physiology and Biophysics, University of Glasgow, Bower Building, Glasgow G12 8QQ, UK

Y.Wang
Laboratory of Plant Physiology and Biophysics, University of Glasgow, Bower Building, Glasgow G12 8QQ, UK

A.Hills
Laboratory of Plant Physiology and Biophysics, University of Glasgow, Bower Building, Glasgow G12 8QQ, UK

H. Griffiths
Plant Sciences, University of Cambridge, Downing Street, Cambridge CB2 3EA, UK

S. Rogers
Computing Science, University of Glasgow, Alwyn Williams Building, Glasgow G12 8QQ, UK

T. Lawson
School of Biological Sciences, University of Essex, Colchester, CO4 3SQ, UK

Yajie Liu
SINTEF Fisheries & Aquaculture, 7465 Trondheim, Norway

Trond W.Rosten
SINTEF Fisheries & Aquaculture, 7465 Trondheim, Norway

Kristian Henriksen
SINTEF Fisheries & Aquaculture, 7465 Trondheim, Norway

Erik Skontorp Hognes
SINTEF Fisheries & Aquaculture, 7465 Trondheim, Norway

Steve Summerfelt
The Conservation Fund, Freshwater Institute, 1098 Turner Road, Shepherdstown, WV, USA

Brian Vinci
The Conservation Fund, Freshwater Institute, 1098 Turner Road, Shepherdstown, WV, USA

Thomas Hahn
Stockholm University, Stockholm Resilience Centre, SE-10691 Stockholm, Sweden

Constance McDermott
University of Oxford, Environmental Change Institute, Oxford OX1 3QY, United Kingdom

Claudia Ituarte-Lima
Stockholm University, Stockholm Resilience Centre, SE-10691 Stockholm, Sweden

Maria Schultz
Stockholm University, Stockholm Resilience Centre, SE-10691 Stockholm, Sweden

Tom Green
Simon Fraser University, Centre for Dialogue, Vancouver, British Columbia, Canada V6B 1L6

Magnus Tuvendal
Stockholm University, Stockholm Resilience Centre, SE-10691 Stockholm, Sweden

Carsten Nesshöver
UFZ-Helmholtz Centre for Environmental Research, Department of Conservation Biology & UFZ Science-Policy Expert Group, Permoser Str. 15, 04318 Leipzig, Germany
UFZ-Helmholtz Centre for Environmental Research, Department of Environmental Politics & UFZ Science-Policy Expert Group, Permoser Str. 15, 04318 Leipzig, Germany

Timo Assmuth
Finnish Environment Institute (SYKE) Environmental Policy Centre, FI-00251 Helsinki, Finland

Katherine N. Irvine
Social, Economic and Geographical Sciences, The James Hutton Institute, Craigiebuckler, Aberdeen AB15 8QH, United Kingdom

Graciela M. Rusch
Norwegian Institute for Nature Research, Terrestrial Ecology Department, Sluppen, 7485 Trondheim, Norway

Kerry A. Waylen
Social, Economic and Geographical Sciences, The James Hutton Institute, Craigiebuckler, Aberdeen AB15 8QH, United Kingdom

Ben Delbaere
ECNC-European Centre for Nature Conservation, 5000 LG Tilburg, The Netherlands

Dagmar Haase
UFZ-Helmholtz Centre for Environmental Research, Department of Computational Landscape Ecology, Permoser Str. 15, 04318 Leipzig, Germany
Humboldt Universität zu Berlin, Institute of Geography, Lab for Landscape Ecology, Rudower Chaussee 16, 12489 Berlin, Germany

Lawrence Jones-Walters
Norwegian Institute for Nature Research, Human Dimension Department, Fakkelgarden, 2624 Lillehammer, Norway

Hans Keune
INBO & Belgian Biodiversity Platform & University of Antwerp, Kliniekstraat 25, 1070 Brussels, Belgium

Eszter Kovacs
Szent István University, Institute of Nature Conservation and Landscape Management, Páter K. u. 1., H-2100 Gödöllő, Hungary
Environmental Social Science Research Group (ESSRG), Rómer Flóris u. 38, Budapest H-1024, Hungary

Kinga Krauze
European Regional Centre for Ecohydrology of the Polish Academy of Sciences, Tylna 3, 90-364 Lodz, Poland

Mart Külvik
Estonian University of Life Sciences, Institute of Agricultural and Environmental Sciences, Kreutzwaldi 1, Tartu 51014, Estonia

Freddy Rey
Univ. Grenoble Alpes, Irstea, UR EMGR, 2 rue de la Papeterie, BP 76, 38402 Saint-Martin-d'Hères, France

Jiskavan Dijk
Norwegian Institute for Nature Research, Terrestrial Ecology Department, Sluppen, 7485 Trondheim, Norway

Odd Inge Vistad
Norwegian Institute for Nature Research, Human Dimension Department, Fakkelgarden, 2624 Lillehammer, Norway

Mark E. Wilkinson
Environmental and Biochemical Sciences, The James Hutton Institute, Craigiebuckler, Aberdeen AB15 8QH, United Kingdom

Heidi Wittmer
UFZ-Helmholtz Centre for Environmental Research, Department of Environmental Politics & UFZ Science-Policy Expert Group, Permoser Str. 15, 04318 Leipzig, Germany

Brendan P.Walsh
Intelligent Efficiency Research Group (IERG), School of Engineering, University College Cork, Cork, Ireland

Sean N. Murray
Intelligent Efficiency Research Group (IERG), School of Engineering, University College Cork, Cork, Ireland

D.T.J.O' Sullivan
Intelligent Efficiency Research Group (IERG), School of Engineering, University College Cork, Cork, Ireland

Geoffrey P. Hammond
Department of Mechanical Engineering, University of Bath, Bath BA2 7AY, United Kingdom

Institute for Sustainable Energy and the Environment (I•SEE), University of Bath, Bath BA2 7AY, United Kingdom

Áine **O'Grady**
Department of Mechanical Engineering, University of Bath, Bath BA2 7AY, United Kingdom

Muhammad Waseem Ahmad
BRE Centre for Sustainable Engineering, School of Engineering, Cardiff University, Cardiff CF24 3AA, United Kingdom

Monjur Mourshed
BRE Centre for Sustainable Engineering, School of Engineering, Cardiff University, Cardiff CF24 3AA, United Kingdom

David Mundow
SMS Plc – Energy & Utilities Division, Prennau House, Copse Walk, Cardiff Gate Business Park, Cardiff CF23 8XH, United Kingdom

Mario Sisinni
SMS Plc, Via Gaudenzio Ferrari 21, 21047 Saronno, Italy

Yacine Rezgui
BRE Centre for Sustainable Engineering, School of Engineering, Cardiff University, Cardiff CF24 3AA, United Kingdom

Sheila M.W. Reddy
Office of the Chief Scientist, The Nature Conservancy, 334 Blackwell St., Suite 300, Durham, NC 27701, USA

Robert I. McDonald
Office of the Chief Scientist, The Nature Conservancy, 4245 N. Fairfax Dr, Suite 100, Arlington, VA 22203, USA

Alexander S. Maas
Office of the Chief Scientist, The Nature Conservancy, 334 Blackwell St., Suite 300, Durham, NC 27701, USA
Department of Agricultural and Resource Economics, Colorado State University, Clark B-320, Ft. Collins, CO 80523, USA

Anthony Rogers
Office of the Chief Scientist, The Nature Conservancy, 334 Blackwell St., Suite 300, Durham, NC 27701, USA
Nicholas School of the Environment, Duke University, Durham, NC 27708, USA
Pew Charitable Trust, 901 E. St., NW, Washington, DC 20004-2008, USA

Evan H. Girvetz
Office of the Chief Scientist, The Nature Conservancy, 1917 1st Avenue, Seattle, WA 98101, USA
International Center for Tropical Agriculture, CGIAR, Kasarani Rd., ICIPE Complex, Nairobi, Kenya

Jennifer Molnar
Office of the Chief Scientist, The Nature Conservancy, 4245 N. Fairfax Dr, Suite 100, Arlington, VA 22203, USA

Tim Finley
Texas Operations, The Dow Chemical Company, 2301 Brazosport Rd., APB Building, Freeport, TX 77541, USA

Gená Leathers
Texas Operations, The Dow Chemical Company, 2301 Brazosport Rd., APB Building, Freeport, TX 77541, USA

Johnathan L. DiMuro
Global EH&S and Sustainability, The Dow Chemical Company, 2020 Dow Center, Midland, MI 48674, USA

PREFACE

Environmental Accounting was usually used as a subset of accounting, aiming to incorporate both economic and environmental information. But in this book, environmental accounting was considered as an environmental tool when environmental research took economic effect into consideration. It is widely used when referring to natural resources studies, environmental cost evaluation, environmental fees determination and ecology assessment. This book presented an in-depth research about how the environmental and ecological fields were combing the economic factors. Topics in this book included the relationship between environmental science and policy, taking account factor into the environmental modelling process, carbon footprint accounting, *et al.* Environmental Accounting is also considered an important tool in studies from a perspective of energy, which is also included in this book.

Editor

1 ACCOMMODATING CONSENSUS AND DIVERSITY IN ENVIRONMENTAL KNOWLEDGE PRODUCTION: ACHIEVING CLOSURE THROUGH TYPOLOGIES IN IPBES

Jasper Montana

Department of Geography, University of Cambridge, Downing Place, CB2 3EN, UK

ABSTRACT

How can a diversity of perspectives be accommodated in scientific and political consensus on environmental issues? This paper adopts a science and technology studies (STS) approach to examine how the pursuit of consensus-based knowledge and diverse participation, as seemingly contradictory commitments, have been converted into practice in the Intergovernmental Platform on Biodiversity and Ecosystem Services (IPBES). Through a series of negotiations, these commitments have been translated into a set of situated practices that now dominate this expert panel. Consensus has been achieved through the pursuit of closure, in which meetings of expert and administrator groups produce texts, tables and images that stabilise ostensibly collective decisions. Within this framework, diverse perspectives have been accommodated through the production of typologies, such as lists of comparable options, which allow for the coexistence and commensurability of a range of knowledges and experts. However there is a politics to typologies, which requires specific attention to how decisions are made (deliberation), who participates in them (participation), and the extent to which these participants are representative of broader knowledge and policy communities (representation). While the potential of typologies to accommodate consensus and diversity offers the hope of realising 'unity in diversity' for both environmental knowledge and policy, recognising the politics of their production is important for more equitable processes of environmental governance.

Keywords: Consensus, Diversity, Environmental knowledge, IPBES, Typology, Unity in diversity

INTRODUCTION

In environmental governance, there is a growing tension between the pursuit of scientific and political consensus, and the recognised need to open up governance processes to diverse participants and worldviews. In the production of environmental knowledge, the inclusion of a wider range of experts and knowledges is not only considered important for traditional forms of capacity building, but also for attaining more equitable environmental outcomes (Castree et al., 2014; Mooney et al., 2013; Turnhout et al., 2016). While many initiatives now recognise the need for wider participation, approaches for achieving diversity in practice are still in experimental stages (e.g. Clark et al., 2016; Palsson et al., 2013; van der Hel, 2016). In light of this challenge, this paper asks: How can consensus-based knowledge and policy processes take account of divergent perspectives?

This paper examines the case of the Intergovernmental Platform on Biodiversity and Ecosystem Services (hereafter IPBES, or 'the Platform'), which is an international expert panel for biodiversity established through the United Nations system. Influenced by the perceived success of the Intergovernmental Panel on Climate Change (IPCC), the foundation of IPBES was catalysed by an ambition to bring greater unity to biodiversity knowledge. Billed as an 'IPCC for biodiversity' it was argued that an intergovernmental structure would allow biodiversity science "to evolve

[…] towards greater unity and integration." (Loreau et al., 2006: 246) In the following years, the IPCC became an important model for emulation. In fact, the formal rules of IPBES agreed in 2012 reflected many features of the climate panel, including: an intergovernmental structure; highly formalised procedures that governed its functions, structures and processes; and a system of documentation based on the circulation of meeting reports, decision documents and drafted texts between groups of administrators, experts and peer reviewers.

During this period, expectations of what constituted good environmental knowledge were also changing. Although premised on the success of the long-standing climate panel, the negotiations around IPBES opened up opportunities to scrutinise the perceived failings of the IPCC. Notably, the climate panel had been criticised for having limited participation from developing countries (Hajer, 2012; Hulme and Mahony, 2010), was lacking in the meaningful integration of diverse disciplinary perspectives (Ford et al., 2012; Godal, 2003; IAC 2010), and prioritised a framing of the climate as a 'global kind' that privileged top-down forms of governance (Hulme, 2010; Jasanoff, 2010). In designing IPBES, it was also recognised that biodiversity was materially different to climate, encompassing highly dynamic and diversely distributed socioecological relationships that needed attention at multiple scales. Similarly, the need to account for the rights and agency of indigenous peoples and local communities, which had been raised in the Convention on Biological Diversity (see for example, Reimerson, 2013), also prefigured new approaches to knowledge production (Turnhout et al., 2012). These emergent perspectives were by no means universally held amongst the architects of IPBES, but were none the less integral to a call for greater inclusivity, in which a wider group of voices were to be welcomed into the process.

To examine how the pursuit of consensus and diversity were converted into practice in the case of IPBES, this paper draws on theoretical and methodological approaches from science and technology studies (STS). This field of scholarship has drawn attention to how science shapes, and is shaped by, the settings in which it is produced (Jasanoff, 2004). In particular, it has shown that the precise mechanisms through which knowledge production takes place reflect the commitments of the institutional and political cultures in which they are situated (Jasanoff,

2005; Miller, 2008). The structure of IPBES as an ostensibly global organisation in the United Nations system, for example, predisposes the Platform to produce what have been considered 'global kinds of knowledge' (Turnhout et al., 2016). However, previous scholarship has also shown that such commitments translate into practice in different ways in different settings. Even broadly prevalent commitments, such as that of objectivity, become enacted through practices that are highly context dependent (Jasanoff, 2011). In the case of IPBES, the commitments to consensus-based knowledge and diverse participation were therefore translated into their own situated practices. In this paper, I empirically examine this process and identify the emergence of one particular solution – the typology – as a means of achieving closure in heterogeneous settings. While typologies facilitate coexistence and commensurability in environmental knowledge production, they also bring the politics of participation to the fore. When inclusion in a typology is dependent on being involved in their negotiation, the questions of who participates and how in environmental knowledge production becomes increasingly important.

CASE STUDY AND METHODS

This research is based on the case study of IPBES, an international expert institution for biodiversity (www.ipbes.net). IPBES was formally established in 2012 with the mandate to "strengthen the science-policy interface for biodiversity and ecosystem services" (IPBES, 2012). The Platform has an intergovernmental framework, which brings together an international cast of over 1000 experts and 120 governments to produce its first work program between 2014 and 2018 (see overview in Montana, 2016). It has four broad functions: to conduct assessments on the state of knowledge on biodiversity; build international capacity across knowledge and policy communities; catalyse the development of policy support tools; and support new knowledge generation. The Platform is intended to produce a range of outputs, including assessment reports, methodological guides, participatory processes, and online catalogues. IPBES was the result of around a decade of discussions, workshops, and formal intergovernmental negotiations that took place both inside and outside of the United Nations system (see account in Vadrot, 2014b).

This research applied qualitative methods to collect and analyse data from interviews, participant observation, and official documents between December 2013 and February 2016. Interviews for this research were conducted with both IPBES experts (n = 12) and administrators (n = 5). Experts were defined as those selected for the IPBES work program (anonymised as E1 - E12). Administrators were defined as those working in the secretariat, technical support units, Bureau and Multidisciplinary Expert Panel (anonymised as A1 - A5). Administrators worked in collaboration to oversee and coordinate the IPBES process and are grouped accordingly here. Interviews were semi-structured and conducted in person or via Skype. Interview data was complemented by participant observation at IPBES meetings, including three Plenary meetings (Antalya, Turkey in 2013; Bonn, Germany in 2015; Kuala Lumpur, Malaysia in 2016), one expert group meeting (Ushuaia, Argentina in 2015), and a joint meeting of the Multidisciplinary Expert Panel, Bureau, and three task forces (Bonn, Germany in 2015). This was complemented by background research during a four-month institutional placement with the IPBES secretariat from January until April 2015. Some data was also extracted from official documents. Collected data was analysed using computer-assisted thematic analysis (Atlas.ti) using an iterative three-pass coding process. Quotations provided in this paper are representative of the coded themes.

INSTITUTIONALISING CONSENSUS

Although an ambition for forging consensus and increasing diverse participation were articulated in the lead up to the establishment of IPBES, the precise mechanisms through which these commitments would be translated into practice were subject to a series of negotiations. In the intergovernmental Plenary of IPBES, the institutional conditions of knowledge production were negotiated in the form of operating principles, rules and procedures. These documents set the scope of the IPBES work program and established expert groups to carry it through to completion. They defined rules that would guide the selection of experts and specified the procedures for peer review. These 'rules on paper' sketched out the rough framework within which IPBES would carry out its work, allowing the subsidiary bodies, secretariat and expert groups to interpret and operationalize them into what would later become the Platform's 'rules in use' (to draw on the language of 'new institutionalism', Young, 2002). From this perspective, the institutionalisation of consensus was the result of a series of negotiations that took place across the Platform's formal structures.

IPBES emerged as an outcome of the United Nations system, in which consensus-based decision making and state sovereignty are central principles of operation. It is perhaps no wonder, then, that the rules prescribed consensus as the chief decision-making framework of its intergovernmental Plenary. Defined by tradition, consensus is achieved through the absence of formal objections, rather than the outcome of a majority vote or unanimous agreement (UNEP, 2007). This provides, according to an administrator in IPBES, "an environment where people are basically owning and being part of whatever is being produced" (Interviewee A5, June 2015). Driven by this motive, consensus was institutionalised as "a principle that runs across the whole operations of IPBES" (Interviewee A3, May 2015). However, unlike the Plenary, the administrative and expert groups of IPBES did not have strict rules that governed how decision making should take place. Frameworks of deliberation instead emerged as local cultures dependent on the particular styles and preferences of those involved. While meetings of the formal administrative bodies (i.e. the Multidisciplinary Expert Panel and Bureau) worked towards consensus in accordance with United Nations tradition, decision making in other expert groups was much more variable. As one of the administrators explained, expert groups "are all a little bit different. It gets down into the level of individuals and characters. And then different set ups." (Interviewee A5, June 2015) Yet, despite this heterogeneity, the administrative and expert groups of IPBES had a recognisable common project that tied them together. Ultimately, they had been charged with the completion of the IPBES work programme, which meant: developing and following the rules of the Platform; coming to forms of collective agreement; and completing the various deliverables that they had been mandated to produce. In fulfilling this commitment, the administrators and experts met, discussed, debated, negotiated and ultimately produced texts, tables and images. This practice of *putting on paper* stabilised group decisions and provided a means of achieving temporary closure.

The attainment of closure, as a form of consensus, did not necessarily mean that member states, expert groups and administrators explicitly agreed. Rather, closure in IPBES allowed for the temporary stabilisation of decisions, which were characterised by three dimensions: negotiation, collective ownership, and the production of a written record (Fig. 1.).

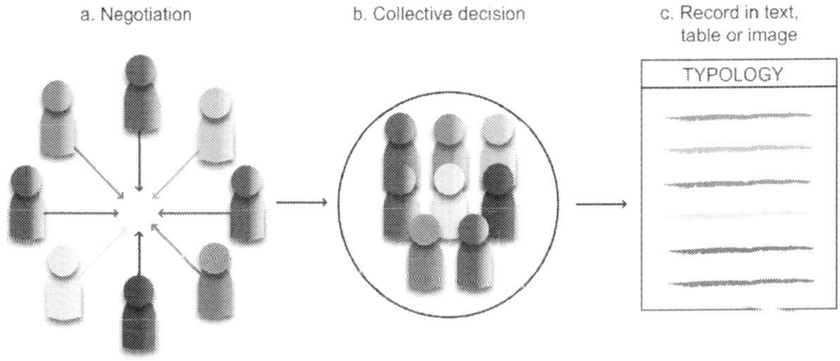

Figure 1. The three dimensions of closure in IPBES: a) the negotiation between divergent perspectives; b) the formation of an ostensibly collective decision; and c) the production of a documented record of the decision.

Firstly, closure in IPBES was reached through a process of negotiation. This was most explicitly recognisable in the Plenary, but was also carried out across the administrative and expert groups. As one of the administrators explained:

Of course, when the experts work they are negotiating in a way also. Trying to bring different perspectives in from different disciplines. (Interviewee A3, May 2015)

The experts and administrators participating in the process originated from diverse backgrounds and therefore brought competing perspectives into their work. Despite this, productive negotiations were driven by the need to work together to complete the work programme: an ambition that was widely accepted as a requirement of participation in the IPBES process.

Secondly, negotiations consistently moved towards a conclusion that was seen as a collective decision. Authors were expected to take ownership of the content of their reports, as one of the participating experts explained:

- any publication that has multiple authors, anything expressed in that multiple author publication is, I guess, the views of all the authors. And you don't agree or sign off on a manuscript

> or a draft report, agreeing for it to be published unless you are happy with the content and any messages that are contained therein. (Interviewee E8, March 2015)

Reflecting on their experience at an early author meeting, another of the IPBES experts noted that: "we ended up having everyone having bought into this and driving it, so that was the nice thing about the meeting." (Interviewee E1, December 2014) This process of achieving joint ownership of decisions was central to achieving closure in IPBES.

Finally, once disputes had been settled and compromises had been found, the collectively negotiated positions were put on paper. This process of producing documentation has previously been recognised as central to the Platform's commitment to transparency (Granjou et al., 2013), but it also functions as a critical device for stabilising consensus across the Platform. The production of texts, tables and images, from the lists of selected experts to the annual budget, are representative of these moments of closure in the Plenary, administrative or expert group meetings.

In IPBES, the pursuit of closure became a collective venture that was underpinned by a lively operational discourse: consensus was described as a problem to be 'solved'; expert groups were given 'marching orders'; 'time-bound' activities were delegated to 'task forces'; and the Multidisciplinary Expert Panel was 'mandated' to complete 'work programme deliverables'. It was this drive to demonstrate success and relevance that Kovács and Pataki (2016: 133) characterise as a "need for speed" in their critical evaluation of the IPBES process. This operational discourse established a logic in which success could be measured by the completion of products and the passing of decisions, which were often greeted with enthusiastic applause during Plenary meetings. As such, the production of texts, tables and images to stabilise ostensibly collective decisions was highly effective. However, the drive towards closure also came into conflict with other established commitments in the work of the Platform. In particular, the ambition to include a greater diversity of experts and knowledges in its work.

NORMALISING DIVERSITY

Like consensus, the precise practices through which a commitment to diversity was institutionalised in IPBES were determined through a series of negotiations. The IPBES operating principles, agreed intergovernmentally in 2012, set out that the Platform should carry out its work:

taking into account the need for different disciplines and types of knowledge, gender balance, and effective contribution and participation by experts from developing countries (IPBES, 2012: 4).

Following this, the approval of the IPBES conceptual framework in 2013 provided a framework that made an "explicit, formal incorporation of knowledge systems other than western science" (Díaz et al., 2015). These documents were important stabilising devices for diversity in the Platform, however they required further interpretation and operationalization. While the Plenary defined disciplines, knowledges, gender, and world regions as desirable categories of diversity, the means through which difference would be determined was not specified in exact terms. Deferred to the administrators of the Platform, new systems of classification emerged that would allow experts and knowledges to be divided, counted and reported in the expert selection process. In 2015, for example, experts were divided up by their gender, their world region, their area of expertise, and the source of their nomination (see categories in Table 1.) Expert identity was no longer simply defined by eminent authority, but derived through new markers of expertise based on embodied lived experience in different social worlds. Throughout expert selection, these categories provided a device for numerically tallying diversity across the Platform. Expert groups were evaluated against negotiated ideals of balance, and the result – a list of selected experts – represented yet another written product of closure.

Table 1. System of expert categorisation as applied during IPBES expert selection in April 2015

Type of nomination	World region	Gender	Discipline or practice expertise
• Government • Stakeholder	• Africa • Asia-Pacific • Eastern Europe • Latin America and Caribbean • Western Europe and Other	• Woman • Man	• Economics • Freshwater • Indigenous and Local Knowledge • Marine • Natural sciences and conservation • Practitioners • Social sciences • Terrestrial • Other expertise

However, the system of expert selection had important features that set it apart from other forms of closure. Embedded within the process were records of both a collective decision and the accommodation of diversity (see examples in Fig. 2.). Each individual selected according to this system was clearly recognised as an expert by the Platform. In this sense, each expert was equivalent to any other expert. Yet, at the same time, each expert was anything but equivalent, and was defined by clear markers of difference that the Platform carefully accounted for. Seen as a whole, each expert list therefore encompassed both commonalities and desirable differences amongst experts. A previous analysis of the composition of the Multidisciplinary Expert Panel, for example, has shown how categorisation allowed for this expert group to be broken up into different units of difference, encompassing gender, disciplines and world regions, allowing diversity to be metrically compared between different groups (Montana and Borie, 2016). In other words, the expert groups themselves can be understood as typologies that both accommodate a consensus around who should be considered an expert in the IPBES process and the inclusion of the diverse categories of embodied experience that are also considered important. Typologies such as this became prevalent across the entire IPBES process.

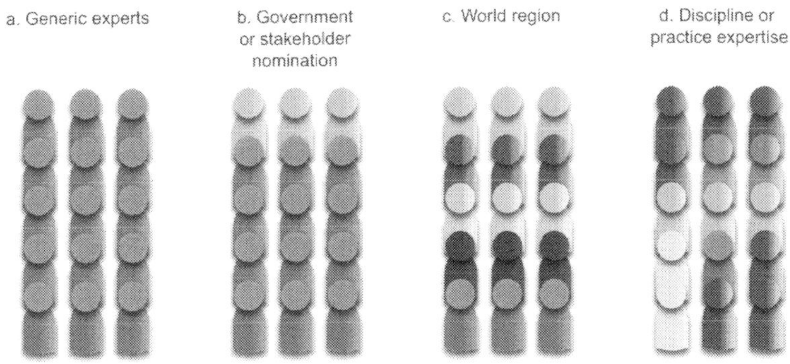

Figure 2. Characterisations of the diversity embedded in the expert groups of IPBES, illustrating how diversity was characterised by the Platform. A single expert group could be perceived as: a) generic experts; b) representatives of either government or stakeholder nominations; c) representatives of different world regions; or d) representatives of different disciplines or domains of practice.

PRODUCING TYPOLOGIES

Throughout the work of IPBES, typologies have become important devices for accommodating consensus and diversity in the production of environmental knowledge. During the production of the methodological assessment on biodiversity models and scenarios, which brought together over 80 different experts between October 2014 and February 2016, typologies loomed large in the report writing process. Confronted with divergent perspectives at one of the author meetings on this report, one of the coordinating lead authors commented that:

- we all have a different idea about what a model is and we probably need a top-down decisions about how they are defined. None of us will probably like it, but that's what we need (Field Notes, March 2015)

In response, another of the authors at the meeting suggested: "or, we need a typology of models" (Field Notes, March 2015). Presented in this way, typologies appear as powerful devices for accommodating difference without resorting to the top-down impositions of 'global kinds of knowledge'. In fact, the final methodological assessment report adopted in early 2016 developed numerous typologies in the form of tables, which reflected a range of "illustrative and non-exhaustive examples" (IPBES, 2016). More broadly, similar typologies have emerged as solutions to divergent perspectives across IPBES on issues ranging from the values to the scales of biodiversity (see Table 2.). Even the conceptual framework of the Platform, which explicitly incorporates multiple framings of biodiversity, can be understood as a typology. Identified in Borie and Hulme (2015: 494) as "a solution to create an agreement out of disagreement, to create a consensus out of dissensus", the use of colour coding to identify different knowledge systems, from natural science to indigenous and local knowledge, provided a powerful device for allowing their co-existence in the conceptual framework. Typologies have therefore been used frequently in IPBES as a means of reaching closure while acknowledging diversity, and in doing so allowing those involved to achieve the mandate of completing the IPBES work programme.

Table 2. Some illustrative examples of typologies in IPBES

Source of typology	**Category of difference**	**Kinds of difference**
Conceptual framework (Díaz et al., 2015)	Knowledge systems	• Scientific knowledge • Other knowledge systems
Preliminary guide to the diverse conceptualisation of values (IPBES, 2015b: 3)	Values	• Non-anthropocentric • Anthropocentric
Preliminary guide to the diverse conceptualisation of values (IPBES, 2015b: 5)	Valuation methods	• Biophysical and ecological • Cultural and social • Economic • Public health • Holistic, Indigenous, and local knowledge-based
Guide to assessment across all scales (IPBES, 2015a: 27)	Spatial scales for assessments	• Global • Regional • Subregional • National

The production of typologies in IPBES is likely to have implications that go beyond just facilitating closure. Typologies both impose categories – making kinds comparable to other kinds – but also erase differences that may later be considered desirable. In their analysis of how categories and standards shape the modern world, Bowker and Star, (2000: 196–203) advanced a critique about the power of these typological categories to impose particular frameworks of behaviour on groups of people, which can have real impacts on real lives. While these practices may be considered broadly benign in a biodiversity expert panel, the impact of the typological approach in shaping how nature is understood and governed should remain a point of reflection for the Platform. For example, Bowker and Star (2000: 116) also noted that the production of typologies "presets the options about the range of possibilities." In this sense, decisions made in IPBES directly constrain the range of possible options available to downstream processes like the Convention on Biological Diversity (CBD). While some might consider that the production of typologies leave the options too wide open for policy makers, and that difficult choices between divergent perspectives should take place in IPBES instead, others are likely to welcome the retention of diversity in the products of the Platform. Rather than being considered the end of the story, the typology framework in IPBES might be better understood as the temporary stabilisation of an underlying politics of knowledge that is often left obscured. And, these politics can be made available for analysis by asking a series of key questions.

THE POLITICS OF TYPOLOGIES

The proliferation of typologies as a means of accommodating the production of consensus-based knowledge while taking account of diversity appears as an exemplary solution in IPBES. At once, it recognises a plurality of perspectives and appeals to the pragmatic logic of getting things done. However, the production of typologies is not always a straightforward process. As the co-chair of the methodological assessment on models and scenarios reported to the Plenary in 2016: "it took a lot of pain and a lot of work to reach a consensus – even in our eighty person team – to produce that typology" (Field Notes, February 2016). This reflection hints at the politics that is inherent to the production of typologies. Politics that should be acknowledged in the further use and analysis of this device in environmental governance. In particular, questions should be raised about how collective decisions are made, who participates in the process, and the extent to which those involved represent the diversity of perspectives that exist outside any given deliberative process? In other words, the process of negotiation itself needs to be opened up to examination.

Building on previous work in the field of STS, I draw on insights from democratic theory to reflect on the importance of taking account of deliberation, participation and representation in the production of typologies (a similar framework was applied to compare the workings of expert advisory bodies in Jasanoff, 2005). In doing so, I argue that sensitivity to these factors should be taken into account if the production of typologies is to lead to more equitable forms of environmental governance. This consideration should be carried out recognising that the pursuit of diverse participation is often motivated by very different factors. The growing push to bring a wider group of experts and knowledges into the production of sustainability and environmental change research, for example, is considered to fulfil a number of possible functions, including: a normative function that increases the legitimacy of the process; a substantive function that strengthens the knowledge base towards more appropriate solutions; and/or an instrumental function that supports more collaborative relationships and joint ownership of the knowledge produced (Blackstock et al., 2007). Regardless of the motivation, inclusion and exclusion of different perspectives deeply influence the kinds of knowledge considered to be relevant in the IPBES process, and ultimately the kinds of political responses that will be seen as appropriate to address biodiversity decline (Turnhout et al., 2016). Reflection on why diverse participation is considered desirable will help elucidate questions regarding where, when and how the politics of knowledge production matter.

How are Decisions Made?

If diverse participation is going to have more than a legitimating effect, it needs to be effective. Different approaches to communicating information and making decisions can greatly impact the effectiveness of participation in deliberative processes (Fung, 2006). More than simply being present in negotiations, effective participation requires approaches to discussion and decision making that support the active inclusion of diverse actors. In global expert panels, like IPBES, effective participation often relies upon those participating having knowledge and experience of the local cultures of negotiation, as much as the subject matter under discussion (Jasanoff and Martello, 2004: 347). For example, the speed with which IPBES has carried out its work has previously been identified as a major limitation in establishing the knowledge and experience required amongst all participants for equitable and effective participation (Kovács and Pataki, 2016). Ensuring that deliberative frameworks do not preference some actors over others, or providing adequate capacity building support to acculturate newcomers, would help overcome the situation described by an IPBES expert following one of the first author meetings, in which: "there are a few people who are actually driving [the report writing process] and there are other people who are just running along" (Interviewee E1, December 2014). In these cases, diversity risks becoming as much performative as it is practical, providing only the normative function of legitimation, which this author described as having some authors "just there to make up numbers" (Interviewee E1, December 2014).

Who Participates?

If typologies provide a means of accommodating diversity, then it matters what kind of diversity is included in the process. In political systems, the premise of participatory democracy recognises the value of bringing different voices into deliberations (Fung, 2006). Based on this assertion, the significance of participation in IPBES was perhaps best demonstrated through the influence of the Bolivian delegation in specific negotiations, which arguably drove the inclusion of indigenous and local knowledge systems in both the IPBES conceptual framework and rules of procedure (Borie and Hulme, 2015; Vadrot, 2014a). However, while voluntary participation of each member state in the Plenary was protected by the Platform's rules, the administrative and expert groups of IPBES were necessarily finite in size. As such, despite the apparent willingness to open up to diverse forms of knowledge and expertise in IPBES, the strict rules that governed who can and can't participate in these groups inevitably left many potential perspectives outside of negotiations. Furthermore, a lack of engagement with the process further exacerbated the problems of achieving diverse participation. The recognised absence of social scientists in the early stages of the IPBES work programme (see, for example, Larigauderie et al., 2016) provided one notable illustration of how limited participation might hinder attempts to accommodate diversity in knowledge production. Participation clearly matters in the production of knowledge. As one of the co-chairs of the methodological assessment reflected: "no matter how much the individuals involved come into this process keen to leave their particular perspectives at the door, it is almost impossible to do that completely." (Interviewee E12, June 2015) Yet, with finite author groups in global expert panels, there will always be a requirement for participants to bring in perspectives from others on the outside.

Is Representation Sufficient?

To counter the concern of limited participation, proponents of processes, like IPBES, might argue that diversity can be represented through a sub-set of selected participants. In such a case, concern should be less focused on the inclusion and exclusion of individuals, and more on how included individuals bring the perspectives of others into negotiations with them. In political systems, the premise of political representation refers to the relationship through which political actors advocate and act on the behalf of larger groups (Heywood, 2013: 197). Reflecting this approach, the management of knowledge and expertise in IPBES can also be understood to conform to a representative framework. However, it has been noted that the tradition of consensual and cumulative knowledge production most suited to a system of representative participation is not uniform across domains of knowledge (Obermeister, In Press). While it may be reasonable to assume that the inclusion of a natural scientist will ensure that a breadth of natural scientific knowledge is included in an assessment, it is less clear if the diverse range of pluralist approaches in the social sciences or indigenous knowledge systems can be similarly accommodated through a small sub-set of representative participants. In considering the design and analysis of knowledge production processes, attention should be given to the different way in which selected participants establish accountability to the wider communities of research and practice that they claim to represent. Beyond the peer review process in IPBES, which has targeted a small set of scientists and practitioners that are already engaged with the process, the consideration of how to strengthen the accountability of its knowledge production to communities outside of the intergovernmental Plenary merits further attention.

CONCLUSION

The adoption of typologies as means of reaching closure in IPBES has allowed the diversity of knowledges and experts that have been brought into the Platform to be detectable in its outputs. Although this paper has focused primarily on the production of the typologies, how they are received in the various policy forums once they leave the IPBES process is of comparable, if not greater, importance. Presented with a menu of diverse options, negotiators in the CBD or elsewhere are not bound to take account of the diversity presented to them. Rather, they may choose to disassemble these typologies in favour of single 'global kinds', which have a tendency to prescribe top-down management frameworks and reinforce hegemonic norms (Hulme, 2010; Turnhout et al., 2016). The inclusion of indigenous and local knowledge systems, for example, provides an opportunity for these alternative forms of knowledge to be carried into political forums for consideration. However, a lack of capacity in interpreting these knowledge systems could easily lead them to being disregarded, despite their inclusion in typologies. How typologies that encompass new configurations of knowledges are received into policy forums is a prime area for further research.

In considering the production and reception of typologies, it is notable that typologies, which make differences in perspective explicit, commensurable and co-existent, appeal to a different kind of environmental politics. From the diverse conceptualisation of the values of nature, to the multi-faceted criteria for the identification of experts, typologies facilitate a harmonisation of approaches that reflects a commitment to *unity in diversity*. The notion that both unity and difference can meaningfully co-exist has provided inspiration for natural and political philosophers for centuries. While in the nineteenth century the naturalist, von Humboldt (1860), wrote of unity in diversity as a definitive quality of Nature itself, the concept persists today as a motto and political ideal of the European Union. In less explicit terms, a commitment to unity in diversity can be recognised across the domains of environmental science and policy. The negotiation of co-management arrangements that accommodates, and values, the different stewardship systems of the First Nations of Canada and the Canadian state, provides one illustrative example (see Houde, 2007). Similarly, the differentiated commitments of greenhouse gas reductions from different nations that functioned as a key characteristic of the 2015 Paris Agreement (see Rajamani, 2016), acknowledged that different actors can each take different action, while essentially all doing the same thing: namely responding to climate change. The ability for collective action, while accepting difference, therefore opens up wider possibilities for environmental policy.

In the production of environmental knowledge, a discourse of unity that embraces diversity has already inspired a new wave of thought and activity (Clark et al., 2016; Cornell et al., 2013; Mace, 2013; Mauser et al., 2013; van der Hel, 2016). Harnessed effectively, these experiments in environmental knowledge production could lead to a reimagining of the concept of consensus itself. Here, consensus might no longer simply reflect the certainty of hegemonic science, or the pursuit of the lowest common denominator in political settings. After all, it has been acknowledged that the unification of nature through universalist sciences is unlikely to lead to the unification of politics (Latour, 2004). Instead, the pursuit of consensus in environmental governance might be reimagined and refocused to embrace diversity in knowledge and politics, and become at ease with the continuation and coexistence of difference.

Funding

This work was supported by the Economic and Social Research Council, UK [University of Cambridge DTC].

ACKNOWLEDGEMENTS

The author would like to thank Bill Adams, Josie Chambers, Chris Sandbrook and attendees at the Science and Democracy Network 2016 Annual Meeting for comments and guidance.

REFERENCES

1. Blackstock et al., 2007 K.L. Blackstock, G.J. Kelly, B.L. HorseyDeveloping and applying a framework to evaluate participatory research for sustainability Ecol. Econ., 60 (2007), pp. 726-742
2. Borie and Hulme, 2015 M. Borie, M. HulmeFraming global biodiversity: IPBES between mother earth and ecosystem services

Environ. Sci. Policy, 54 (2015), pp. 487-496

3. Bowker and Star, 2000 G.C. Bowker, S.L. StarSorting Things Out: Classification and Its Consequences MIT Press, Cambridge, MA (2000)
4. Castree et al., 2014 N. Castree, W.M. Adams, J. Barry, D. Brockington, B. Buscher, E. Corbera, D.Demeritt, R. Duffy, U. Felt, K. Neves, P. Newell, L. Pellizzoni, K. Rigby, P. Robbins, L. Robin, D.B.Rose, A. Ross, D. Schlosberg, S. Sorlin, P. West, M. Whitehead, B. WynneChanging the intellectual climate Nat. Clim. Change, 4 (2014), pp. 763-768
5. Clark et al., 2016 W.C. Clark, L. van Kerkhoff, L. Lebel, G.C. GallopinCrafting usable knowledge for sustainable development Proc. Natl. Acad. Sci., 113 (2016), pp. 4570-4578
6. Cornell et al., 2013 S. Cornell, F. Berkhout, W. Tuinstra, J.D. Tàbara, J. Jäger, I. Chabay, B. de Wit, R.Langlais, D. Mills, P. Moll, I.M. Otto, A. Petersen, C. Pohl, L. van KerkhoffOpening up knowledge systems for better responses to global environmental change Environ. Sci. Policy, 28 (2013), pp. 60-70
7. Díaz et al., 2015 S. Díaz, S. Demissew, C. Joly, W.M. Lonsdale, A. LarigauderieA rosetta stone for nature's benefits to people PLoS Biol., 13 (2015)
8. Ford et al., 2012 J. Ford, W. Vanderbilt, L. Berrang-FordAuthorship in IPCC AR5 and its implications for content: climate change and indigenous populations in WGII Clim. Change, 113 (2012), pp. 201-213
9. Fung, 2006 A. FungVarieties of participation in complex governance Public Administration Review (2006), pp. 66-75 December 2006
10. Godal, 2003 O. GodalThe IPCC's assessment of multidisciplinary issues: the case of greenhouse gas indices Clim. Change, 58 (2003), pp. 243-249
11. Granjou et al., 2013 C. Granjou, I. Mauz, S. Louvel, V. TournayAssessing nature? The genesis of the intergovernmental platform on biodiversity and ecosystem services (IPBES) Sci. Technol. Soc., 18 (2013), pp. 9-27
12. Hajer, 2012 M.A. HajerA media storm in the world risk society:

enacting scientific authority in the IPCC controversy (2009–10) Crit. Policy Stud., 6 (2012), pp. 452-464

13. Heywood, 2013 A. HeywoodPolitics (4th ed.), Palgrave Macmillan, Hampshire; New York, NY (2013)
14. Houde, 2007 N. HoudeThe six faces of traditional ecological knowledge: challenges and opportunities for Canadian co-management arrangements Ecol. Soc., 12 (2007), p. 34
15. Hulme and Mahony, 2010 M. Hulme, M. MahonyClimate change: what do we know about the IPCC? Prog. Phys. Geogr., 34 (2010), pp. 705-718
16. Hulme, 2010 M. HulmeProblems with making and governing global kinds of knowledge Glob. Environ. Change, 20 (2010), pp. 558-564
17. IAC, 2010 IACClimate Change Assessments: Review of the Processes and Procedures of the IPCC InterAcademy Council, Alkmaar, The Netherlands (2010) IPBES, 2012
18. IPBESFunctions, Operating Principles and Institutional Arrangements of the Intergovernmental Science-Policy Platform on Biodiversity and Ecosystem Services, Intergovernmental Science-Policy Platform on Biodiversity and Ecosystem Services (2012)
19. IPBES, 2015a IPBESGuide on the Production and Integration of Assessments from and Across All Scales. IPBES/4/INF/9 Intergovernmental Science-Policy Platform on Biodiversity and Ecosystem Services, Bonn, Germany (2015)
20. IPBES, 2015b IPBESPreliminary Guide Regarding Diverse Conceptualization of Multiple Values of Nature and Its Benefits, Including Biodiversity and Ecosystem Functions and Services. IPBES/4/INF13 Intergovernmental Science-Policy Platform on Biodiversity and Ecosystem Services, Bonn, Germany (2015)
21. IPBES, 2016 IPBESSummary for Policymakers of the Assessment Report of the Methodological Assessment of Scenarios and Models of Biodiversity and Ecosystem Services
22. S. Ferrier, K.N. Ninan, P. Leadley, R. Alkemade, L. Acosta-Michlik, H.R. Akçakaya, L. Brotons, W.Cheung, V. Christensen, K.H. Harhash, J. Kabubo-Mariara, C. Lundquist, M. Obersteiner, H.Pereira, G. Peterson, R. Pichs-Madruga, N.H. Ravindranath, C. Rondinini, B. Wintle (Eds.), Intergovernmental Science-Policy

Platform on Biodiversity and Ecosystem Services, Bonn, Germany (2016)

23. Jasanoff and Martello, 2004 S. Jasanoff, M.L. MartelloKnowledge and governance S. Jasanoff, M.L. Martello (Eds.), Earthly Politics: Local and Global in Environmental Governance, MIT Press, Cambridge MA.; London (2004), pp. 335-350
24. Jasanoff, 2004 S. JasanoffOrdering knowledge, ordering society S. Jasanoff (Ed.), States of Knowledge: The Co-production of Science and Social Order., Routledge, London (2004)
25. Jasanoff, 2005 S. JasanoffDesigns on Nature: Science and Democracy in Europe and the United States Princeton University Press, Princeton; Oxford (2005)
26. Jasanoff, 2010 S. JasanoffA new climate for society Theory Cult. Soc., 27 (2010), pp. 233-253
27. Jasanoff, 2011 S. JasanoffThe practices of objectivity in regulatory science
28. C. Camic, N. Gross, M. Lamont (Eds.), Social Knowledge in the Making, University of Chicago Press, Chicago (2011), pp. 307-337
29. Kovács and Pataki, 2016 E.K. Kovács, G. PatakiThe participation of experts and knowledges in the Intergovernmental Platform on Biodiversity and Ecosystem Services (IPBES) Environ. Sci. Policy, 57 (2016), pp. 131-139
30. Larigauderie et al., 2016 A. Larigauderie, M. Stenseke, R.T. WatsonBiodiversity assessments: IPBES reaches out to social scientists Nature, 532 (2016), p. 313
31. Latour, 2004 B. LatourPolitics of Nature: How to Bring the Sciences into Democracy Harvard University Press, Cambridge, MA (2004)
32. Loreau et al., 2006 M. Loreau, A. Oteng-Yeboah, M.T.K. Arroyo, D. Babin, R. Barbault, M. Donoghue, M.Gadgil, C. Häuser, C. Heip, A. Larigauderie, K. Ma, G. Mace, H.A. Mooney, C. Perrings, P. Raven, J. Sarukhan, P. Schei, R.J. Scholes, R.T. WatsonDiversity without representation Nature, 442 (2006), pp. 245-246
33. Mace, 2013 G. MaceEcology must evolve Nature, 503 (2013), pp. 191-192
34. Mauseretal.,2013W.Mauser,G.Klepper,M.Rice,B.S.Schmalzbauer,

H. Hackmann, R. Leemans, H.MooreTransdisciplinary global change research: the co-creation of knowledge for sustainability Curr. Opin. Environ. Sustain., 5 (2013), pp. 420-431

35. Miller, 2008 C.A. MillerCivic Epistemologies: constituting knowledge and order in political communities Sociol. Compass, 2 (2008), pp. 1896-1919
36. Montana and Borie, 2016 J. Montana, M. BorieIPBES and biodiversity expertise: regional, gender and disciplinary balance in the composition of the interim and 2015 multidisciplinary expert panel Conserv. Lett., 9 (2016), pp. 138-142
37. Montana, 2016 J. MontanaHow IPBES works: the functions, structures and processes of the Intergovernmental Platform on Biodiversity and Ecosystem Services C-EENRG Working Papers Centre for Environment, Energy and Natural Resource Governance, University of Cambridge, Cambridge (2016), pp. 1-23
38. Mooney et al., 2013 H.A. Mooney, A. Duraiappah, A. LarigauderieEvolution of natural and social science interactions in global change research programs Proc. Natl. Acad. Sci., 110 (2013), pp. 3665-3672
39. Obermeister, 2016 N. Obermeister, in press From dichotomy to duality: addressing interdisciplinary epistemological barriers to inclusive knowledge governance in global environmental assessments Environmental Science and Policy.
40. Palsson et al., 2013 G. Palsson, B. Szerszynski, S. Sörlin, J. Marks, B. Avril, C. Crumley, H. Hackmann, P. Holm, J. Ingram, A. Kirman, M.P. Buendía, R. WeehuizenReconceptualizing the 'anthropos' in the anthropocene: integrating the social sciences and humanities in global environmental change research Environ. Sci. Policy, 28 (2013), pp. 3-13
41. Rajamani, 2016 L. RajamaniAmbition and differentiation in the 2015 Paris agreement: interpretative possibilities and underlying politics Int. Comp. Law Q., 65 (2016), pp. 493-514
42. Reimerson, 2013 E. ReimersonBetween nature and culture: exploring space for indigenous agency in the convention on biological diversity Environ. Polit., 22 (2013), pp. 992-1009
43. Turnhout et al., 2012 E. Turnhout, B. Bloomfield, M. Hulme,

J. Vogel, B. WynneConservation policy: listen to the voices of experience Nature, 488 (2012), pp. 454-455

44. Turnhout et al., 2016 E. Turnhout, A. Dewulf, M. HulmeWhat does policy-relevant global environmental knowledge do? The cases of climate and biodiversity Curr. Opin. Environ. Sustain., 18 (2016), pp. 65-72
45. UNEP, 2007 UNEPMultilateral Environmental Agreement: Negotiator's Handbook (2nd ed.), University of Joensuu, Joensuu, Finland (2007)
46. Vadrot, 2014a A.B.M. VadrotThe epistemic and strategic dimension of the establishment of the IPBES: epistemic selectivities at work Innovation, 27 (2014), pp. 361-378
47. Vadrot, 2014b A.B.M. VadrotThe Politics of Knowledge and Global Biodiversity Routledge, London (2014)
48. van der Hel, 2016 S. van der HelNew science for global sustainability? The institutionalisation of knowledge co-production in future earth Environ. Sci. Policy, 61 (2016), pp. 165-175
49. von Humboldt, 1860 A. von Humboldt Cosmos: a Sketch of a Physical Description of the Universe, vol. 1, Harper and Brothers Publishers, New York (1860)
50. Young, 2002 O.R. YoungThe Institutional Dimensions of Environmental Change: Fit, Interplay, and Scale MIT Press, Cambridge, MA; London (2002)

2

INTEGRATED VALUATION OF A NATURE-BASED SOLUTION FOR WATER POLLUTION CONTROL. HIGHLIGHTING HIDDEN BENEFITS

Camino Liquete[a], Angel Udias[a], Giulio Conte[b], Bruna Grizzetti[a], Fabio Masi[b]

[a] European Commission, Joint Research Centre (JRC), Directorate D - Sustainable Resources, Via Enrico Fermi 2749, 21027 Ispra, VA, Italy

[b] IRIDRA Srl, Via La Marmora 51, Florence 50121, Italy

ABSTRACT

In this study we assess multiple benefits (environmental, social and economic) provided by a multi-purpose green infrastructure (a series of constructed wetlands surrounded by a park) in a peri-urban area, and compare it with the alternative grey infrastructure and with the previous situation (a poplar plantation). We apply a multi-criteria analysis as a basis for integrated valuation. We address specific policy needs (strategic objectives) for the local territorial planning in the implementation of the EU Water Framework Directive. The analysis is used retrospectively (*ex post* evaluation) but our results could also be used prospectively to appraise new proposals of constructed wetlands under similar circumstances.

The results reflect that the green infrastructure performs equal or even better than the grey infrastructure alternative for water purification and flood protection, it has a similar cost, and it provides additional benefits (like wildlife support and recreation). The most preferred alternative is the green infrastructure, followed by the grey infrastructure and the poplar plantation.

This study demonstrates (a) the effectiveness of investments on nature-based solutions, (b) the potential of green infrastructures for delivering a broad range of ecosystem services, and (c) the utility of integrating different value systems and stakeholders› viewpoints to support environmental decision-making.

Keywords: Sustainable urban drainage systems, Constructed wetland, Ecosystem services, Integrated valuation, Multi-criteria analysis, Water management

INTRODUCTION

Natural ecosystems are hypothesized to provide viable (cost-efficient and effective) solutions to tackle numerous societal challenges such as climate change, disaster prevention, sustainable cities and water resource management. In the present EU policy context, the use of natural ecosystems as smart solutions is promoted by several strategies. The EU Biodiversity Strategy[1] have set specific policy targets for maintaining and enhancing ecosystems and their services by establishing green infrastructures and restoring degraded ecosystems. Green infrastructures[2] are considered to provide multiple benefits contributing to achieve the objectives of several policies, including climate change and environmental policies, disaster risk management, health and consumer policies and the Common Agricultural Policy. Specifically for water policy, the recent Blueprint to safeguard Europe›s water resources[3] indicated that green infrastructures and nature-based solutions, such as natural water retention measures, can greatly contribute to the provisioning of ecosystem services and should be adopted as measures in the implementation of the Water Framework Directive and the Flood Directive through the territorial planning. Therefore there is a great interest in investing in nature-based solutions assuring multiple ecosystem services. But how to measure the effectiveness of these measures and how to account for the multiple benefits they provide? These are the two key questions we want to address in the present study, based on a real case application.

Nature-based solutions are defined as actions inspired by, supported by or copied from nature that help societies address a variety of environmental, social and economic challenges in sustainable ways (DG Research and Innovation, 2015). Other definitions highlight the contribution of well-managed and diverse ecosystems to enhance human resilience and sustainable development, thus focusing on ecosystem services. For instance, Maes and Jacobs (2015) define nature-based solutions as *any transition to a use of ecosystem services with decreased input of non-renewable natural capital and increased investment in renewable natural processes*. Eggermont et al. (2015) differentiate three types of nature-based solutions that share the aim of improving the delivery of a range of ecosystem services: solutions of no or minimal intervention in ecosystems, interventions in managed ecosystems and landscapes that look for sustainability and multi-functionality, and creation or deep modification of ecosystems usually to build green or blue infrastructures. The case study shown in this paper corresponds to the third type. Any of these visions implies that maintaining and enhancing natural capital and ecosystem services is of crucial importance. In Europe, investing in nature-based solutions can lead to wide socio-economic benefits, provision of jobs, and low-carbon technology innovations, that is, to sustainable economy and development as envisaged by the EU Horizon 2020 vision (Maes and Jacobs, 2015). Some of the nature-based engineered solutions already used in urban planning and water management (e.g. green roofs, bio-infiltration rain gardens, vegetation in street canyons) have demonstrated to be more efficient, cost-effective, adaptable, multi-purpose and long-lasting than the so-called 'grey infrastructure' alternatives (e.g. Gill et al., 2007; Pugh et al., 2012; Ellis, 2013; Flynn and Traver, 2013; Raje et al., 2013).

In order to support the implementation of innovative nature-based solutions in environmental management and land use planning, valuation becomes essential. Valuation can refer to monetisation (assessing a monetary value) or to an estimation of worth or importance (Dendoncker et al., 2014). In this case, valuing for sustainability and for environmental decision-making requires to account for ecological, social and economic aspects, which are considered the three pillars of integrated valuation

(Boeraeve et al., 2014; Dendoncker et al., 2014; Gómez-Baggethun et al., 2014). One of the possible methodologies to achieve value integration is multi-criteria analyses (MCA). MCA is a framework for exploring and ranking the performance of alternative decision options according to multiple objectives (Belton and Stewart, 2002; Hajkowicz and Collins 2007). It can combine a wide assortment of information (e.g. qualitative and quantitative) and opinions. The approach has been largely applied for water resource management (Hajkowicz and Collins, 2007). MCA establishes preferences between options (usually identifying the most preferred option) by reference to an explicit set of objectives for which it has established measurable criteria (Department for Communities and Local Government, 2009). Their aim is to simplify handling complex information to take difficult decisions in a consistent way.

In this study we assess the benefits of a multi-purpose nature-based solution for water pollution control in a peri-urban area located in Gorla Maggiore (northern Italy), using an ecosystem service approach and applying an integrated valuation based on MCA for local water management. This solution is compared with the alternatives "doing nothing" and with the construction of a conventional grey infrastructure. This case study gives an example of integrating different value systems and stakeholders› viewpoints, thus providing hands-on guidance for integrated valuation in ecosystem service assessments linked to (water) decision-making.

STUDY AREA: THE ALTERNATIVES

The study area is located in Gorla Maggiore, a small municipality in northern Italy (Fig. 1). Gorla Maggiore is one of the case studies of the EU FP7 project OpenNESS (http://www.openness-project.eu/). This project has 27 case studies across Europe to test practical solutions that integrate the concepts of natural capital and ecosystem services into land, water and urban management. In particular, the main objectives of the Gorla Maggiore case are:

- To investigate all the benefits that a neo-ecosystem could provide in terms of ecosystem services (water purification, flood regulation, natural habitat, recreation).

- To compare the green infrastructure (water park) with other conventional grey infrastructures and with the previous situation (a private poplar plantation).
- To integrate the ecosystem service approach in the decision-making process and in river basin management plans, through the direct involvement of the stakeholders.

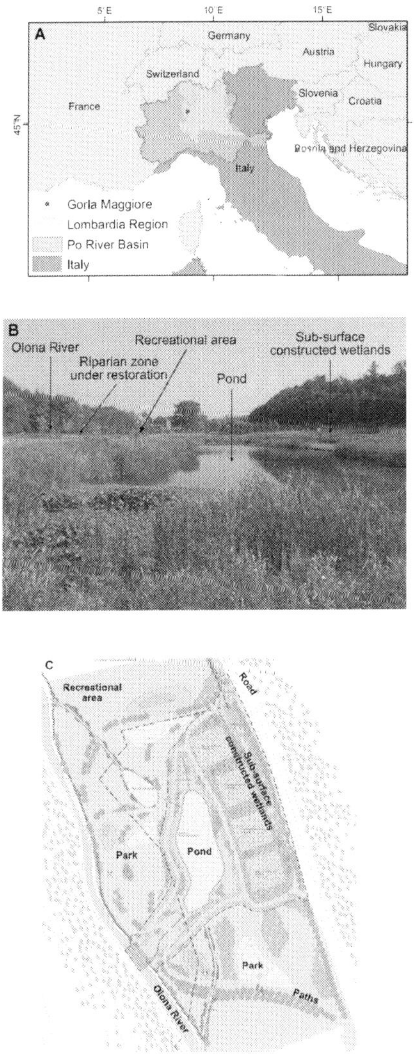

Figure 1. Location, illustration and map of the study area with the present nature-based solution.

Gorla Maggiore hosts an innovative nature-based solution to treat water pollution from the adjacent urban area. It consists of a green infrastructure that was specifically designed to treat the water discharged into the Olona River by the Combined Sewer Overflow (CSO), i.e. the excess flow of mixed sewage and rainwater that cannot be treated in the waste-water treatment plant during heavy rain events. These overflows are relatively common in the site (70 events were registered between March and August 2014, Masi et al., 2016) and contain not only storm water but also untreated human and industrial waste, toxic materials, and solids.

Since constructed wetlands are starting to be considered as an eco-suitable technology to treat CSO (Meyer et al., 2013), the construction of the Gorla Maggiore water park has been funded by the regional government and a private foundation (Regione Lombardia and Fondazione Cariplo) as a trial to test the feasibility of constructed wetlands to treat the CSO. The treatment of the overflows during rainy events is a critical issue in the Lombardy Region, since there are several thousands of CSOs that contribute significantly to the overall pollution load to surface water. To tackle the problem a regional law (R.R. n.3 from 24 March 2006), compliant with the EU Water Framework Directive, limits the pollutant load discharged by the CSO.

In this paper we compare and value the benefits provided by three different alternatives in Gorla Maggiore. The alternatives represent the most realistic options for water pollution control in the site facing the pressure from the urban area:

1. The present Gorla Maggiore water park, called here the green infrastructure. The water park is a constructed ecosystem of about 9 ha built on the Olona River bank during 2011–2012. It includes (a) a pollutant removal area composed of a grid, a sedimentation tank and four vertical sub-surface flow constructed wetlands; (b) a multipurpose area with a surface flow constructed wetland or pond with multiple roles, such as pollution retention (secondary and tertiary treatment), buffer tank for flood events, maintenance of biodiversity and recreational area; and (c) a recreational park with restored riparian trees, green open space, walking and cycling paths

and some services (e.g. picnic table, toilets, bar) maintained by a voluntary association (http://www.calimali.org/).

2. The standard solution used to treat waters from CSOs: a first-flush and a buffer tank, called here grey infrastructure. It comprises an underground first flush storage tank of over 1000 m^3, from which the water can be pumped to the wastewater treatment plant within the 48 h after moderate rainfalls; and a dry retention pond for second flush (open-air), where the excess water during heavy rainfall can be stored until naturally drained. The size and technical characteristics of the grey infrastructure correspond to the legal requirements in Regione Lombardia.
3. Keeping the previous situation or "doing nothing": the poplar plantation. It replicates the previous private use of 4.2 ha for productive forestry, in particular for young poplars that are relatively common along the Po valley.

MATERIALS AND METHODS

We applied a multi-criteria analysis (MCA) based on the analytic hierarchy process (AHP), a well-established methodology developed in the 1970's (Saaty, 1980) that deals with multi-criteria decision-making and allows for the participation of multiple stakeholders (including individual and group preferences). We combine the MCA with an ecosystem services approach since most of the criteria identified are actually ecosystem services. AHP is a useful tool based on mathematical and psychological fundamentals to analyse complex decisions, involving multiple stakeholders and multiple alternatives. It applies a hierarchical structure that facilitates the definition of priorities and preferences in decision-making (Saaty, 1980). The application of the AHP to get a functional MCA for Gorla Maggiore involved the following steps: 1) identify the problem and structure it as a hierarchy; 2) assess the different alternatives; 3) pairwise compare and judge the elements of the hierarchy; 4) estimate the overall priority values to rank the alternatives; and 5) run a sensitivity analysis of weights.

Structure the Problem

Stakeholders

To structure the problem the view of different stakeholders is fundamental. Stakeholders are actors (individuals, groups or organizations) whose interests may be affected, positively or negatively, by a specific public decision or policy. Three main groups of stakeholder associated with the problem and defined in the preliminary research were involved in this study case, totalling 19 participants:

- Water management institutions (local & regional governments, environmental regulators and water management authorities).[4] These managers or representatives from institutions are responsible for the implementation of environmental or water regulations at river basin, regional or local scales.
- Scientific & technical experts (university professors, consultants & smes, scientists).[5] These technicians are some researchers involved in the Openness project, the SME in charge of building and monitoring the green infrastructure or other university professors and consultants specialised in grey infrastructures.
- Local actors & ngos (local experts, citizens' associations, environmental associations, municipalities).[6] These local stakeholders are assumed to represent the local views.

The choice of stakeholders was designed to cover all groups that may be affected by or involved in the project and those who can take decisions to implement such a project. Thus, the stakeholders cover: (1) the key water managers or decision-makers according to the Italian administrative framework (they have high influence and from low to high interest); (2) experts of both conventional grey infrastructures and green infrastructures to treat CSO that may support and evaluate the different alternatives, together with economic, environmental and statistical experts able to assess the project (in general they have high interest and low influence); and (3) representatives of social groups affected by the project decision and potential users of the area (in this case study they have low influence and high interest, those with low interest refused the invitation to take part of the study).

Problem Hierarchy

The first step in an AHP is to structure the problem as a hierarchy. At the uppermost level is the goal(s) or objective(s) to be achieved, followed by a series of criteria and sub-criteria that contribute to attain the main goal. At the bottom level we find a number of alternatives that should be evaluated with respect to the criteria and sub-criteria. Choosing the appropriate criteria and sub-criteria is the main challenge in a MCA. They are site and context specific, and they should be designated by all the involved stakeholders (Garfì et al. 2011).

Fig. 2 illustrates the final AHP hierarchy defined in Gorla Maggiore. It is the result of some literature review (e.g. Ellis, 1998, Ananda and Herath, 2003, Martin et al., 2007, Chávez et al., 2012, Xu et al., 2014, Kukrety et al., 2013, Kurka and Blackwood, 2013, Petrini et al., 2016), a detailed analysis of the case study, two dedicated meetings with the stakeholders, and further bilateral dialogues with some of the stakeholders (in particular with the technical responsible for the infrastructure in the town hall, a local expert and the regional environmental agency).

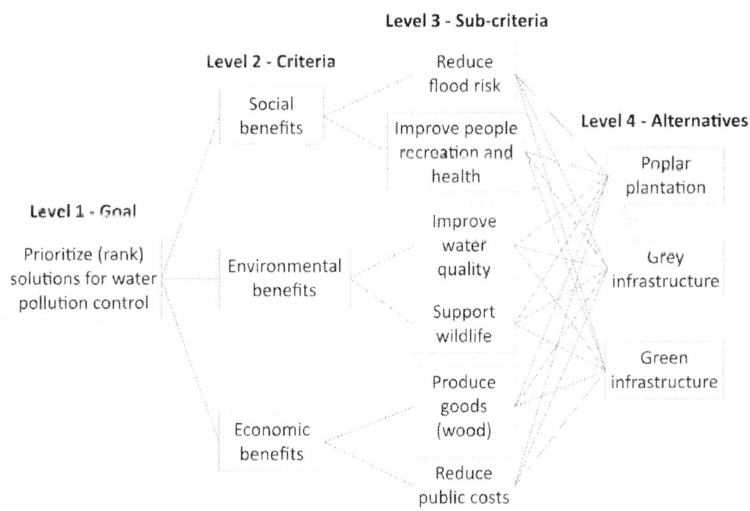

Figure 2. The four levels of hierarchical structure used in this case study. Note that five of the six sub-criteria are ecosystem services.

The social process followed to build the hierarchy included the following steps:

- The main goal of the study was planned by the authors and the regional government (the main funder of the infrastructure) and shared by the rest of stakeholders since the beginning of the project.
- The sub-criteria (basically, the ecosystem services) included in the analysis were selected by all the stakeholders through a dedicated meeting in October 2013 and, hence, they represent the main interests of the stakeholder group. Naming and classification of the sub-criteria were further elaborated by the authors (in an inclusive manner) to avoid ambiguity and double-counting. Then, they were approved by the rest of stakeholders through bilateral discussions and during another meeting in January 2015.
- The selection of and grouping by criteria was an academic exercise performed by the authors. They covered the three pillars of an integrated valuation, but they were neither planned nor discussed within the stakeholders group. That is, the main interests expressed by the stakeholder group naturally covered ecological, social and economic aspects.
- The alternatives were suggested by the authors based on tangible scenarios under the case study context. Their main narratives were shared and accepted by the all the stakeholders during the meeting of October 2013. The technical descriptions were further itemized and modified to include all stakeholders comments after the meeting of January 2015. Similarly, the indicators were proposed by the authors based on suitability and feasibility, and they were revised and in some cases adjusted after the discussion of January 2015. Some of the stakeholders, notably the regional government, the water agency, the municipality and the experts, contributed actively to the design and measurement of indicators and alternatives.

In this case the goal represents the main objective to construct the site, water pollution control, and how to select the best option. The second level includes broader priorities for public policies, actually integrating the concepts of natural capital and ecosystem services into water and urban management. It refers to criteria dealing with the environmental, social and economic benefits. The third level of the hierarchy had originally 11 sub-criteria that were further discussed and grouped into the final 6 sub-criteria of Fig. 2 which represent the main areas of concern. It is recommended that the number of elements in a level do not exceed 10 (Kukrety et al., 2013). Keeping a reduced number of sub-criteria helps avoiding confusion due to a large number of comparisons and, consequently, decreases the uncertainty of the process. Five of the six sub-criteria are benefits or ecosystem services provided by the Gorla Maggiore water park (water purification, flood protection, maintenance of natural habitats, recreation and wood extraction) while the sixth one is an analysis of public costs. Each of the sub-criteria is then measured against a number of indicators shown in Table 1.

Table 1. Criteria, sub-criteria and indicators selected for this case study. For a complete description of the indicators see Section 3.2.

Criteria	Sub-criteria	Indicators
Social benefits	Reduce flood risk	Peak flow reduction Reduction of flooding down-stream
	Improve people recreation and health	• No. of visitors/users • Frequency of visits
Environmental benefits	Improve water quality	• Load reduction of dissolved organic carbon • Load reduction of nitrogen
	Support wildlife	• Expert judgment about biodiversity • Landscape diversity
Economic benefits	Produce market goods	• Value of wood production
	Reduce public costs	• Total construction costs • Total maintenance costs

Finally, the fourth and bottom level of the hierarchy includes the alternatives, in terms of water pollution control implementations, that might be more adequate for the success of the public policies. The alternatives, already described in Section 2, represent the previous situation or "doing nothing" (a poplar plantation), the conventional solution used to treat waters from CSOs (the grey infrastructure), and the present Gorla Maggiore water park (the green infrastructure).

Alternatives Assessment

The performance of each alternative under the six sub-criteria was assessed based on the indicators of Table 1. As already explained in Section 3.1.2 (the steps of the social process), the indicators were proposed by the authors based on their relevance and feasibility to be measured under the three alternatives, and they were further refined through an iterative process with the stakeholders. Actually, some of the stakeholders were instrumental to monitor the site and perform the assessments, or to provide feedback on the measured data. Some of the indicators are evaluated with a quantitative scale (e.g. cost) and other with a qualitative scale (e.g. expert judgment about biodiversity).

The assessment of alternatives took one year of monitoring and research (roughly the year 2014). Due to the small scale and the specificity of the case study, each of the indicators has a specific quantification method, as follows:

- Peak flow reduction (%): this indicator estimates the reduction of the maximum peak flow of the hydrograph generated by a rain event with return time of 10 years. The hydrograph is derived from the design data of the CSO-constructed wetland with a runoff model. The value for the grey infrastructure was calculated according to the Lombardia Region guidelines for buffer tanks. The green infrastructure value was estimated from the modelling simulations performed in Rizzo et al. (2015), which account both the pond (buffer tank) and the vertical flow bed lamination effects. No lamination capacity has been assumed for the poplar plantation since the CSO was directly discharged to the river before the construction of artificial wetlands.
- Reduction of flooding downstream (m^3): reduction of the water volume flowing downstream from a rain event with return time of 10 years. The methodology followed to measure this indicator is the same of the previous one.
- Load reduction of dissolved organic carbon (t/yr): amount of dissolved organic carbon avoided to be discharged in the river, contributing to the improvement of the river water quality. The load reduction for the green infrastructure was directly

measured in 68 registered CSO events from February 2014 to February 2015 using a mass balance approach (Masi et al., 2016). The load reduction for the grey infrastructure is derived from the same mass balance approach based on the first flush volume requested by the Lombardia region law, equal to 989 m^3 for the CSO of Gorla Maggiore, and assuming no load reduction capability of the buffer tank. No removal efficiency has been assumed for the poplar plantation, since the CSO was directly discharged to the river in the years before the construction of artificial wetlands.

- Load reduction of nitrogen (t/yr): amount of nitrogen avoided to be discharged in the river, contributing to the improvement of the river water quality. The value is expressed in tonnes of N-NH_4^+ per year. The methodology followed to measure this indicator is the same of the previous one.
- Expert judgment about biodiversity: a biologist and an ecologist assessed qualitatively the expected community diversity and richness in a monoculture plantation with standard farming procedures (for the poplar plantation), in a managed grassland (for the grey infrastructure), and in the set of constructed wetlands (green infrastructure). Biological sampling of the site (macroinvertebrates, macrophytes, birds and amphibians) supported the assessment of the green infrastructure. This indicator has 5 categories: very low, low, medium, high and very high.
- Landscape diversity: This is the Shannon›s diversity index of the habitats of each alternative estimated with the software ArcGIS and Fragstat. For the poplar plantation and the green infrastructure, habitat mapping was based on existing satellite images of the site. Habitats of the grey infrastructure were inferred from similar infrastructures and adapted to the site.
- No. of visitors/users and frequency of visits: The number and frequency of local visits in the water park was estimated through a mail survey distributed in Gorla Maggiore that got 71 responses. It is considered that there is no access to the private land under the poplar plantation scenario. The grey

infrastructure is assumed to have less visits than the green infrastructure due to the lack of biodiversity and related educational facilities, but the surrounding recreational park can still attract visits. These indicators have 5 categories from very low to very high.

- Value of wood production: The area of poplar plantation was mapped from satellite images while the average regional plant density, productivity, mass volume, wood price and investment costs were derived from the literature (e.g. Istituto di Sperimentazione per la Pioppicoltura, 2001; Fondazione Lombardia per l'Ambiente, 2008;Nervo et al., 2011). We estimated the net profit from the wood harvest being ca. 255 euro per ha and year, and assumed a 20 years rotation period. The other alternatives have no wood production.
- Total construction costs and total maintenance costs: The actual construction and maintenance costs (for a 20 years lifespan) of the green infrastructure were reported by the funders, the town hall and the construction company. The costs of the grey infrastructure were estimated from other existing infrastructures by the construction company. For this study the poplar plantation has no public costs, even if under the regional legislation "doing nothing" to treat the CSO would lead to an infraction and a fine for the municipality.

Pairwise Comparisons

A crucial step in the AHP methodology is the pairwise comparison judgments among the elements at one level of the hierarchy. The AHP comparison is usually performed in each level of the hierarchy (e.g. sub-criteria) linked to the previous level (e.g. criteria), but due to time-constraints during the meetings and taking advantage of the clarity of the attributes, we preferred to compare only the sub-criteria of level 3. The question posed was: Regarding the implementation of water pollution control solutions, which sub-criteria do you think should be given more importance? How much more?

The pairwise comparison survey was carried out during the meeting in January 2015 in the headquarters of the regional government (Regione Lombardia), Milan (Italy). The sub-criteria, indicators and alternatives were technically described to the participants. Then each participant was asked to indicate the relative preference of each sub-criteria over another, based on either the point of view of the public or private institution they were representing, or on their personal judgment (perception). Qualitative (verbal) comparisons among the 15 combinations (all possible discrete combinations of 6 sub-criteria) were converted into quantitative values by using a numerical scale of integers ranging from 1 to 5 (Table 2).

Table 2. Scale to represent the intensity of preference

Intensity of importance	Definition (verbal scale)	Explanation
1	Equal importance	Two sub-criteria contribute equally to the objective
2	Slightly preferred	Experience and judgment slightly favor one sub-criteria over another
3	Preferred	Experience and judgment strongly favor one sub-criteria over another
4	Strongly preferred	A sub-criteria is favored very strongly over another; its dominance is demonstrated in practise
5	Extremely more important	The evidence favoring one sub-criteria over another is of the highest possible order of affirmation

Each one of the comparison or judgment matrices assumes the form:

$$A = \begin{bmatrix} a_{11} & a_{12} & \cdots & a_{1n} \\ a_{21} & a_{22} & \cdots & a_{2n} \\ \vdots & \vdots & \cdots & \vdots \\ a_{n1} & a_{n1} & \cdots & a_{nn} \end{bmatrix}$$

where a_{ij} represents the pairwise comparison rating for attribute i and attribute j. Given the reciprocal property of the matrix, if $a_{ij}=x$, then $a_{ji}=1/x$ where $x\neq0$ (i.e. if sub-criteria i has one of the above numbers assigned to it when compared with sub-criteria j, then j has the reciprocal value when compared with i). Mathematically, the principal eigenvector for each matrix, when normalised, becomes the vector of priorities for that matrix (Saaty, 1980). This method also allows controlling potential inconsistent responses (Kukrety et al., 2013). The consistency ratio (CR) is calculated based on properties of reciprocal matrices (Saaty, 1980). Values of CR≤0.1 are considered as acceptable. The consistency ratio equal or below 0.1 was checked for all judgments, getting 18 valid responses from the 19 stakeholders present in the exercise. This means that one response out of 19 was considered inconsistent and was discarded. The individual judgments (decision matrices shown in the supplementary material of this paper) were then aggregated by stakeholders' group and all together, calculating the geometric means (Saaty and Vargas, 2012).

Synthesizing Priorities

The pairwise comparison judgments are used together with the alternatives assessment to develop overall priorities for ranking the alternatives. The overall priority values are calculated by multiplying the priority of an element (in this case the sub-criteria weights) by the value(s) attributed to it (in this case a normalised average of their indicators). The products are then summed across each branch of the hierarchy using a simple additive function.

To integrate all these values some normalisation and weights are necessary. First, the input data for each alternative (indicators) must be transformed to a comparable scale, since they measure attributes in different scales, intensities and type of data (both numeric and categorical). All indicators are transformed to relative scale using an "ideal" mode, where the "ideal" alternative (or maximum score) is the best performant value of each indicator. The scores range from zero to one. We assume a linear distribution of values for this normalisation (minimum-maximum). When two indicators exist per sub-criteria, both are weighted equally (50%) to estimate an average score per sub-criteria. This decision is based on expert opinion, considering that all indicators have similar representativeness and it would be less biased to consider them equally. The attributes were considered too technical to ask for weights of indicators to the stakeholders. The only exception is the sub-criteria 'public costs' where, based on their magnitude, the weight of the indicators 'construction costs' and 'maintenance costs' was 95% and 5% respectively.

The preliminary results were calculated in the stakeholders meeting of January 2015 and were shown to all participants. These global priorities helped synthetize information, rank alternatives and challenge thinking in the case study.

Sensitivity Analysis

In the context of the MCA presented in this study, sensitivity analysis reveals which criteria weights are most likely to change the outcome in selecting an alternative water pollution control, to investigate the extent to which disagreements between people may affect the final overall decision. Sensitivity analyses are typical run by holding all the variables (weights in this case) constant in a model, except for the one being tested, and varying that variable over its likely range of values. Such tests are referred to as one-way or one-at-a-time sensitivity analyses. In this case, since the weights have a linear relationship with the final scores of each alternative, a simple and very effective approach is iteratively combine the extreme weighs (maximum and minimum) proposed by the stakeholders and analyse what would be the final performance of each alternative under the new weights combination. Considering six sub-criteria, the number of combinations is 2^6–64.

Another possible source of uncertainty is the impact of changes on the value of indicators (i.e. on the assessment of alternatives). However, the level of uncertainty of the assessment is considered very low, since one of the key characteristics of this study is that the infrastructures were already built (i.e. this is an *ex post* assessment) and most of the indicators could be quantified based on existing data. In particular, the poplar plantation existed before, the green infrastructure could be monitored and the grey infrastructure could be observed and extrapolated from other sites.

RESULTS AND DISCUSSION

Integrated Valuation of Alternatives

The three alternatives envisaged for the MCA and described in Section 2 (i.e. the poplar plantation or "doing nothing", the conventional grey infrastructure, and the presently existing green infrastructure) are scored according to the establish criteria, sub-criteria and indicators of Fig. 2 and Table 1. In general, the poplar plantation has the lowest values in all sub-criteria but the wood exploitation (Table 3), where the net benefits would be private. Still it has more habitats' diversity than the grey infrastructure. One may argue that the poplar may reduce flooding and pollution to a certain extent, but the magnitude would be negligible compared with ad-hoc infrastructures, especially because the CSO would not flow to the poplar. This option involves "doing nothing" to treat the CSO, which may actually have negative public cost effects (in the form of penalties) in the near future linked to the non-compliance with the regional water legislation. We did not include it in the public cost scores due to lack of available quantifications.

Table 3. Assessment of alternatives, also called performance matrix. The percentage of change between alternatives is calculated with respect to the maximum value per indicator, with a negative sign indicating a decrease from the original situation to the new one.

	Alternatives			**Changes between alternatives (%)**	
Indicators	Poplar	Grey	Green	From poplar to grey	From poplar to green
Peak flow reduction (%), return time of 10yr	0	80	86	93	100
Reduction of flooding downstream (m3), return time of 10yr	0	8100	8900	91	100
Load reduction of dissolved organic carbon (t/yr)	0	9.5	11.7	81	100
Load reduction of nitrogen (t/yr)	0	0.2	0.4	50	100
Expert judgment about biodiversity	low	low	high	0	50
Landscape diversity (Shannon's diversity index)	1.89	1.85	2	−2	6
No. of visitors/users	very low	medium	high	50	75
Frequency of visits	very low	medium	high	50	75
Value of wood production (profit from harvest in EUR)	21420	0	0	−100	−100
Total construction costs (EUR)	0	844750	900000	94	100

Total maintenance costs per 20yr expected lifespan (EUR)	0	27824	29590	94	100

The grey and the green infrastructures have a similar performance in the water-related indicators, with a slight advance of the green infrastructure, more important for pollution reduction indicators where the green option performs significantly better than the grey one (Masi et al., 2016). Construction and maintenance costs are quite similar, in this case resulting slightly more expensive in the green option. All these data come from the existing experience either in Gorla Maggiore or in similar conditions and, thus, can be considered as precise estimations. The foremost difference among the two infrastructures is their performance in terms of wildlife support and recreation. The green infrastructure provides substantially larger benefits as natural habitat and green space for recreation. The surface area of the grey infrastructure could be covered by managed grass, which is not a very diverse habitat, and it could retain muddy sediments and odours after the heavy rains.

The weighting of sub-criteria provided by the stakeholders' group (Table 4) was relatively homogeneous, but still we can appreciate some patterns. Based on the personal weighting factors attributed to each sub-criteria during the pairwise comparison, there seems to be three main interest groups: managers/institutions, technicians, and local stakeholders (see Section 3.1.1).

Table 4. Descriptive statistics of the weights assigned to each sub-criteria by the stakeholders (%).

	Individual scores or weights				
Sub-criteria	Minimum	Maximum	Mean	Median	St. Dev.
Flooding	7.1	36.5	22.1	22.0	12.9
Pollution	8.7	38.3	26.9	26.4	20.3
Wildlife	10.8	35.3	21.4	20.7	14.3
Recreation	6.2	33.2	13.7	11.8	7.1
Wood	3.2	6.8	4.8	4.4	3.6
Costs	4.5	26.6	11.2	8.4	5.9

Managers tend give more importance to the costs (even if to a moderate extent) and to water-related benefits, especially pollution control for which the infrastructure was build, while they give less importance to recreation (Fig. 3). Technicians value similarly flood protection and pollution control and provide the less weight to wood production and public costs. Surprisingly, local stakeholders put the highest relevance in wildlife support. They are also the ones attaching the highest weight to recreation (Fig. 3).

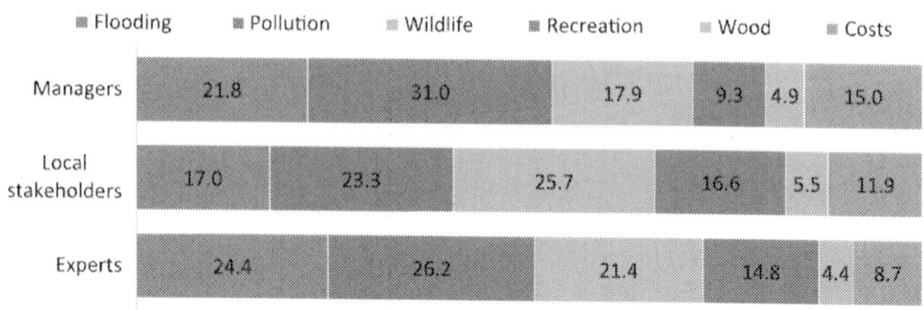

Figure 3. Preference elicitation of the sub-criteria by stakeholders' groups (%). (For interpretation of the references to color in this figure legend, the reader is referred to the web version of this article.)

We have tested the MCA model with the three groups of stakeholders, getting very similar results for the prioritisation of options. Here we just present the integrated results from all the participants. In general, the higher weight was assigned to the criteria of water pollution control, as it is the general scope of the infrastructure. Reducing the flooding risk and providing natural habitat scored very similar as second most important criteria. Then, in decreasing order, the stakeholders considered important recreation, the reduction of public costs and wood exploitation.

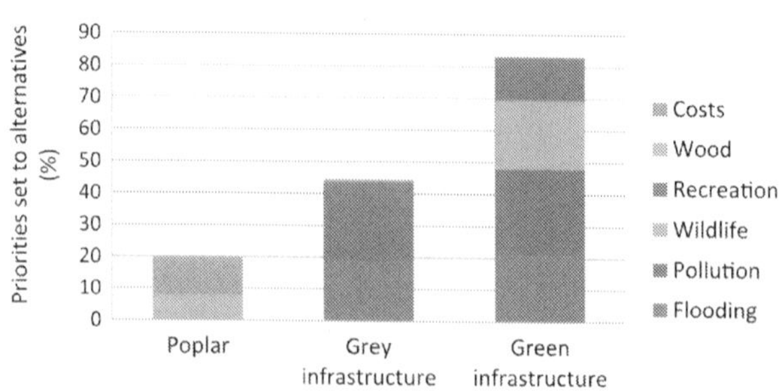

Figure 4. Comparison of alternatives based on the results of the MCA (score in %). (For interpretation of the references to color in this figure legend, the reader is referred to the web version of this article.)

The best option under the MCA is the green infrastructure, followed by far by the grey infrastructure and finally the poplar plantation (Fig. 4). The combination of weights and scores highlights that the largest differences between options are linked to pollution control and natural habitat.

Results of the Sensitivity Analysis

The results of the one-at-a-time sensitivity analysis performed for the global weights proposed by the stakeholders are shown in Fig. 5 and in Supplementary material. It shows how the final score of each of the alternatives varies according to changes in the weights of each of the sub-criteria considered, using the minimum and maximum weights given by all stakeholders. For example, it is observed that the final score of the poplar alternative has almost no variation related to safety, water quality or recreation. This is due to the zero or near-zero utility value of this alternative in relation to those sub-criteria. Something similar happens with the green alternative and the sub-criteria wood production and public costs. These sub-criteria have little effect on the final score of the alternative because it is the most expensive or the less (wood) productive one.

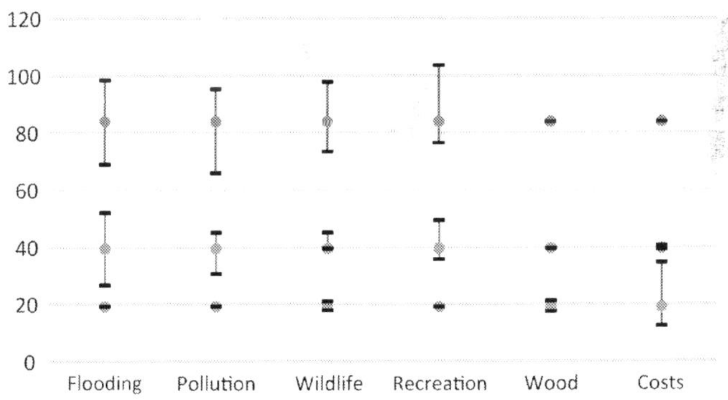

Figure 5. Variation of the final alternative priority (score in %) depending on each single sub-criterion. The central values (green circles for the green infrastructure, grey circles for the grey infrastructure and orange circles for the poplar plantation) point to the global score shown in Fig. 4. Error bars represent the influence of minimum and maximum weights. (For interpretation of the references to color in this figure legend, the reader is referred to the web version of this article.)

Regarding a higher dimension sensitivity analysis (more than one variable at a time), we compared the maximum and minimum scores for the three alternatives changing a single sub-criterion at a time while keeping the rest stable. Fig. 6 shows the final score of each alternative under the 64 combinations generated. We observe that the best alternative is always the green infrastructure. Its minimal score (65.8, referring to changes in the weight of water quality) remains significantly higher than the maximum score of the grey infrastructure (52.2, varying the importance of flood protection). Only once (see the combination 4 of Fig. 6), when taking the public costs as the most relevant factor, the poplar plantation beat the green infrastructure as the best option. This result is logical if we analyse the values of the performance matrix, where the green infrastructure has systematically (with two little exceptions) the highest punctuation regardless of the weights assigned to the criteria. The total score of the poplar alternative varies following a saw tooth form which depends on the value weight of the public costs sub-criterion. These peaks are not revealed in the grey and green alternatives where this sub-criterion has a minor influence.

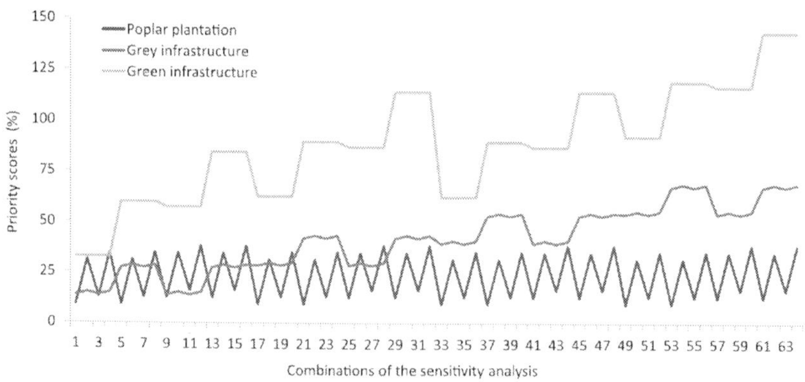

Figure 6. Ranking of alternatives under the 64 combinations of a high dimension sensitivity analysis. The priority scores set the priorities of the three alternatives in each combination of values of sub-criteria. Combination 1 assigns the weights of all the sub-criteria to their minimum value while combination 64 assigns all weights to their maximum value. The scores show a rising trend with the increasing weights (especially for flood protection, water quality, recreation and wildlife support). This rising trend is greater for the green alternative. (For interpretation of the references to color in this figure legend, the reader is referred to the web version of this article.)

Benefits from the Green Infrastructure

Since the green infrastructure (i.e. the constructed wetlands and wet retention pond with a surrounding riparian park) is consistently identified as the best option for the case study, we deepen on the analysis of the socio-economic and environmental benefits it provides. The Gorla Maggiore water park is here analysed under an ecosystem service approach which, together with the MCA, allows evaluating its multiple benefits. In particular the ecosystem services quantified are water purification, flood regulation, natural habitat and recreation.

To analyse the actual water purification, two automatic sampling stations allowed to measure the pollution (organic matter, through the chemical oxygen demand, and ammonia concentration) entering and leaving the constructed wetlands during each CSO event. The system shows average removal efficiencies of 72–96% which confirms a satisfactory performance for water pollution control of CSO, especially in the vertical sub-surface flow constructed wetlands (Masi et al., 2016). The average load reduction is shown in Table 3.

Related to flood control, the mathematical model simulates the unsaturated water flow in the vertical sub-surface constructed wetlands' beds (Richards equation), the height of the ponding layer above those wetlands, and the height of the free water surface constructed wetland (mass balance equations). Modelling results show that the pond multipurpose area acts as a buffer tank with good lamination also obtained at high return time (return time 10 years, maximum flow rate 3.4 m^3s^{-1}, volume 11,497 m^3s^{-1}, duration 4.8 h). The free water surface constructed wetland works as a buffer tank to store 71% of the influent volume, with a less but not negligible role in lamination of vertical sub-surface flow beds (11%). Also the peak flow (86.2% of reduction) and the outflow duration (27.3 times higher than the CSO event duration) are satisfactorily managed during the 10 year return time events. The system is an effective bio-retention technology to manage the urban runoff towards the river (Rizzo et al., 2015). Regarding downstream flood protection, the effect of the Gorla Maggiore water park seems negligible at the catchment level (at large scale) due to the relative small volume of the infrastructure related to the Olona water discharge. Still, the combined effect of multiple similar infrastructures along the Olona River should be investigated, since the accumulated retention could be important.

The maintenance of healthy biota populations and habitats, usually acting as nursery or shelter areas, is crucial for wildlife support. Two field surveys run during 2014 indicate a high capacity of the constructed wetlands and surrounding park to provide wildlife support. Numerous macroinvertebrate taxa were identified in the pond, among which some Odonata (dragonflies and damselflies) that could be part of the Red List of Threatened Species. The pond also seems to play a crucial role in the survival of local amphibian and to support bird populations. One priority species for the EU Birds Directive was identified on the site. The littoral zone of the pond is characterized by dense beds of emergent macrophytes covering ca. 10% of the surface. Attached to them, floating-leaved species cover ca. 15%. These introduced plants contribute to the further purification of the water outflowing from the reed bed system and provide substantial benefits in relation to microhabitats for fauna and an aesthetic appearance to the whole system.

In terms of recreation, the water park has several facilities such as information panels, well-maintained paths and toilets. The accessibility from the town is excellent and it is the only green open space in Gorla Maggiore. People can practise a wide range of activities including educational, sportive and leisure activities. According to the responses collected in our questionnaire from the local residents, the park is highly visited by people living at close distance (less than 1.5 km) and it is primarily used for walking/dog walking or for sightseeing/enjoying nature. On average, each respondent has visited the park twice per month, usually accompanied by 1–2 persons. Other less practised activities in decreasing importance are running or biking, watching wildlife, educating children to nature, playing with kids, sunbathing and picnicking (Reynaud et al., 2016). The level of appreciation of the users of the park is between moderate and high.

Impacts and Transferability of the Results

The ecosystem services delivered by wetlands have been recognized and valued as a nature-based solution for watershed management issues (ten Brink et al., 2012). The main finding from this research is that the green infrastructure (constructed wetlands and park) performs equal or even better than the grey alternative for water purification and flood protection, it has a similar cost, and it provides additional benefits (wildlife support and recreation) specially valued by the local residents and stakeholders. This is relevant for regional water management, where biodiversity and recreation are not usually taken into account for water resources decision-making. This is in line with a similar study that analyses nature-based and collaborative solutions to future water scarcity (Reddy et al., 2015). They also find, with the use of ecosystem service valuation methods and MCA, that nature-based solutions may be cost-effective and cost-competitive for businesses, while providing public and ecosystem benefits (Reddy et al., 2015).

In this case study the MCA is used retrospectively, to evaluate an already built infrastructure (*ex post* evaluation) but our results could also be used prospectively to appraise new proposals of constructed wetlands under similar circumstances. In particular, AHP is a tool that can help decision-makers translate stakeholders' desires and expectations into beneficial public policies or decisions. Our findings can be particularly useful for similar situations: small municipalities aiming to treat their CSO as requested, for instance, by the EU Water Framework Directive. New alternatives (variants of green and grey infrastructures) can be explored based on this experience to more accurately choose the type of installation convenient in each context. Our methods and results could offer practical tools to use the concept of ecosystem services, to select the best option between a multi-purpose green infrastructure and grey alternatives, to apply a cost-benefit analysis (with further economic analysis), and to communicate with the stakeholders and local community, increasing the awareness about the benefits provided by new or restored ecosystems.

In this case study, the environmental quality of the Olona river falls under the Po river basin management plan responsibility of the "Autorita di Bacino del Fiume Po". However, the land and water resource planning is under the authority of the "Regione Lombardia". This regional government has adopted River Contracts as local strategic planning instruments in different sub-basins, among which the Olona-Bozzente-Lura,[7]involving local authorities and institutional stakeholders most of which are represented in our stakeholders group. The most recent programme of actions was adopted in 2014.[8] The strategic objectives of the River Contract are: 1) reduction of water pollution; 2) reduction of flood risk; 3) restoration of landscape, environmental and urban systems relative to river corridors; and 4) sharing of information and knowledge on water. Thus, the ecosystem services approach and the MCA developed in this study address the policy needs (strategic objectives) for the local territorial planning in the implementation of the Water Framework Directive.

CONCLUSIONS

Water-related projects usually tackle complex problems and provide a series of "unmeasurable" benefits, most of which can be assessed from an ecosystem service perspective. They also involve multiple interested actors, from governments to NGOs or lay citizens. MCA allows to evaluate and integrate, both quantitative and qualitatively, a set of crucial criteria with or without monetary value attached. The use of an ecosystem services' approach linked to a MCA for water projects (like the one shown in this paper) may lead to more informed and better decisions. In particular, the case study shown in this article contributed to raise the awareness of the stakeholders on the significance of aquatic habitats and encourage them to promote or implement policies that protect freshwater ecosystems together with people's wellbeing.

In general, the nature-based solution constructed in Gorla Maggiore contributes to: (1) mitigate some water-related issues of the municipality, potentially increasing the ecological status of the Olona River; (2) maintain and even improve biodiversity in the area; and (3) contribute to the residents' livelihood through recreational and educational services.

The application of the *ex post* MCA led to the identification of the green infrastructure as the best among three alternatives. The costs of the works are significant, but do not exceed those of a traditional grey infrastructure.

More broadly, the MCA highlighted the multi-functionality of the green infrastructure, accounting for multiple benefits of interest for the stakeholders. The involved actors learnt about a possible way to structure the decision process and about their own and others› perspectives. Hence, MCA brings a degree of structure, analysis and openness to complex decision that go beyond the practical reach of cost-benefit analysis.

At the same time, by integrating several criteria in the decision analysis (ecological, social and economic), the approach allowed the integration of different policies relevant to the territorial planning. This is of great interest for the regional authorities that are involved in the development and implementation of the River Basin Management Plans under the EU Water Framework Directive.

This study illustrates the effectiveness and multi-functionality of constructed wetlands and restored riparian zones for watershed management, and its capacity to deliver a broad range of ecosystem services. It also demonstrates the utility of integrating different value types (environmental, social and economic) and stakeholders› viewpoints to support environmental decision-making.

Acknowledgements

This research has been funded by the EU FP7 project OpenNESS (Grant agreement no. 308428).

REFERENCES

1. Ananda and Herath, 2003 J. Ananda, G. Herath The use of analytic hierarchy process to incorporate stakeholder preferences into regional forest planning For. Policy Econ., 5 (2003), pp. 13-26
2. Belton and Stewart, 2002 V. Belton, T. Stewart Multiple criteria decision Analysis. An Integrated Approach, Springer, US (2002), p. 372, 10.1007/978-1-4615-1495-4
3. Boeraeve et al., 2014 F. Boeraeve, N. Dendoncker, S. Jacobs, E. Gómez-Baggethun, M. DufrêneHow (not) to perform ecosystem service valuations: pricing gorillas in the mist Biodivers. Conserv. (2014), 10.1007/s10531-014-0796-1
4. ten Brink et al., 2012 P. ten Brink, L. Mazza, T. Badura, M. Kettunen, S. WithanaNature and its Role in the Transition to a Green Economy (A TEEB report) United Nations Environment Programme (2012), p. 59
5. Chávez et al., 2012 M.D. Chávez, P.B.M. Berentsen, A.G.J.M. Oude LansinkAssessment of criteria and farming activities for tobacco diversification using the analytical hierarchical process (AHP) technique Agric. Syst., 111 (2012), pp. 53-62
6. Dendoncker et al., 2014 N. Dendoncker, H. Keune, S. Jacobs, E. Gómez-BaggethunInclusive ecosystem services valuation S.

Jacobs, N. Dendoncker, H. Keune (Eds.), Ecosystem Services: Global Issues, Local Practices, Elsevier, New York (2014), pp. 3-12

7. Department for Communities and Local Government, 2009 Department for Communities and Local GovernmentMulti-criteria analysis: a manual Communities and Local Government, Publications, London (2009), p. 165
8. DG Research and Innovation, 2015 DG Research and InnovationTowards an EU research and innovation policy agenda for nature-based solutions & re-naturing cities Final Report of the Horizon 2020 Expert Group on 'Nature-Based Solutions and Re-Naturing Cities' (full version), Publications Office of the European Union, Luxembourg (2015), p. 70
9. Eggermont et al., 2015 H. Eggermont, E. Balian, J.M.N. Azevedo, V. Beumer, T. Brodin, J. Claudet, B.Fady, M. Grube, H. Keune, P. Lamarque, K. Reuter, M. Smith, C. van Ham, W.W. Weisser, X. Le RouxNature-based solutions: new influence for environmental management and research in Europe Gaia, 24 (4) (2015), pp. 243-248
10. Ellis, 1998 F. EllisHousehold strategies and rural livelihood diversification J. Dev. Stud., 35 (1) (1998), pp. 1-38
11. Ellis, 2013 J.B. EllisSustainable surface water management and green infrastructure in UK urban catchment planning J. Environ. Plan. Manag., 56 (1) (2013), pp. 24-41
12. Flynn and Traver, 2013 K.M. Flynn, R.G. TraverGreen infrastructure life cycle assessment: a bio-infiltration case study Ecol. Eng., 55 (2013), pp. 9-22
13. Fondazione Lombardia per l'Ambiente, 2008 Fondazione Lombardia per l'Ambiente, 2008. Progetto Kyoto Lombardia (volume di presentazione dei risultati della ricerca finanziata da Regione Lombardia, Fondazione Lombardia per l'Ambiente, APAT e ERSAF Lombardia e promossa da Regione Lombardia, Fondazione Lombardia per l'Ambiente e ARPA Lombardia). Available at: 〈www.kyotolombardia.org〉
14. Garfietal.,2011 M.Garfi,L.Ferrer-Martí,A.Bonoli,S.TondelliMulti-criteria analysis for improving strategic environmental assessment

of water programmes. A case study in semi-arid region of Brazil J. Environ. Manag., 92 (2011), pp. 665-675

15. Gill et al., 2007 S. Gill, J. Handley, R. Ennos, S. PauleitAdapting cities for climate change: the role of the green infrastructure J. Built Environ., 33 (1) (2007), pp. 115-133
16. Gómez-Baggethun et al., 2014 Gómez-Baggethun, E., MartínLópez, B., Barton, D., Braat, L., Saarikoski, H., Kelemen, García-Llorente, M., van den Bergh, E.J., Arias, P., Berry, P., Potschin, M., Keene, H., Dunford, R., Schröter-Schlaack, C., Harrison, P., 2014. State-of-the-art report on integrated valuation of ecosystem services. EU FP7 OpenNESS Project Deliverable 4.1., European Commission FP7
17. Hajkowicz and Collins, 2007 S. Hajkowicz, K. CollinsA review of multiple criteria analysis for water resource planning and management Water Resour. Manag., 21 (9) (2007), pp. 1553-1566
18. Istituto di Sperimentazione per la Pioppicoltura, 2001 Istituto di Sperimentazione per la Pioppicoltura, 2001. Criteri oggettivi per la valutazione del pioppeto maturo.
19. Kukrety et al., 2013 S. Kukrety, S. Jose, J.R.R. AlavalapatiExploring stakeholders' perceptions with analytic hierarchy process—a case study of red sanders (*Pterocarpussantalinus* **L.) restoration in India** Restor. Ecol., 21 (6) (2013), pp. 777-784
20 Kurka and Blackwood, 2013 T. Kurka, D. BlackwoodParticipatory selection of sustainability criteria and indicators for bioenergy developments Renew. Sustain. Energy Rev., 24 (2013), pp. 92-102
21. Maes and Jacobs, 2015 J. Maes, S. JacobsNature-based solutions for Europe›s sustainable development Conserv. Lett. (2015), 10.1111/conl.12216
22. Martin et al., 2007 C. Martin, Y. Ruperd, M. LegretUrban stormwater drainage management: the development of a multicriteria decision aid approach for best management practices Eur. J. Oper. Res., 181 (2007), pp. 338-349
23. Masi et al., 2016 F. Masi, A. Rizzo, R. Bresciani, G. ConteConstructed wetlands for combined sewer overflow treatment: ecosystem services at Gorla Maggiore, Italy Ecol. Eng. (2016), 10.1016/j.ecoleng.2016.03.043
24. Meyer et al., 2013 D. Meyer, P. Molle, D. Esser, S. Troesch, F. Masi,

U. DittmerConstructed wetlands for combined sewer overflow treatment—comparison of German, French and Italian approaches Water, 5 (2013), pp. 1-12

25. Nervo et al., 2011 G. Nervo, D. Coaloa, L. Vietto, A. Giorcelli, G. AllegroCurrent situation and prospects for European poplar culture: the role of Italian research Disert. Terc. Congr. Int. De. Sali. En. Argent. (2011)
26. Petrini et al., 2016 M.A. Petrini, J.V. Rocha, J.C. Brown, R.C. BispoUsing an analytic hierarchy process approach to prioritize public policies addressing family farming in Brazil Land Use Policy, 51 (2016), pp. 85-94
27. Pugh et al., 2012 T.A.M. Pugh, A.R. MacKenzie, J.D. Whyatt, C.N. HewittEffectiveness of green infrastructure for improvement of air quality in urban street canyons Environ. Sci. Technol., 46 (14) (2012), pp. 7692-7699
28. Raje et al., 2013 S. Raje, R. Kertesz, K. Maccarone, K. Seltzer, M. Siminari, P. Simms, B. Wood, J.SansaloneGreen infrastructure design for pavement systems subject to rainfall-runoff loadings Transp. Res. Rec., 2358 (2013), pp. 79-87
29. Reddy et al., 2015 S.M.W. Reddy, R.I. McDonald, A.S. Maas, A. Rogers, E.H. Girvetz, J. North, J.Molnar, T. Finley, G. Leathers, J.L. DiMuroFinding solutions to water scarcity: incorporating ecosystem service values into business planning at the Dow Chemical Company's freeport, TX facility Ecosyst. Serv., 12 (2015), pp. 94-107
30. Reynaud et al., 2016 Reynaud, A., Lanzanova, D., Liquete, C., Grizzetti, B., 2016. Going green? Economic valuation of a multipurpose water infrastructure in Northern Italy. Workshop On Non-Market Valuation, 20-21st June 2016, Bordeaux, France
31. Rizzo et al., 2015 Rizzo, A.Boano, F.Revelli, R.Ridolfi, L.Bresciani, R.Masi, F., 2015. Additional ecological services of CSO-CW besides water treatment: modelling CSO-CW behaviour for urban runoff management. In: Proceedings of the 6th International Symposium on Wetland Pollutant Dynamics and Control (WETPOL), Constructed Wetland Association and Cranfield University, York, United Kingdom.

32. Saaty, 1980 T.L. SaatyThe Analytic Hierarchy Process McGraw-Hill International, New York (1980)
33. Saaty and Vargas, 2012 T.L. Saaty, L.G. VargasModels, Methods, Concepts & Applications of the Analytic Hierarchy Process (2nd ed.), Springer Science+Business Media, New York (2012), p. 346
34. Xu et al., 2014 W. Xu, N. Khoshroo, H. Bjornlund, Y. YinEffects of "Grain for Green" reforestation program on rural sustainability in China: an AHP approach to peasant consensus of public land use policies Stoch. Environ. Res. Risk Assess., 28 (4) (2014), pp. 867-880

3 MODELLING WATER USE EFFICIENCY IN A DYNAMIC ENVIRONMENT: AN EXAMPLE USING ARABIDOPSIS THALIANA

S. Vialet Chabrand[a], J.S.A. Matthews[a], O. Brendel[b], M.R. Blatt[c], Y.Wang[c], A.Hills[c], H. Griffiths[d], S. Rogers[e], T. Lawson[a]

[a] School of Biological Sciences, University of Essex, Colchester, CO4 3SQ, UK

[b] EEF, INRA, Université de Lorraine, F-54280 Champenoux, France

[c] Laboratory of Plant Physiology and Biophysics, University of Glasgow, Bower Building, Glasgow G12 8QQ, UK

[d] Plant Sciences, University of Cambridge, Downing Street, Cambridge CB2 3EA, UK

[e] Computing Science, University of Glasgow, Alwyn Williams Building, Glasgow G12 8QQ, UK

ABSTRACT

Intrinsic water use efficiency (W_i), the ratio of net CO_2 assimilation (A) over stomatal conductance to water vapour (g_s), is a complex trait used to assess plant performance. Improving W_i could lead in theory to higher productivity or reduced water usage by the plant, but the physiological traits for improvement and their combined effects on W_i have not been clearly identified. Under fluctuating light intensity, the temporal response of g_s is an order of magnitude slower than A, which results in rapid variations in W_i. Compared to traditional approaches, our new model scales stoma behaviour at the leaf level to predict g_s and A during a diurnal period, reproducing natural fluctuations of light intensity, in order to dissect W_i into traits of interest. The results confirmed the importance of stomatal density and photosynthetic capacity on W_i but also revealed the importance of incomplete stomatal closure under dark conditions as well as stomatal sensitivity to light intensity. The observed continuous decrease of A and g_s over the diurnal period was successfully described by negative feedback of the accumulation of photosynthetic products. Investigation into the impact of leaf anatomy on temporal responses of A, g_s and W_i revealed that a high density of stomata produces the most rapid response of g_s but may result in lower W_i.

Keywords: Stomatal conductance, Photosynthesis, Dynamics, Diurnal, Intrinsic water use efficiency

Nomenclature

A	Net CO_2 assimilation
A_G	Gross CO_2 assimilation
R_d	Mitochondrial respiration
g_s	Stomatal conductance to water vapour
g_m	Mesophyll conductance to CO_2
g_b	Boundary layer conductance to water vapour
g_t	Total conductance to CO_2
W_i	Intrinsic water use efficiency
C_a	Atmospheric CO_2 concentration
C_i	CO_2 concentration in the intercellular airspaces
C_c	CO_2 concentration at the site of carboxylation
a	Stomatal pore area
a_s`	Steady state target of stomatal pore area
a_{min}	Minimum stomatal pore area
a_{max}	Maximum stomatal pore area

α_L Slope of the relationship
θ_L Curvature factor of the curve
k_i Time constant for an increase in a
k_d Time constant for an decrease in a
L Percentage of efficiency
SD Stomatal density
D Diffusivity of water in air
V Molar volume of air
L Depth of stomatal porc
P_a The atmospheric pressure
S_a Factor representing the influence of the rate of accumulation of sugars
S_e Factor representing the influence of the rate of export of sugars
Vc_{max} Maximum Rubisco activity
J_{max} Maximum electron transport activity
A Proportion of light absorbed by PSII

INTRODUCTION

In order to meet the projected demand for cereal production by 2050, crop yields must improve by 1.16–1.31% each year; however, current estimates are well below this required rate [1]. The primary determinant of crop yield is the cumulative rate of photosynthesis over the growing season and is determined by the ability of the plant to capture light and CO_2, use this energy to convert the CO_2 to biomass, and how much of this biomass ends in usable yield. Improving photosynthetic efficiency is recognised as an important but unexploited avenue to increase yield potential in crop plants [2]. Increasing photosynthetic efficiency is accompanied by a higher CO_2demand, which can be limited by the resistance of CO_2 diffusion into the leaf. Any attempt to decrease this resistance greatly increases the water loss by transpirationfrom the leaf.

Photosynthetic productivity is linked to water consumed by the plant and often measured as water use efficiency (WUE). WUE can be defined at different scales of time and space and, at the leaf level, it is often assessed as the ratio of CO_2 fixed per unit of H_2O transpired (E). Intrinsic water use efficiency (W_i) is defined when stomatal conductance to water vapour (g_s) is used instead of E. The use of g_s to describe the stomatal control on the rate of E facilitates the comparisons between different leaves and environmental conditions. The photosynthetic capacity of the leaf determines the net CO_2 assimilation (A) as a function of the variation in the micro-climate surrounding the leaf. Over the diurnal period, A is mainly determined by the irradiance absorbed by the leaf and the limitation of CO_2 imposed by stomatal control. Under field conditions, environmental variables that affect both photosynthesis and stomatal behaviour are rarely constant. For example, light intensity (and spectral quality) alters in time scales of seconds to hours to which A and g_s must respond. The temporal response of A and g_s to a fluctuating environment are asynchronous, with g_s response often an order of magnitude slower than A, which results in rapid variations of W_i. Thus, it is important when describing the kinetic response of W_i to use an approach that considers responses by A and g_s simultaneously.

Intrinsic water use efficiency (W_i) is dependent on the anatomy (e.g. stomatal size and density) and the physiology (e.g. behaviour) of stomata as well as the leaf biochemistry (e.g. activity of the Calvin cycle), all of which interact to determine the kinetics of CO_2 and H_2O gaseous exchange between the leaf and atmosphere. The dynamic nature of the interactions between the different components that determine W_i are not fully understood and need to be addressed if we are to successfully improve both A and W_i under dynamic field conditions.

It is possible to conceptualise the inherent complexity of gas exchange over a fluctuating light regime through modelling, which will improve our understanding of the W_i response by simulating a number of gas exchange scenarios (e.g. changes in light intensity and humidity) that would normally be difficult to assess in a reasonable amount of time using experimental approaches. Current models focus on predicting $g_s$in steady state [3] and cannot be used to infer the impact of stomatal behaviour on Aor W_i under dynamic conditions. Although temporal responses of g_s have previously been described using a dynamic model [4,5], the relationship between stomatal response and leaf level gas exchange was not clearly described. We propose to use a model that will take into consideration the anatomy and physiology of stomata to more accurately represent the stomatal control of W_i.

To scale stomatal responses to leaf level g_s, the two most important stomatal characteristics are aperture and density [6–8]. A high stomatal density does not necessarily result in a higher g_s as stomata ultimately control their aperture depending on the guard cell responses to the external (e.g. light intensity) and internal (e.g. mesophyll demand for CO_2) stimuli [9]. To link stomatal behaviour to leaf level gas exchange responses, we propose a 'big stoma' approach that consists of simulating the response of one stoma that is representative of the heterogeneous response of many stomata and scaling the response to the leaf level. This approach was incorporated in an enhanced version of the multi compartments model described by Noe and Giersch [10] to predict A and W_i. Scaling up the dynamic of the stomatal response to the leaf level, with the improved model for CO_2 diffusion inside the leaf, will help to dissect W_i into traits of interest and predict potential gains in W_i.

The objective of this study was to develop a new model combining our most recent knowledge of kinetics in stomatal behaviour and photosynthesis to describe the temporal response of W_i over the course of a day with natural dynamic variations in irradiance. All the parameters of the model described here incorporate a trait of interest for W_i and were adjusted using Bayesian inference. The model was validated using a dataset with a different irradiance pattern to assess the predictive power of the model. A sensitivity analysis was finally performed to show the interaction among the parameters and display the potential gain in W_i in the case of one or two parameters changing. We used the output of the model to understand how temporal responses in g_s impacts A and W_i.

MATERIAL AND METHODS

Dynamic Modelling of Photosynthesis and Stomatal Conductance

The model essentially consists of four differential equations describing the diffusion of CO_2 between different compartments represented by the atmosphere, the intercellular air spaces and the photosynthetic tissues (Fig. 1). The exchanges between these compartments are dependent on the stomatal aperture and the resistance of diffusion in the mesophyll cells. In addition, the model took into account the limitation of photosynthesis and stomatal aperture that appeared during a period of light.

$$\begin{cases} \dfrac{da}{dt} = \dfrac{a_s - a}{k_i} if a < a_s \\ \dfrac{da}{dt} = \dfrac{a_s - a}{k_d} if a \geq a_s \end{cases} \tag{1}$$

Figure 1. Schematic of the diffusion of gas exchange between the leaf and atmosphere compared to the conceptual version used in the model. The total conductance to water vapour (g_t) controlling the gradient of H_2O from leaf to atmosphere was composed by the stomatal conductance (g_s) and the boundary layer conductance (g_b). The gradient of CO_2 from the atmosphere to the site of carboxylation was also dependent on mesophyll conductance (g_m). The H_2O concentrations inside and outside the leaf were represented by e_s and e_a respectively. The CO_2 concentrations in three compartments were considered: in the atmosphere (C_a), in the intercellular airspaces (C_i) and at the sites of carboxylation (C_c). The different compartments of the leaf were represented as a standardized volume defined by the thickness of the intercellular airspace (d_a) and the photosynthetic tissues (d_p).

The first differential equation (Eq. (1)) described the temporal variations of the stomatal pore area (a) with a_s the steady state target followed by a and two time constants, k_i and k_d, for an increase or a decrease of a respectively. Considering the spatial heterogeneity of the stomatal response, a top-down approach was used, signifying that the model simulated the response of one stoma representative of the sum of the individual stomatal responses and scaled it to leaf level instead of trying to integrate the response of each stoma.

The steady state target of a (a_s) as a function of the light intensity (PPFD) was predicted using a non-rectangular hyperbola [4]:

$$a_s = [a_{min} + \frac{\alpha_L PPFD + (a_{max} - a_{min}) - \sqrt{\alpha_L PPFD + (a_{max} - a_{min})^2 - 4\theta_L \alpha_L PPFD(a_{max} - a_{min})}}{2\theta_L}] \cdot L \tag{2}$$

with a_{min} and a_{max} the minimum and maximum stomatal pore area, α_L the slope of the relationship, θ_L the curvature factor of the curve and L the percentage of efficiency (see below).

$$\frac{dC_i}{dt} = [g_t(C_a - C_i) - g_m(C_i - C_c)]\frac{RT_l}{d_a P_a} \quad (3)$$

Eq. (3) described the variation of the CO_2 concentration in the intercellular airspaces (C_i) with C_a the atmospheric CO_2 concentration and C_c the CO_2 concentration at the sites of carboxylation.

The resistance of CO_2 diffusion from the air to the leaf was represented by the total conductance to CO_2 (g_t):

$$g_t = 1/\left(1.6/g_s + 1.37/g_b\right) \quad (4)$$

with g_b the boundary layer conductance to water vapour estimated in the gas exchange chamber.

The stomatal conductance to water vapour (g_s) was derived from a using the equation of Dow et al. [7]:

$$g_s = \frac{SD \cdot D \cdot a}{V\left(l + \frac{\pi}{2}\sqrt{\frac{a}{\pi}}\right)} \quad (5)$$

with SD the stomatal density, D the diffusivity of water in air, V the molar volume of air and l the depth of the stomatal pore.

The resistance of CO_2 diffusion in the mesophyll cells (g_m) was calculated following the equations described by Von Caemmerer et al. [11] and the values given for Arabidopsis thaliana. As the temperature of the leaf was regulated during the experiment, g_m was considered stable.

The last part of the equation $\frac{RT_l}{d_a P_a}$ was used to convert the fluxes (μmol m^{-2} s^{-1}) between the two compartments (leaf and atmosphere) into concentrations of CO_2(μmol mol^{-1}) contained in the volume represented by the intercellular airspaces (See the Appendix in Noe and Giersch [10] for more information), with R the gas constant, T_l the leaf temperature, P_a the atmospheric pressure, and d_a the depth of the intercellular airspaces (d_a = leaf thickness (m) • airspaces (%)).

$$\frac{dC_c}{dt} = [g_m(C_i - C_c) + R_d - A_G L]\frac{RT_l}{d_p P_a} \quad (6)$$

Eq. (6) described the variation of C_c with R_d the mitochondrial respiration during the day, A_G the gross CO_2 assimilation and L the efficiency coefficient of a_s and A_G (see below). The gross assimilation was calculated using C_c at the current state of the solver and the equations of Farquhar et al. [12]. The net CO_2 assimilation A was calculated at the end of the simulation as: $A = A_G - R_d$. In the last part of the equation d_a is replaced with d_p representing the depth of the photosynthetic tissues (d_p = leaf thickness (m) •[1 − airspaces (%)]).

$$\frac{dL}{dt} = S_e(1 - L) - S_a A_G L \tag{7}$$

The last equation (Eq. (7)) described the negative feedback of sugar accumulation on a_s (Eq. (2)) and A_G during a period of light, with S_a being a factor representing the influence of the rate of accumulation of sugars, and S_e a factor representing the influence of the rate of export of sugars.

Originally in Noe and Giersch [10], this equation only described the negative feedback of sugar export on A. The equation was modified to allow recovery of the response of A_G in the absence of light($Se(1-L)$) and tested using the diurnal data sets described below. The initial simulations showed that the model was able to reproduce the slow decrease of A during the day but not the similar decrease observed for g_s. Eq. (2) was therefore modified to include a decrease in g_s during the diurnal period similar to that for A.

Solving Differential Equations

The differential equations were solved using the function lsodes from the R package deSolve (v1.12). The initial absolute tolerance was set to 0.01 for all the outputs and decreased by 10 each time the solver failed to converge. The initial values of the differential equations were chosen based on the observed data. For the pore area, the initial aperture was calculated by solving Eq. (5) for the first recorded g_s.

Parameterization of the Model

Parameter values known to impact the diurnal variation of gas exchange were adjusted using the Bayesian inference (see below) to fit the observed data. During the diurnal period, gross CO_2 assimilation (A_G) was determined by maximum Rubisco activity (Vc_{max}), maximum electron transport activity (J_{max}) and the proportion of light absorbed by PSII (α). The day respiration in presence of light (R_d) was used to calculate the net CO_2 assimilation (A). The stomatal conductance to water vapour (g_s) was determined as described above by a_{min}, a_{max}, α_L and k_i/k_d. The diurnal variation of A and g_s were linked by the negative feedback of sugar export described by S_a and S_e.

Parameter values known to remain unchanged during the diurnal period of measurement or that could be assigned a priori were treated as constants in the model, to facilitate the adjustment, and were chosen from the literature or measured on Arabidopsis thaliana (see below). The Michaelis constants and the temperature responses of the Farquhar model were parameterized using values from Walker et al. [13]. The thickness of the leaf was set to 200 μm [14,15] and the ratio of airspace in the leaf was set to 0.255 [14,16]. The value of θ_L in equation 2 was chosen at 0.7 by analogy with the similar equation describing the electron transport rate [17]. The boundary layer conductance (g_b) and atmospheric pressure (P_{atm}) were set to the values provided by the LI-6400XT: 9.29 mol m^{-2} s^{-1} and 101.8 kPa respectively. Estimates of g_b are computed as a function of leaf area and fan speed (LI-6400XT Instruction Manuals, page 3–93). The parameter values describing the membrane properties required to estimate g_m followed that of Von Caemmerer et al. [18]. Stomatal density (SD) was assessed on plants grown under the same environmental conditions, on both faces of the leaf (section 1.9). As the model simulated the average response of both faces, the densities of both faces were added and set to 400 mm^{-2}. In equation 5, 5.89 μm was used for the depth of the stomatal pore (l) [7], the diffusivity of water in air at 25 °C (D) was $24.2e^{-6}$ m^2 s^{-1}, the molar volume of air at 25 °C (V) was 0.02446533 m^3 mol^{-1}[19].

Bayesian Inference

The parameter values were estimated using Bayesian inference by comparing the model outputs to recorded gas exchange data [20]. The inference was performed on two objective functions as A and g_s were adjusted simultaneously and in this case, there was a risk of bimodal posterior distributions. Monte Carlo Markov Chain (MCMC) using inter-chain adaptation were previously able to sample from bimodal distributions and this technique was thus used to adjust our model (INCA, [21,22]). To remove any adaptive effect, the final algorithm used to sample the posterior distributions was the random walk metropolis (RWM) with the covariance matrix estimated previously with INCA as an input. The use of Bayesian Inference allowed us to use prior information available in the literature and shown in Table 1. The results of the inference were posterior estimates describing the probability density of the possible parameter values considering the error of prediction and the sensitivity of the model for these parameters and their inter-correlation. The posteriors were represented using quantiles of the posterior distribution at 2.5% and 97.5%, also called credible intervals (similar to a confidence interval). The medians were also shown as the most representative set of parameter values. Two chains with 30,000 iterations were thinned every 2 samples and tested for different criteria: the acceptance rate greater than 15% (to achieve a rapid mixing of the chain), the Effective Sample Size (ESS) > 100 (which is usually enough to describe 95% probability intervals) and the convergence using Monte Carlo Standard Error (MCSE) less than 6.27% (which allows the true mean to be within 5% of the area under a Gaussian distribution around the estimated mean; [23]). The algorithms used to perform the Bayesian inference were implemented using the R software (R Core Team, 2015, v3.2.2).

Table 1. Parameters of the dynamic gas exchange model adjusted with Bayesian inference: mean and standard deviation (SD) of the prior values estimated from the literature (see footnotes) and quantiles at 2.5%, 50% and 97.5% of the credible interval of each posterior distribution estimated using Bayesian inference.

	unit	Prior		Posterior		
		Mean	SD	2.5%	50%	97.5%
a_{min} [1]	μm^2	3.56	0.2	4.03	4.16	4.27
a_{max} [2]	μm^2	100	10	72.15	93.24	113.87
αL	$\frac{\mu m^2}{\mu molm^{-2}s^{-1}}$	0	10	0.742	0.767	0.793
k_i	s	100	1e4	1290	1430	1580
k_d	s	100	1e4	463	541	627
Vc_{max} [3]	$\mu molm^{-2}s^{-1}$	65.5	22.6	89.8	99.3	128.8
J_{max} [3]	$\mu molm^{-2}s^{-1}$	102.4	21.9	180.5	186.4	193
R_d	$\mu molm^{-2}s^{-1}$	0	10	0.01	0.14	0.37
α	$\frac{\mu mol(electron)}{\mu mol(photon)}$	0	1	0.177	0.183	0.192
S_e		0	10	$0.263e^{-6}$	$8.486e^{-6}$	$32.904e^{-6}$
S_a		0	10	$0.207e^{-6}$	$0.280e^{-6}$	$0.385e^{-6}$

1. Parameter estimated using the observed stomatal conductance under dark conditions.
2. Parameter estimated from Dow et al. [7]
3. Parameter estimated by averaging published values for Col-0 by Flexas et al. [42], Tholen et al. [14], Heckwolf et al. [43], Cousins et al. [44], Sade et al. [45].

The accuracy of the model outputs compared to the observed data was assessed using the Root Mean Square Error (RMSE):

$$RMSE = \sqrt{\frac{\sum_{t=1}^{n}\left(x_{obs,t} - x_{mod,t}\right)^2}{n}} \qquad (8)$$

where x_{obs} and x_{mod} represents the observed and modelled values to be compared, and n the number of observations.

Sensitivity Analysis

The sensitivity analysis was performed using the same data set and parameter values estimated previously using Bayesian inference. Each parameter value was altered by ±50%, the impact on the model outputs was summarized by using the daily mean of Aand g_s divided by the corresponding values obtained using the original parameters values. This resulted in a factor centred on 1 representing the positive or negative impact of the parameter alterations on A and g_s (e.g. a factor of 1.1 corresponds to an increase of 10% in the original parameter value).

Plant Material

Arabidopsis thaliana ecotype Col-0 plants were grown in a controlled environment under fluctuating light conditions provided by a Heliospectra LED light source (Heliospectra AB, Göteborg, Sweden). The fluctuating light regime was reconstructed from natural variations in light intensity recorded during a day in July 2013 at the University of Essex, UK (Fig. 2a) assuming a constant spectral distribution. The growth environment was maintained at a relative humidity of 55–65%, a temperature of 21–22 °C and a CO_2 concentration of 400 µmol mol^{-1}. Plants were well watered, and positioning under the Heliospectra light was altered daily. Measurements were taken on the youngest fully expanded leaf.

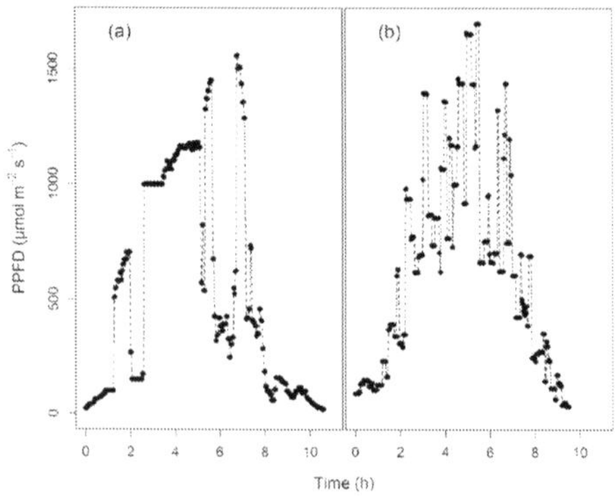

Figure 2. Light regimes a recorded from natural variations at the University of Essex, UK, July 2013, and b simulated using a Gaussian altered with random variations.

Model Simulations of Step Change of Light

Using the set of parameter values previously estimated, the model was used as a simulator to assess the diversity of the temporal response of g_s after a step change of light. The step change was performed by initially setting the light to an intensity of 100 μmol m^{-2} s^{-1} for 30 min followed by a step increase in intensity to 1000 μmol m^{-2} s^{-1} for 2 h. All other environmental conditions were kept constant, with the [CO_2] maintained at 400 μmol mol^{-1} and a leaf temperature of 25 °C. Parameter values of SD, k_i and α_L were increased or decreased by a factor of 2 to illustrate their impact on temporal responses of g_s, A and W_i. The rate of change of g_s was estimated using a linear regression on the first 10 min of the variation. The impact of the temporal response of g_s on A and W_i was characterised by calculating the percentage of variation of the mean A and W_i during the 2 h after the step change of light.

Leaf Gas Exchange

Photosynthesis gas exchange parameters (A and g_s) were measured using a Li-Cor 6400XT portable gas exchange system (Li-Cor, Lincoln, Nebraska, USA), with dew point and vapour pressure deficit maintained via a Li-Cor 610 portable dew point generator. Throughout the measurement cuvette, conditions were maintained at a CO_2 concentration of 400 ppm (corresponding to ambient growth), leaf temperature of 25

°C, and a leaf to air water vapour pressure deficit of 1 (±0.2) kPa. The largest fully expanded mature leaf was used, with A and g_s allowed to stabilize under the controlled cuvette conditions for a minimum of 30 min. At this point, the automatic 12 h light program (mirroring that of the growth conditions, Fig. 2a) was started, with Aand g_s recorded every 2 min. W_i was calculated as A/g_s. For the model validation, the same measurement conditions were used in the cuvette with the exception of the light that reproduced a different pattern of light intensity (Fig. 2b).

Stomatal Density

Following the dental impression methods of Weyers & Johansen (1985), negative impressions of the ab- and ad-axial leaf surface were made using Xantoprene Polysiloxane precision material (Heraesus Kulzer Ltd). Positive impressions were produced by coating nail varnish on the dry polymer and used to count the number of stomata in 9 field of view using an Olympus BX60 light microscope (Olympus Europa, Southend-On-Sea, UK). The area of view was measured using an eyepiece graticule to express the number of stomata by mm^{-2}. Stomatal impressions of the leaf surface were taken on the same leaves as gas exchange measurements were conducted.

RESULTS

Bayesian Inference

The two chains (describing the posterior distributions) resulting from the Bayesian inference successfully converged and were used to calculate the credible intervals of each parameter shown in Table 1. The credible interval of a_{min} was significantly higher and not overlapping the prior distribution, which was not the case for a_{max}, suggesting that the observed data were informative for a_{min} but not for a_{max}. The model estimated α_L precisely, despite the fact that the prior distribution covered a large range of possible values, suggesting that the model was particularly sensitive to this parameter. The time constants, k_i and k_d, displayed a strong asymmetry with values almost 3 times higher for k_i compared to k_d. Vc_{max} and J_{max} values were sensibly higher than the prior means estimated from the literature, however the prior distribution and the credible interval overlapped, suggesting that these values were still in the range observed in the literature.

The medians of the posterior estimates shown in Table 1 were used to represent the most representative prediction of the model (Fig. 3a–f). Under a diurnal fluctuating light regime, our dynamic model accurately described g_s (Fig. 3b, RMSE: 0.014 mol m^{-2} s^{-1}; Suppl. Fig. S1a, R^2:0.984) and A (Fig. 3c, RMSE: 0.6 μmol m^{-2} s^{-1}; Suppl. Fig. S1b, R^2:0.993). Consequently, the rapid variations of W_i were also accurately reproduced (Fig. 3e, RMSE: 2.58 μmol mol^{-1}; Suppl. Fig. S1c, R^2:0.97). By adjusting the model to the observed g_s, our 'big stoma' approach allowed us to predict the average behaviour of the stomata with a variation of the pore area from 4 to 14 μm^2 (Fig. 3a). Under these environmental conditions A and g_s displayed a decrease in efficiency of approximately 8% at the end of the diurnal period (Fig. 3d), which was described by S_e and S_a. The model seemed to be less sensitive to S_e as the credible interval was much larger than the one of S_a. The model successfully describes the variation of C_i but also predicts the variations of C_c that showed a larger variation compared to C_i due to low g_m (Fig. 3f).

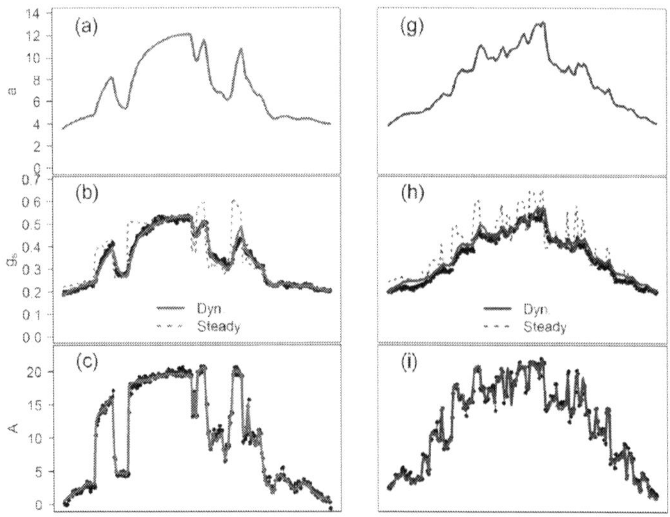

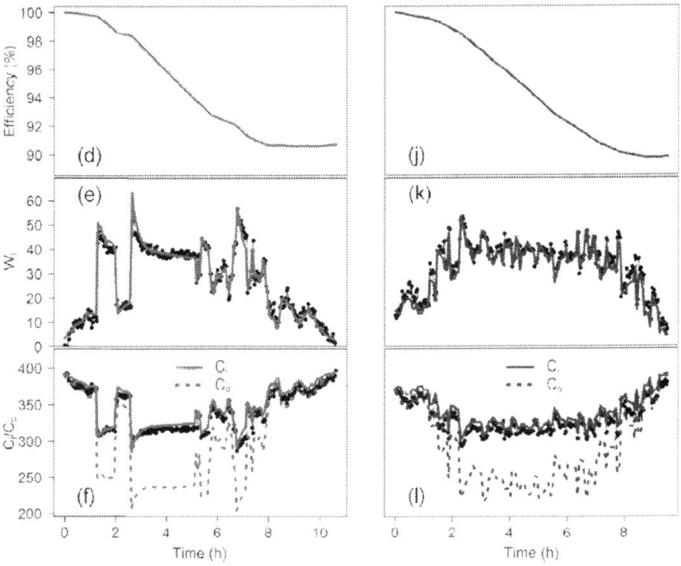

Figure 3. Modelled (solid & dashed lines) and observed (black circles) diurnal variations in gas exchange under two different light regimes (Fig. 2). a–f Modelled data (red lines) were fitted to observations (black circles) using Bayesian inference to estimate the most credible set of parameter values under the light regime described in Fig. 2a. g-l Modelled data (blue lines) were simulated using the same set of parameter values and the light regime describe in Fig. 2b. The red and blue dashed line in b and h represented the steady state g_s target (Steady) if conditions remained constant, whilst solid lines represent the dynamic outputs. (For interpretation of the references to colour in this figure legend, the reader is referred to the web version of this article.)

Model Validation

To validate the model, the same plant and set of parameter values (Table 1) were used to predict A, g_s and W_i under a different light regime (Fig. 2b). Our dynamic model was able to accurately predict the observed gas exchanges (Fig. 3g-l): g_s (Fig. 3h, RMSE: 0.023 mol m^{-2} s^{-1}; Suppl. Fig. S1d, R^2:0.983), A (Fig. 3i, RMSE: 0.69 μmol m^{-2} s^{-1}; Suppl. Fig. S1e, R^2:0.988) and W_i (Fig. 3k, RMSE: 2.95 μmol mol^{-1}; Suppl. Fig. S1f, R^2:0.961). Compared to the rapidity of the stomatal response (k_i: 24 min and k_d: 9 min), the light variations were too fast (on average a change every 8 min) to observe the complete stomatal response and, as a consequence, a followed the general tendency of the light intensity

(Fig. 3g). The efficiency of A and g_s during the diurnal period showed a decrease of around 10% similar to that of the estimate during the other diurnal period (Fig. 3j).

Sensitivity Analysis

A sensitivity analysis of the model was undertaken to illustrate the impact of varying individual parameters on A and g_s (Fig. 4). The parameters that displayed the most sensitivity for g_s were (ordered by importance): SD, a_{min}, α_L, k_i, and S_a. For A, the parameters were (ordered by importance): α, J_{max}, Vc_{max}, g_m, SD, a_{min}, α_L and S_a, showing the interdependence of A and g_s. Even if the model was sensitive to the variation of a parameter, this variation would not necessarily improve W_i. For example, Vc_{max} only displayed variation in A that resulted in a decrease in W_i suggesting that under the conditions used here A is not limited by CO_2 diffusion. The greatest impact on both A and g_s was observed when SD was decreased, while increasing photosynthetic capacity (Vc_{max} and J_{max}) by 50% only improved A by 10% under the environmental conditions used here.

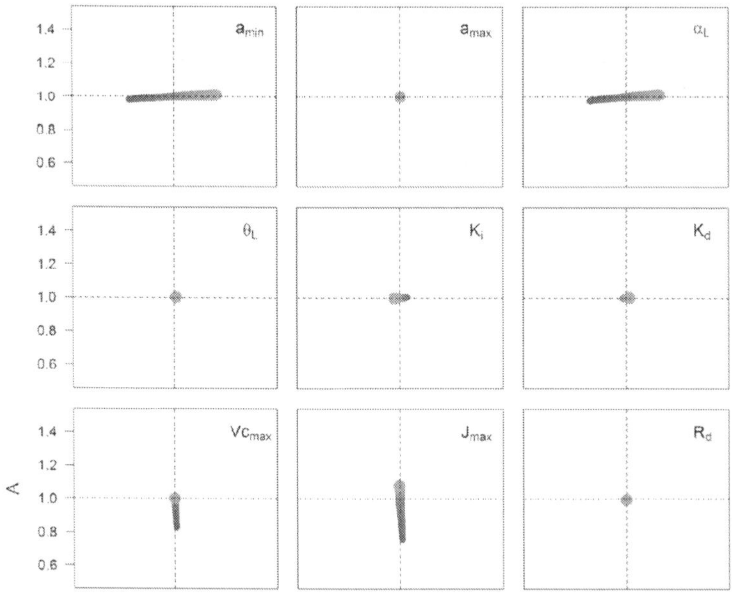

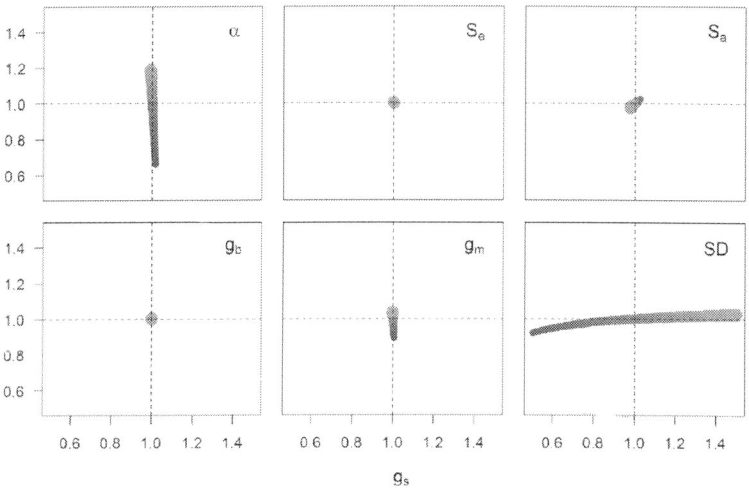

Figure 4. Relative variation of the daily mean net assimilation (A) and daily mean stomatalconductance to water vapour (g_s) using parameter values ranging ±50% around the estimated parameters from Table. 1. The size and the colour of the circles are proportional to the deviation of the parameter values (small and red circles correspond to decreased parameter values whilst large green circles correspond to increased parameter values). The length and direction are relative to the impact of the parameter on A and g_s.The simulated daily mean of A and g_s for each parameter set was divided by the mean values of A and g_s determined using the parameters in Table 1. The (1,1) coordinate showed similar daily mean values of A and g_s compared to values simulated using the parameter set described in Table 1. (For interpretation of the references to colour in this figure legend, the reader is referred to the web version of this article.)

Impact of stomatal characters on g_s, A and W_i temporal responses

In order to determine the impact of stomatal characters on the temporal response of g_s, A and W_i, the influence of the rapidity of the stomatal response (k_i) and the steady states g_s (SD, α_L) to a step increase in light intensity were examined (Fig. 5). The initial slope of the g_s response displayed large differences (e.g. 0.58–0.29 mol m^{-2} s^{-2}, Fig. 5b) even if the rapidity of increasing stomatal aperture did not change (Fig. 5a). The different temporal responses of g_s driven by SD, k_i and α_L limited A by 8.8% (Fig. 5c), 2% (Fig. 5g) and 6.1% (Fig. 5k) respectively. The impact on W_i was due to a larger decrease in g_s rather than A following the step increase in light intensity, with an increase of W_i by 82% (Fig. 5d), 7.4% (Fig. 5h) and 19.2% (Fig. 5l) respectively.

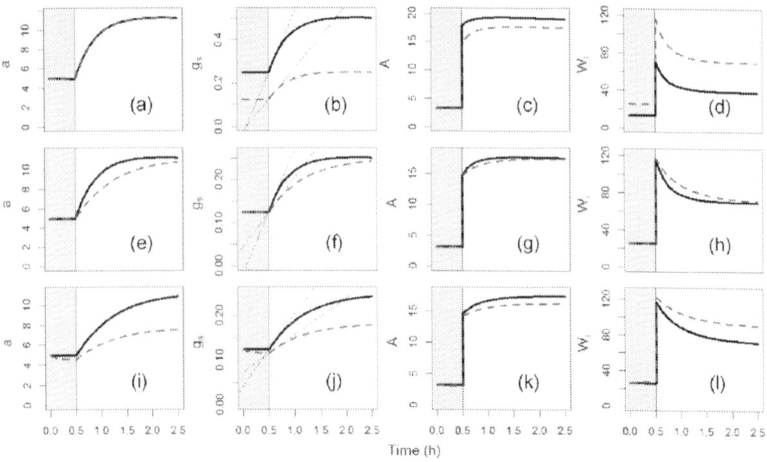

Figure 5. Simulation of the temporal response of a, g_s, A and W_i with different alterations on SD, k_i, and α_L under a step change of light from 100 (shaded area) to 1000 (white area) μmol m^{-2} s^{-1}. a-d Impact of a variation of the stomatal density (SD) from 400 (black line) to 200 mm^{-2} (red dashed line). e-h Impact of a variation of the time constant for stomatal aperture (k_i) from 1430 (black line) to 2860s (red line). i-l Impact of a variation of the slope of the relationship between the pore area and the light (α_L) from 0.77 (black line) to 0.38 $\mu m^2/\mu mol m^{-2} s^{-1}$(red line). The dotted lines on (b, f, j) represent the rates of g_s variation for the first 10 min of the step change estimated by linear regression. (For interpretation of the references to color in this figure legend, the reader is referred to the web version of this article.)

DISCUSSION

Over different diurnal light regimes, our new dynamic model was able to incorporate the rapid fluctuations in light and accurately predicted g_s, A and W_i. To our knowledge, this is the first attempt to model stomatal behaviour and scale it up to leaf level to predict gas exchange under dynamic light regimes. This 'big stoma' approach to model g_s provides a direct estimation of the rapidity of the stomatal response, whereas most current models consider only an instantaneous response [3]. Moreover, the simulated average pore area (a) predicted by the model was in a similar range to those previously reported in the literature [24,25].

Previous studies have attempted to directly model temporal responses of g_s[4,5], but our approach models the speed of increasing pore area rather than the speed of g_s increase, and is therefore more mechanistic and biologically relevant when describing the dynamics of gas exchange. Although we have validated the model using responses to fluctuations in light intensity, other environmental variables (e.g relative humidity) could be included in the future and would greatly improve the predictive power of the model.

Our model also incorporates negative feedback on A and g_s that revealed a 10% decrease in both by the end of the diurnal period. Noe and Giersch [10] suggested that the accumulation of sugars from photosynthesis could negatively regulate A. The decrease in g_s and therefore subsequent decrease in guard cell turgor, could be related to the decrease of A[26–28]. Such mechanisms for negative feedback are not well documented but are important for breeders, as any increase in photosynthetic capacity could lead to a great decrease in efficiency over the diurnal period. By accurately predicting diurnal gas exchange and dissecting W_i into the key parameters controlling dynamic responses, we believe our model will provide an important tool for future breeding strategies to identify targets for improved W_i under different environmental conditions.

Previous studies have shown spatial heterogeneity in stomatal distribution and behaviours [29–32] that have not been included in the model. Instead, the model predicted the behaviour of a stoma that is representative of the sum of individual stomatal responses over the entire leaf surface. This simplification did not appear to impact on the quality of the predictions and it is difficult to assess the temporal and spatial heterogeneity in stomatal response; therefore, further investigations would be required to determine if and how such variations would impact model outputs.

Stomatal density (SD) was shown to be the most sensitive stomatal parameter for improving W_i, with decreases in SD resulting in a substantial reduction in g_s but minimal impact on A. This is in agreement with the work of Franks et al. [33], who showed significant variation of W_i in Arabidopsis mutants with different SD. The Bayesian inference and the sensitivity analysis revealed that the model was not sensitive to changes in a_{max}, as this value was not reached at any point over the diurnal period. In contrast, a_{min} and α_L were important determinants of W_i but are rarely reported or examined in this context in the literature. The model predicts that, under dark conditions, a_{min} is statistically greater than zero resulting in significant nocturnal transpiration, which has been reported previously [34] and has implications for overall plant water use. As SD cannot change over the diurnal period, α_L could be an important determinant of plant water use by adjusting the magnitudes of change in g_s as a function of the environment.

Although the coefficient of light absorption by PSII (α) was an important determinant of A and W_i in the field, this parameter does not show a large diversity; therefore, Vc_{max}, J_{max} and g_m are the most promising targets for improving A. The model predicts that increasing g_m could lead to similar improvements in A and W_i[35] as those observed when Vc_{max} and J_{max} were increased through manipulating enzymes associated with the Calvin cycle (e.g. [36]).

The sensitivity analysis performed here should be seen as a tool to determine the key parameter(s) for improving W_i under specific environmental conditions (e.g. fluctuating light intensity). Indeed, the parameter values estimated with the Bayesian inference are specific for this study and the variation around the determined parameter may change between individuals and/or species. For example, an individual with different photosynthetic capacity (Vc_{max}, J_{max}) or SD will exhibit different thresholds of A limitations (Fig. 5c) and the results of the sensitivity analysis may also vary. To generalize the sensitivity analysis, ranges of biologically possible parameter values and the interactions between them (e.g. relationship between size of the pore and SD) should be used; however, to our knowledge, this information is not yet fully available in the literature.

Our model simulations clearly showed that the temporal response of g_s is the product of several parameters (SD, α_L and k_i/k_d), and the rapidity of the stomatal response cannot be interpreted by comparing the initial slope of variation in g_s, as is often reported in the literature [37,38]. Changes in the initial slope of the g_s response are correlated with SD, α_L and k_i/k_d and therefore are not only indicative of the speed of changes in stomatal aperture. In terms of gas diffusion, stomatal size constrains the maximum pore area but the observed pore area also depends on the sensitivity of the guard cell (α_L) responses to light intensity. As a result, it is theoretically possible to have large stomata that even at high light retain a small pore area. Considering the relationship between stomatal size and density [39], our results suggest that having a high density of small stomata will result in fast changes in g_s but also in a high average g_s value through the diurnal period, decreasing W_i.

The model simulations also revealed large variations in W_i (>80%) when the parameter values controlling the temporal response of g_s were altered. Indeed, the decrease in A due to greater limitation of CO_2 diffusion was minimal compared to the improvement in W_i. The importance of the temporal response of g_s was highlighted previously [40,41], but not the importance of the additive effects of parameter interactions on A and W_i as revealed by our model. Therefore, our model could be used to find the optimal parameter set that would decrease g_s without impacting Aand maximizing W_i to direct future breeding and research programmes to identify plants with reduced water usage, but the same or greater productivity.

CONCLUSION

We assessed and validated a new dynamic model of leaf gas exchange that accurately predicts g_s, A, and W_i under fluctuating light regimes such as those experienced by plants in the field. The model uses a unique framework that takes into account leaf anatomy, biochemistry and the physiological responses of stomata on W_i. The model revealed important negative feedback controls on A and g_s towards the end of the diurnal period that resulted in decreases in A with implications for overall plant productivity. Importantly, the model enabled W_i to be dissected into key parameters such as stomatal sensitivity to light (α_L) and minimal pore area under dark conditions (a_{min}) that have previous been neglected and could provide new and realistic targets for future improvements in crop water use efficiency.

ACKNOWLEDGEMENTS

This work supported SV-C through a BBSRC grant BB/1001187_1 to TL. NERC funding is acknowledged for PhD studentship to JSAM (Env-East DTP). MRB and YW were supported by BBSRC grants BB/L001276/1 and BB/L019205/1.

SUPPLEMENTARY DATA

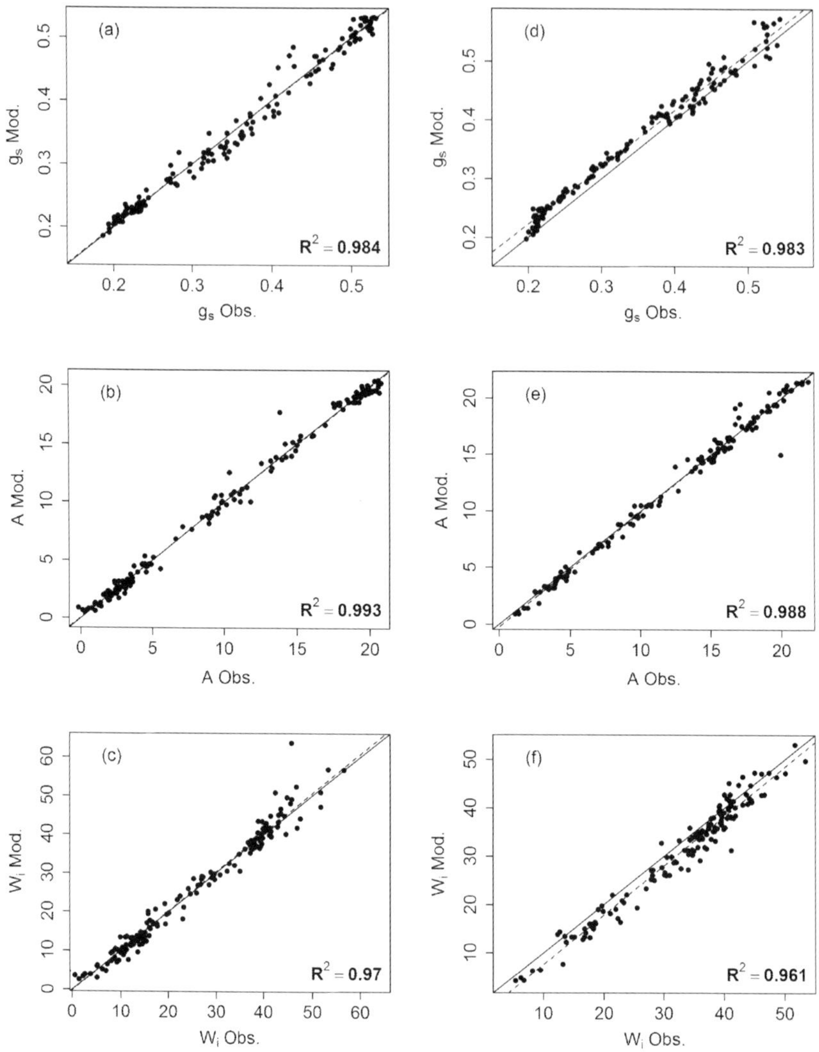

Figure 1: Modelled (Mod.) vs observed (Obs.) stomatal conductance to water vapour (g_s, mol m^{-2} s^{-1}), net CO_2 assimilation (A, μmol m^{-2} s^{-1}) and intrinsic water use efficiency (W_i, μmol(CO_2)/mol(H_2O)) measured under: **a-c** light regime described in Fig. 2a for the model parameterization, **d-f** light regime describe in Fig. 2b for the model validation. Solid lines and dashed lines represented the 1:1 lines and linear regressions respectively. The coefficient of determination (R^2) was derived from the linear regression.

REFERENCES

1. A.J. Hall, R.A. Richards, Prognosis for genetic improvement of yield potential and water-limited yield of major grain crops, Field Crop. Res. 143 (2013) 18–33, http://dx.doi.org/10.1016/j.fcr.2012.05.014.
2. X.-G. Zhu, S.P. Long, D.R. Ort, Improving photosynthetic efficiency for greater yield, Annu. Rev. Plant Biol. 61 (2010) 235–261, http://dx.doi.org/10.1146/ annurev-arplant-042809-112206.
3. G. Damour, T. Simonneau, H. Cochard, L. Urban, An overview of models of stomatal conductance at the leaf level, Plant Cell Environ. (2010), http://dx. doi.org/10.1111/j.1365-3040.2010.02181.x.
4. M.U.F. Kirschbaum, L.J. Gross, R.W. Pearcy, Observed and modelled stomatal responses to dynamic light environments in the shade plant Alocasia macrorrhiza, Plant Cell Environ. 11 (1988) 111–121, http://dx.doi.org/10. 1111/1365-3040.ep11604898.
5. S. Vialet-Chabrand, E. Dreyer, O. Brendel, Performance of a new dynamic model for predicting diurnal time courses of stomatal conductance at the leaf level, Plant Cell Environ. 36 (2013) 1529–1546, http://dx.doi.org/10.1111/pce. 12086.
6. H.T. Brown, F. Escombe, Static diffusion of gases and liquids in relation to the assimilation of carbon and translocation in plants, Philos. Trans. R. Soc. B Biol. Sci. 193 (1900) 223 291, http://dx.doi.org/10.1098/rstb.1900.0014.
7. G.J. Dow, J.A. Berry, D.C. Bergmann, The physiological importance of developmental mechanisms that enforce proper stomatal spacing in Arabidopsis thaliana, New Phytol. (2013), http://dx.doi.org/10.1111/nph. 12586.
8. P. Lehmann, D. Or, Effects of stomata clustering on leaf gas exchange, New Phytol. 207 (2015) 1015–1025.
9. T. Lawson, J.I.L. Morison, Stomatal function and physiology, in: Evol. Plant Physiol, Elsevier, 2004, pp. 217–242, http://dx.doi.org/10.1016/B978- 012339552-8/50013-5.
10. S.M. Noe, C. Giersch, A simple dynamic model of photosynthesis in oak leaves: coupling leaf conductance and photosynthetic carbon fixation by a variable intracellular CO2 pool, Funct. Plant Biol. 31 (2004) 1195–1204, http://dx.doi. org/10.1071/FP03251.

11. S. von Caemmerer, J.R. Evans, Temperature responses of mesophyll conductance differ greatly between species, Plant Cell Environ. 38 (2015) 629–637, http://dx.doi.org/10.1111/pce.12449.
12. G.D. Farquhar, S. von Caemmerer, Modelling of photosynthetic response to environmental conditions, in: Physiol. Plant Ecol. II, Springer, Berlin, Heidelberg, 1982, pp. 549–587, http://dx.doi.org/10.1007/978-3-642-68150- 9 17.
13. B. Walker, L.S. Ariza, S. Kaines, M.R. Badger, A.B. Cousins, Temperature response of in vivo Rubisco kinetics and mesophyll conductance in Arabidopsis thaliana: comparisons to Nicotiana tabacum, Plant Cell Environ. 36 (2013) 2108–2119, http://dx.doi.org/10.1111/pce.12166.
14. D. Tholen, C. Boom, K. Noguchi, S. Ueda, T. Katase, I. Terashima, The chloroplast avoidance response decreases internal conductance to CO2 diffusion in Arabidopsis thaliana leaves, Plant Cell Environ. 31 (2008) 1688–1700, http://dx.doi.org/10.1111/j.1365-3040.2008.01875.x.
15. N. Wuyts, C. Massonnet, M. Dauzat, C. Granier, Structural assessment of the impact of environmental constraints on Arabidopsis thaliana leaf growth: a 3D approach, Plant Cell Environ. 35 (2012) 1631–1646, http://dx.doi.org/10. 1111/j.1365-3040.2012.02514.x.
16. K.A. Pyke, J.L. Marrison, R.M. Leech, Temporal and spatial development of the cells of the expanding first leaf of Arabidopsis thaliana (L.) Heyn, J. Exp. Bot. 42 (1991) 1407–1416.
17. S. Von Caemmerer, Steady-state models of photosynthesis, Plant Cell Environ. 36 (2013) 1617–1630, http://dx.doi.org/10.1111/pce.12098.
18. S. von Caemmerer, J.R. Evans, Temperature responses of mesophyll conductance differ greatly between species, Plant Cell Environ. 38 (2015) 629–637, http://dx.doi.org/10.1111/pce.12449.
19. G.H. Riechers, H.G. Jones, Plants and microclimate, Ecology 65 (1984) 1701, http://dx.doi.org/10.2307/1939155.
20. M. Van Oijen, J. Rougier, R. Smith, Bayesian calibration of process-based forest models: bridging the gap between models and data, Tree Physiol. 25 (2005) 915–927 http://www.ncbi.nlm.nih.gov/pubmed/15870058.

21. R.V. Craiu, J. Rosenthal, C. Yang, Learn from thy neighbor: parallel-chain and regional adaptive MCMC, J. Am. Stat. Assoc. 104 (2009) 1454–1466, http://dx. doi.org/10.1198/jasa.2009.tm08393.
22. A. Solonen, P. Ollinaho, M. Laine, H. Haario, J. Tamminen, H. Järvinen, Efficient MCMC for climate model parameter estimation: parallel adaptive chains and early rejection, Bayesian Anal. 7 (2012) 715–736, http://dx.doi.org/10.1214/ 12-BA724.
23. J.M. Flegal, M. Haran, G.L. Jones, Markov Chain Monte Carlo: can we trust the third significant figure? Stat. Sci. 23 (2008) 250–260, http://dx.doi.org/10. 1214/08-STS257.
24. H. Kaiser, L. Kappen, Stomatal oscillations at small apertures: indications for a fundamental insufficiency of stomatal feedback-control inherent in the stomatal turgor mechanism, J. Exp. Bot. 52 (2001) 1303–1313, http://dx.doi. org/10.1093/jexbot/52.359.1303.
25. H. Kaiser, L. Kappen, In situ observation of stomatal movements and gas exchange of Aegopodium podagraria L. in the understorey, J. Exp. Bot. 51 (2000) 1741–1749, http://dx.doi.org/10.1093/ jexbot/51.351.1741.
26. T. Lawson, S. Lefebvre, N.R. Baker, J.I.L. Morison, C.A. Raines, Reductions in mesophyll and guard cell photosynthesis impact on the control of stomatal responses to light and CO2, J. Exp. Bot. 59 (2008) 3609–3619, http://dx.doi. org/10.1093/jxb/ern211.
27. F.A. Busch, Opinion: the red-light response of stomatal movement is sensed by the redox state of the photosynthetic electron transport chain, Photosynth. Res. 119 (2014) 131–140, http://dx.doi. org/10.1007/s11120-013-9805-6.
28. T. Azoulay-Shemer, A. Palomares, A. Bagheri, M. Israelsson-Nordstrom, C.B. Engineer, B.O.R. Bargmann, et al., Guard cell photosynthesis is critical for stomatal turgor production, yet does not directly mediate CO2- and ABA-induced stomatal closing, Plant J. 83 (2015) 567–581, http://dx.doi.org/ 10.1111/tpj.12916.
29. J.D.B. Weyers, T. Lawson, Heterogeneity in stomatal characteristics, Adv. Bot. Res. (1997) 317–352, http://dx.doi.org/10.1016/S0065-2296(08)60124-X.
30. H.G. Jones, Use of thermography for quantitative studies of spatial and temporal variation of stomatal conductance over leaf

surfaces, Plant Cell Environ. 22 (1999) 1043–1055, http://dx.doi.org/10.1046/j.1365-3040.1999. 00468.x.

31. K.A. Mott, I.E. Woodrow, Modelling the role of Rubisco activase in limiting non-steady-state photosynthesis, J. Exp. Bot. (2000) 399–406 (51 Spec No) http://www.ncbi.nlm.nih.gov/pubmed/10938848.
32. J.D. West, D. Peak, J.Q. Peterson, K.A. Mott, Dynamics of stomatal patches for a single surface of Xanthium strumarium L. leaves observed with fluorescence and thermal images, Plant Cell Environ. 28 (2005) 633–641, http://dx.doi.org/ 10.1111/j.1365-3040.2005.01309.x.
33. P.J. Franks, T.W. Doheny-Adams, Z.J. Britton-Harper, J.E. Gray, Increasing water-use efficiency directly through genetic manipulation of stomatal density, New Phytol. 2 (2015), http://dx.doi.org/10.1111/nph.13347.
34. M.A. Caird, J.H. Richards, L.A. Donovan, Nighttime stomatal conductance and transpiration in C3 and C4 plants, Plant Physiol. 143 (2006) 4–10, http://dx. doi.org/10.1104/pp.106.092940.
35. J. Flexas, A. Díaz-Espejo, M.A. Conesa, R.E. Coopman, C. Douthe, J. Gago, et al., Mesophyll conductance to CO2 and Rubisco as targets for improving intrinsic water use efficiency in C3 plants, Plant Cell Environ. (2015), http://dx.doi.org/ 10.1111/pce.12622.
36. A.J. Simkin, L. McAusland, L.R. Headland, T. Lawson, C.A. Raines, Multigene manipulation of photosynthetic carbon assimilation increases CO2 fixation and biomass yield in tobacco, J. Exp. Bot. 66 (2015) 4075–4090, http://dx.doi. org/10.1093/jxb/erv204.
37. P.L. Drake, R.H. Froend, P.J. Franks, Smaller, faster stomata: scaling of stomatal size, rate of response, and stomatal conductance, J. Exp. Bot. 64 (2013) 495–505, http://dx.doi.org/10.1093/jxb/ers347.
38. J.A. Raven, Speedy small stomata? J. Exp. Bot. 65 (2014) 1415–1424, http://dx. doi.org/10.1093/jxb/eru032.
39. A.M. Hetherington, F.I. Woodward, The role of stomata in sensing and driving environmental change, Nature 424 (2003) 901–908, http://dx.doi.org/10. 1038/nature01843.
40. T. Lawson, M.R. Blatt, Stomatal size, speed, and responsiveness impact on photosynthesis and water use efficiency, Plant Physiol. 164 (2014) 1556–1570, http://dx.doi.org/10.1104/pp.114.237107.

41. L. McAusland, S. Vialet-Chabrand, P. Davey, N.R. Baker, O. Brendel, T. Lawson, Effects of kinetics of light-induced stomatal responses on photosynthesis and water-use efficiency, New Phytol. (2016), http://dx.doi.org/10.1111/nph. 14000.
42. J. Flexas, M.F. Ortuno, ˜ M. Ribas-Carbo, A. Diaz-Espejo, I.D. Flórez-Sarasa, H. Medrano, Mesophyll conductance to CO2 in Arabidopsis thaliana, New Phytol. (2007) 501–511, http://dx.doi.org/10.1111/j.1469-8137.2007.02111.x.
43. M. Heckwolf, D. Pater, D.T. Hanson, R. Kaldenhoff, The Arabidopsis thaliana aquaporin AtPIP1;2 is a physiologically relevant CO2 transport facilitator, Plant J. 67 (2011) 795–804, http://dx.doi.org/10.1111/j.1365-313X.2011. 04634.x.
44. A.B. Cousins, O. Ghannoum, S. Von Caemmerer, M.R. Badger, Simultaneous determination of Rubisco carboxylase and oxygenase kinetic parameters in Triticum aestivum and Zea mays using membrane inlet mass spectrometry, Plant Cell Environ. 33 (2010) 444–452, http://dx.doi.org/10.1111/j.1365-3040. 2009.02095.x.
45. N. Sade, A. Gallé, J. Flexas, S. Lerner, G. Peleg, A. Yaaran, et al., Differential tissue-specific expression of NtAQP1 in Arabidopsis thaliana reveals a role for this protein in stomatal and mesophyll conductance of CO2 under standard and salt-stress conditions, Planta 239 (2014) 357–366, http://dx.doi.org/10. 1007/s00425-013-1988-8.

4 COMPARATIVE ECONOMIC PERFORMANCE AND CARBON FOOTPRINT OF TWO FARMING MODELS FOR PRODUCING ATLANTIC SALMON (SALMO SALAR): LAND-BASED CLOSED CONTAINMENT SYSTEM IN FRESHWATER AND OPEN NET PEN IN SEAWATER

Yajie Liu[a], Trond W.Rosten[a], Kristian Henriksen[a], Erik Skontorp Hognes[a], Steve Summerfelt[b], Brian Vinci[b]

[a] SINTEF Fisheries & Aquaculture, 7465 Trondheim, Norway

[b] The Conservation Fund, Freshwater Institute, 1098 Turner Road, Shepherdstown, WV, USA

ABSTRACT

Ocean net pen production of Atlantic salmon is approaching 2 million metric tons (MT) annually and has proven to be cost- and energy-efficient. Recently, with technology improvements, freshwater aquaculture of Atlantic salmon from eggs to harvestable size of 4–5 kg in land-based closed containment (LBCC) water recirculating aquaculture systems (RAS) has been demonstrated as a viable production technology. Land-based, closed containment water recirculating aquaculture systems technology offers the ability to fully control the rearing environment and provides flexibility in locating a production facility close to the market and on sites where cost of land and power are competitive. This flexibility offers distinct advantages over Atlantic salmon produced in open net pen systems, which is dependent on access to suitable coastal waters and a relatively long transport distance to supply the US market. Consequently, in this paper we present an analysis of the investment needed, the production cost, the profitability and the carbon footprint of producing 3300 MT of head-on gutted (HOG) Atlantic salmon from eggs to US market (wholesale) using two different production systems—LBCC-RAS technology and open net pen (ONP) technology using enterprise budget analysis and carbon footprint with the LCA method. In our analysis we compare the traditional open net pen production system in Norway and a model freshwater LBCC-RAS facility in the US. The model ONP is small compared to the most ONP systems in Norway, but the LBCC-RAS is large compared to any existing LBCC-RAS for Atlantic salmon. The results need to be interpreted with this in mind. Results of the financial analysis indicate that the total production costs for two systems are relatively similar, with LBCC-RAS only 10% higher than the ONP system on a head-on gutted basis (5.60 US$/kg versus 5.08 US$/kg, respectively). Without interest and depreciation, the two production systems have an almost equal operating cost (4.30 US$/kg for ONP versus 4.37 US$/kg for LBCC-RAS). Capital costs of the two systems are not similar for the same 3300 MT of head-on gutted salmon. The capital cost of the LBCC-RAS model system is approximately 54,000,000 US$ and the capital cost of the ONP system is approximately 30,000,000 US$, a difference of 80%. However, the LBCC-RAS model system selling salmon at a 30% price premium is comparatively as profitable as the ONP model system (profit margin of 18% versus 24%, respectively), even though its 15-year net present value is negative and its return on investment is lower than ONP system (9% versus 18%, respectively). The results of the carbon footprint analysis confirmed that production of feed is the dominating climate aspect for both production methods, but also showed that energy source and transport methods are important. It was shown that fresh salmon produced in LBCC-RAS systems close to a US market that use an average US electricity mix have a much lower carbon footprint than fresh salmon produced in Norway in ONP systems shipped to the same market by airfreight, 7.41 versus 15.22 kg CO_2eq/kg salmon HOG, respectively. When comparing the carbon footprint of production-only, the LBCC-RAS-produced salmon has a carbon footprint that is double that of the ONP-produced salmon, 7.01 versus 3.39 kg CO_2eq/kg salmon live-weight, respectively.

Keywords: Salmon, Economics, Carbon footprint, Recirculating aquaculture systems, Net pen aquaculture

INTRODUCTION

Farmed Atlantic salmon (*Salmo salar*) is sold globally in various forms and markets. The US is an important market for farmed Atlantic salmon, estimated to be more than 350,000 MT in 2014 (Marine Harvest ASA, 2014), and has shown steady growth since the late 1980s (USDA ERS, 2015). In 2014 the US market was primarily supplied by salmon produced in Chile (126,820 MT), Canada (47,454 MT) and Norway (26,208 MT) (USDA ERS, 2015). The US production of Atlantic salmon (18,000 MT [2012]) is relatively small in comparison to the amount consumed in the US (NOAA, 2013). Limited access to suitable coastal water areas and rigorous regulations in the US (NOAA, 2013) curtail the opportunity to produce Atlantic salmon in open net pen systems, the industry's preferred and established technology for the on-growing phase of salmon farming in Norway, Canada, and Chile. An alternative technology to open net pen systems for salmon production is land-based, closed containment (LBCC) water recirculating aquaculture systems (RAS) technology (LBCC-RAS). LBCC-RAS technology had been used for production of a limited number of species, like eel, beginning in the 1980s (Heinsbroek and Kamstra, 1990). Developments in LBCC-RAS technology since the 1980s have led to the ability to culture a wide variety of fish species including cold-water salmonids (e.g., Arctic char, rainbow trout, and Atlantic salmon to smolt size) (Summerfelt et al., 2004; Bergheim et al., 2009; Dalsgaard et al., 2013; Kolarevic et al., 2014). Most recently, freshwater aquaculture of Atlantic salmon from eggs to harvestable size of 4–5 kg in a LBCC-RAS facility has been demonstrated as a viable production technology (Summerfelt et al., 2013). Land-based, closed containment water recirculating aquaculture systems technology offers the ability to fully control the rearing environment, exclude parasites and obligate pathogens, and provide flexibility in locating a production facility close to the market and on sites where the cost of land and power are competitive. This control and flexibility offers advantages over Atlantic salmon produced in open net pen systems (ONP), which is negatively impacted by sea lice and dependent on access to suitable coastal waters and a relatively long transport distance to supply the US market. Interest in production of Atlantic salmon using LBCC-RAS technology has led to construction of a number of commercial LBCC-RAS farms (Summerfelt and Christianson, 2014). Although their current supply to the US Atlantic salmon market is just beginning, plans for a number of US-based LBCC-RAS farms for Atlantic salmon have been reported in the trade press. It is therefore of particular interest to compare such different approaches for production of the same seafood to the same market.

The aquaculture production of Atlantic salmon has been estimated to exceed 1,900,000 MT in 2014; global production has increased 428% since 1994 (Marine Harvest ASA, 2014). Open net pen farming in the ocean has been the major technology for the on-growing portion of the production cycle. The technology for ONP farming with large net pen volumes, exceeding 60,000 m^3 in one pen, has proven to be cost- and energy-efficient (Ziegler et al., 2013), leading to commercial success and founding a large global business. However, the growth of the industry has not been without environmental conflicts, especially towards wild Atlantic salmon and Sea Trout (*Salmo Trutta*) where negative impacts on wild populations due to escapees have been suggested (Naylor et al., 2005). Alternative methods for growing salmon in closed containment systems for the whole production cycle have been attempted since the beginning of the 1990s, with no commercial success, either land-based or in floating bags (Liu and Sumaila, 2007). Recently, a new interest for producing Atlantic salmon in closed containment systems has arisen (Summerfelt and Christiansen, 2014). A variety of closed containment systems are being suggested (Rosten et al., 2013), but LBCC-RAS technology seems to have found a particular global interest, with LBCC-RAS farms being planned, built and put into production in Europe, North America, China, and Norway (Summerfelt and Christianson, 2014).

Norwegian-farmed Atlantic salmon is sold as fresh, frozen, filleted, smoked and cured product. Fresh whole salmon is the primary product and accounts for approximately three quarters of the total value of exports (Statistics Norway, 2015). Fresh salmon has the highest export price. Denmark, France and Japan are the biggest export countries, making up of one-third of total Norwegian salmon exports (Statistics Norway, 2015). Norwegian salmon made up approximately 8% of the US salmon market in 2014 (USDA ERS, 2015).

The production cost of Atlantic salmon farming in Norway has been charted annually since 1986. From 2008–2012 the production cost has varied between 21.04 and 22.98 NOK per kilo WFE (Directorate of Fisheries, 2014). It has recently increased due to the high cost of sea lice treatment (Liu and Bjelland, 2014). The relatively low investment cost for open net pen production sites compared to the investment cost for proposed LBCC-RAS farms has historically favored open net pen production. Norway has the lowest production cost per kilo of salmon compared to Canada, Great Britain and Chile due to economies of scale (Marine Harvest ASA, 2014).

The economic viability of intensive LBCC-RAS has been evaluated (Muir, 1981; Gempesaw et al., 1993; Losordo and Westerman, 1994; De Ionno et al., 2006; Timmons and Ebeling, 2010), though these studies have largely focused on specific system designs for a single level of output, and have not identified the capital and operating cost savings which may exist as water treatment processes are optimized and as technologies are scaled appropriately. De Ionno et al. (2006) reported that increasing LBCC-RAS facility capacity, increasing sale price, and decreasing facility capital cost were the most important factors affecting economic viability. These savings can be significant and can contribute to the success or failure of an aquaculture business employing this type of technology.

Environmental assessments of ONP salmon production and distribution have identified feed production as a dominating climate aspect of salmon aquaculture production, closely followed by transportation of the salmon to retailer (Ziegler et al., 2013). A shift into more closed systems includes changes such as: replacing ocean current energy with electricity; more alternative materials in the production facilities; controlling interactions with the surrounding environment; collecting and utilizing nutrients in the biosolids produced by the fish; and placing the production close to the market or independent of oceans. There are several potential environmental tradeoffs in this shift. Feed efficiency is especially important, but also the balance between an increase in energy use in the growout phase versus a reduction in transport distance.

This paper aims to investigate whether domestic US production of Atlantic salmon in a LBCC-RAS farm is competitive when compared to a similarly sized ONP system overseas, using investor relevant keys like return of investment, production cost, market price, and carbon footprint. In this paper we present an analysis of the investment needed, the production cost, the profitability and carbon footprint of Atlantic salmon farming from eggs to US market (wholesale) using two different production systems—LBCC-RAS technology and ONP technology using enterprise budget analysis and calculating the carbon footprint with the LCA method. In our analysis we compare the traditional ONP production system in Norway and a model freshwater LBCC-RAS facility in the US. We model the necessary product prices to obtain profitability with LBCC-RAS, and compare the profitability to a similarly-scaled ONP system and provide a sensitivity analysis for the most important impact factors. In addition, we incorporate a comparison of the carbon footprint of the two systems using an overview of the consumed materials, feed, energy, transport and energy source.

MATERIALS AND METHODS

The feasibility of two commercial-scale farming systems for Atlantic salmon, a LBCC-RAS farm in the US and an ONP farm in Norway, is evaluated through a concept-level design and capital and operational cost analysis for 3300 MT head-on gutted (HOG) production systems. The economic performance is evaluated in detail using an enterprise budget analysis, while the environmental performance is evaluated in detail using attributional life cycle analysis. The ONP system evaluated here was scaled down from the more common large-sized facilities in Norway to fit to the comparable LBCC-RAS system.

Open net Pen System Model

Technical design of the ONP model farm is based upon a biological production plan (i.e., bioplan), data and operational practices obtained from Norwegian salmon farmers. Data and specifications of components are gathered from aquaculture industry suppliers in Norway. The ONP model farm includes concept-level design of floating rings, nets, mooring systems, boats, feed barge systems, camera systems, feed distribution systems and remote power systems. The bioplan, which predicted fish growth and size from smolt to harvestable size, results in two active growout sites, using limitations for fish density of 25 kg/m^3 and maximum allowable biomass of 200,000 fish per unit.

The bioplan for the 3300 MT ONP model farm is based upon average ambient sea temperatures from mid-Norway, stocking with two smolt cohorts per year. The ONP system is assumed to stock a cohort of S1 smolts, average size 100 g, on April 1 and a cohort of S0 smolts, average size 75 g, on August 1. Fish growth and associated feed demand are determined by using specific growth rates (SGR) and feed conversion ratios (FCR) given in feed supplier feeding tables for various fish sizes. Fish growth estimates are reduced by 12% to compensate for handling and treatment of the fish during the production cycle. The overall FCR was set to 1.27 to obtain the average FCR from the last 10 years in Norway (Directorate of Fisheries, 2014). Mortalities for smolt to harvest are set to obtain 16% per generation mortality to comply with a dataset available from mid-Norway (Mattilsynet, 2011).

Land-based Closed Containment Recirculating Aquaculture System Model

Technical design of the LBCC-RAS model farm is based on data developed by The Conservation Fund's Freshwater Institute growout trials of Atlantic salmon, some of which has been reported (Summerfelt et al., 2013). This includes concept-level water recirculation system designs for each fish grouping developed in the bioplan. Each water recirculation system design includes multiple recirculation modules to allow for staging and movement of fish throughout the facility. Concept designs for incubation, fry, smolt, pre-growout, and growout rearing areas, as well as a final purging system, are completed using steady-state mass balance analyses. Design water quality criteria used in the mass balance analyses are based on The Conservation Fund's Freshwater Institute growout trials. Thermal growth coefficients (TGC) are used to predict fish growth for the bioplan for the 3300 MT LBCC-RAS model farm. Thermal growth coefficient values are based on data collected in growout trial data from The Conservation Fund's Freshwater Institute. Additionally FCR, mortality, head-on gutted yield, and other performance indicators, which are used to develop a biological plan are taken from past growout trials (Summerfelt et al., 2013). The FCR (kg/kg) and TGC (1000 $g^{1/3}$/ °C days) are set to vary according to these growout trial data at different life stages; FCR: Fry, 0.75; smolt, 0.90; pre-growout, 1.0; growout 1.1; and TGC: Fry, 1.25; smolt, 1.40; pre-growout, 2.00; growout, 2.30. The overall average FCR based on the individual values is 1.09. A maximum biomass density of 80 kg/$m^3$is used for the biological plan of the LBCC-RAS model farm.

The steady-state feed requirement for the LBCC-RAS model farm is 11,815 kg/day. Water supply required for the entire 3300 MT LBCC-RAS model farm is based on allowing no more than 75 mg/L nitrate-nitrogen at maximum loading in each recirculation system, assuming no passive denitrification within the systems. The amount of water supply needed to maintain this nitrate-nitrogen level in the recirculation systems is calculated to be 7.7 m^3/min, including 1.1 m^3/min for finishing/purging the harvested salmon before slaughter. The resulting water required per feed fed is 803 L/kg feed for the systems that have feeding fish, i.e., all RAS except the purge system. The power requirement for the model farm is 2458 kW, comprised primarily of power required for the water recirculation pumps (2079 kW); the total power required per unit of live weight salmon produced is 5.4 kWh/kg (4.6 kWh/kg for pumping only).

Concept-level design characteristics for each rearing area in both production systems are summarized in Table 1; the inputs required for the two systems are summarized in Table 2; illustrative renderings are shown in Fig. 1. The technical design for each model farm allowed the progression of capital and operating costs for comparison of the two production systems. Cost data used in the development of the concept-level estimates provided here is a combination of industry standard published cost data (Directorate of Fisheries, 2014; Marine Harvest ASA, 2014; RS Means, 2010) and project specific vendor quotations obtained in 2010–2011.

Table 1. Concept-level design characteristics for each rearing system in a 3,300 MT HOG Atlantic salmon land-based closed containment farm (LBCC-RAS) and a 3300 MT HOG open net pen farm (ONP).

Fish Rearing Area	Modules	Units per module	Unit diameter by depth (m × m)	Total Rearing Volume (m^3)	Module Flow Rate (m^3/min)	Total Flow Rate (m^3/min)	Total Makeup Flow Rate (m^3/min)	Maximum Module Feed Rate (kg/day)
LBCC-RAS—fry	1	18	2 by 1.0	57	1.5	1.5	0.08	22.9
LBCC-RAS—smolt	2	4	9 by 2.0	1,018	11.4	22.7	0.19	248.0
LBCC-RAS—pre-growout	3	4	10 by 3.0	2,827	22	66	0.57	549.5
LBCC-RAS—growout	8	5	16 by 4.25	34,180	95	757	5.75	2063.5
LBCC-RAS—final purging	1	2	16 by 4.25	1,709	38	38	1.1	–
ONP—System[a]	2	6	157 by 40	587,000	–[b]	–[b]	–[b]	–

[a] The ONP system is a growout system from smolts to harvestable size. Smolts and harvest/packing of the salmon are modeled to be provided by subcontractors.

[b] The water exchange in the ONP system is dependent upon water current and conditions of the nets (mesh size and fouling).

Table 2. Input factors and assumptions used in the financial analysis of two production models (LBCC-RAS system and ONP system) for a 3300 MT HOG Atlantic salmon farm.

Input factors	ONP system	LBCC-RAS system
Feed (US$/kg)	1.48	1.50
Farm labor (# person)	6	10
Farm labor (US$/person/year)	125,000	45,000
Processing labor (# person)	–	6
Processing labor (US$/person/year)	0.38/kg[a]	37,500
Livestock (US$/smolt or US$/egg) smolt)	1.53	0.30
Electric (US$/kWh)	0.17	0.05
Oxygen (US$/kg)	–	0.20
Wellboat cost (US$/kg[a])	0.92	–
Bicarbonate (US$/kg)	–	0.35
Management (US$/year)	–	500,000
Other operating cost	0.43 US$/kg fish	–
Insurance (US$/kg[a])	0.02	0.02b
Tax level	28%	28%
Equity ratio	30%	40%
Interest loans	3.0%	6.0%

[a] Whole fish weight.

[b] First year is 0.04 US$/kg.

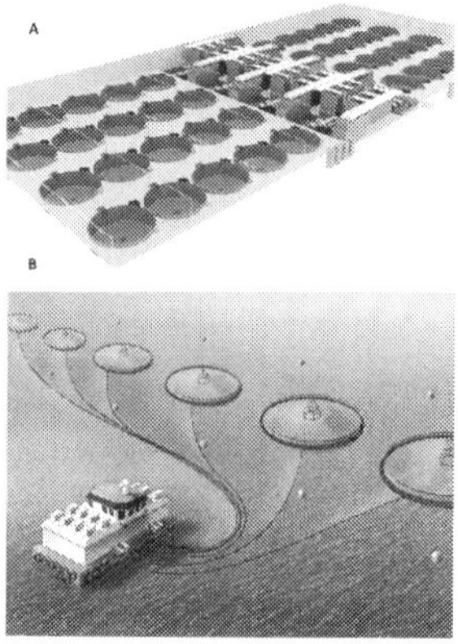

Figure 1. Concept-level renderings of the growout rearing area in a 3300 MT HOG Atlantic salmon LBCC-RAS farm (A) and ONP farm (B).

Economics

Salmon aquaculture is a commercial operation whose purpose to be profitable. The prerequisite for a business to be sustainable is to be profitable in both the short- and long-term and over the investment horizon. The financial performance of these two aquaculture production systems is investigated using an enterprise budget analysis; this allows an assessment of the feasibility and profitability of the two systems. Enterprise budgets, also called production budgets, provide a framework within which all the components of costs and revenues associated with the production of farm products are itemized. The budget is constructed on a production basis, and the assessment is built upon a cash flow analysis. The profitability is calculated based on financial statements such as income statement and balance sheets.

There are a number of well-developed analytical techniques for analyzing profitability (Liu and Sumaila, 2007; Kumar and Engle, 2011). Net present value (NPV) is a commonly used parameter to provide an objective decision of an investment and project. Net present value takes into account the time value of money, and is the difference between the present value of total costs and total revenue over an operational horizon. Positive NPV indicates that an investment is worthwhile. In addition to NPV, other indicators are also used as assessment criteria; these include gross margin, return on investment (ROI), internal rate of return (IRR), payback period, and break-even production and price. Gross margin is expressed as revenue minus variable costs; net income or profit is revenue minus all costs. Return on investment is the rate of return on the initial capital investment and is estimated by profit before taxes divided by the capital investment. Internal rate of return is the discount rate at which net present value of profit is set equal to zero. Breakeven production/ price represents the expected production level and market price at which total sale revenue covers total production costs. Breakeven analysis can inform the conditions necessary for the business to become profitable or to remain in business.

Enterprise Budget

The enterprise budget is estimated based on a total production of 4000 MT wet weight, which is equivalent to 3300 MT of head-on gutted weight. Head-on gutted yield is estimated to be 88% after a 5% loss of weight during final purging for both the ONP and the LBCC-RAS production systems. The estimates of total investment cost and operating cost of each cost item are based on the production system design models and their associated bioplans. The costs include two parts: capital cost and operating cost.

Capital Cost—ONP Model

Capital costs incur at the beginning of the operation, and most of these costs are one-time costs. The capital cost for the 3300 MT ONP model farm is based on information gathered from the Norwegian aquaculture industry, and is thereby considered representative for an ONP farm constructed and operated according to Norwegian laws and regulations (Norway, 2008). The ONP model farm includes 3 licenses and 12 pens, and their associated physical components consisting of floating rings, nets, mooring systems, boats, feed barge systems, camera systems, feed distribution systems and remote power systems. The cost of each item is estimated based on current market price suppliers' command. Compared to estimates reported by Marine Harvest (Marine Harvest ASA, 2014), the capital cost for the ONP model farm is considered representative for a two site ONP farm. We assume that the lifespan of nets and feeding system is 5 years, floating rings is 8 years, camera and power systems is 10 years, and the remainder of the equipment is 20 years. These lifespans are used for calculation of depreciation and replacement cost.

The cost for an ONP farming license in Norway is included in the capital cost estimate for the ONP model farm. The current cost of ONP farming licenses is much higher when compared to license costs of the 1990s (Färe et al., 2005); cost for a license in the current open market is approximately 55 million Norwegian kroners, which is equivalent to 8 million US dollars[1] (Aardal, 2014). The total capital cost of the ONP model farm including licenses at current prices is estimated to be 29.7 million US dollars for a total production of 3300 MT head-on gutted salmon (Table 3).

Table 3. Capital expenses for a 3,300 MT HOG LBCC-RAS and ONP Atlantic salmon farm.

ONP system cost components	Cost (US$)
Licences	23,571,429
Floating rings	1,834,286
Nets	857,143
Moorings	342,857
Boats	1,285,714
Feed barges	1,371,429
Camera systems	214,286
Feed distributors	34,114
Power systems	188,571
Total	29,699,829

LBCC-RAS system cost components	Cost (US$)
RAS Systems	26,640,557
Effluent treatment	3,487,500
Water supply	675,000
Processing	2,112,030
Building	9,426,413
Engineering	5,080,980
Construction management	1,058,538
Bond	254,049
Contingency (10%)	4,848,102
Total	53,583,169

Capital Cost—LBCC-RAS Model

The capital cost of the LBCC-RAS model farm includes all RAS systems, water supply, effluent treatment systems, buildings, engineering services, construction management services, a primary processing facility and general contractor bonding requirements. These components are itemized based on material, equipment, labor and subcontractor services, upon which the costs are estimated. Ten percent contingency is applied to capture uncertainty associated with this level of cost estimation. We assume that the lifespan of materials and equipment is 10 years and the lifespan for buildings and tanks is 20 years. These lifespans are used for calculation of depreciation and replacement cost. The cost of bonding is included as insurance may be required by owners that builders must have for large projects and is typically passed back to the owner. There are currently no comparable license costs for a LBCC-RAS farm in the US. The total capital cost including contingency of the LBCC-RAS model farm is estimated to be 53.6 million US dollars for a total production of 3300 MT head-on gutted salmon (Table 3).

Operating cost—ONP model

The operating cost for the ONP model farm is estimated based on data collected by the Norwegian Directorate of Fisheries (2014) and also Marine Harvest ASA (2014), and are the average costs of the last five years, 2009–2013. Since there are uncertainties associated with these items and the overall cost has increased gradually in the last several years, we applied a 2% increase for the first five year's estimates, and a 3% increase for the remaining year's estimate to account for uncertainties for each cost item. In other words, it is assumed that each cost item will increase 2% for the first five years and 3% for the rest. The operating costs are the average estimates over 15 years. The breakdown of costs is presented in Table 4.

Table 4. Operating expenses for a 3,300 MT HOG LBCC-RAS and ONP Atlantic salmon farm.

Cost item	ONP system		LBCC-RAS system	
	Cost (US$)	Cost (NOK)	Cost (US$)	Cost (NOK)
Feed	2.05	14.34	1.90	13.33
Smolt	0.47	3.30	–	–
Egg	–	–	0.12	0.86
Labor	0.31	2.15	0.52	3.65
Well boat	0.18	1.23	–	–
Health	0.03	0.18	–	–
Electricity	–	–	0.33	2.32
Oxygen	–	–	0.15	1.07

Water treatment	–	–	0.09	0.62
Insurance	0.02	0.16	0.18	1.27
Primary processing	0.43	3.03	0.12	0.83
Transportation	0.25	1.58	–	–
Sales & marketing	0.09	0.60	–	–
Maintenance	0.14	0.99	0.47	3.26
Interest	0.60	4.21	0.65	4.52
Depreciations	0.18	1.28	0.58	4.09
Others	0.33	2.32	0.49	3.45
Total	5.08	35.37	5.60	39.27

Operating Cost—LBCC-RAS Model

The operating cost for the LBCC-RAS model farm is estimated based on the bioplan designed for an annual production of 3300 MT after primary processing. Cost items include feed, oxygen, bicarbonate, electricity, eggs, labor, stock insurance, interest and depreciation. Feed amount and thus cost, is calculated based on the feed required for growth multiplied by feed conversion ratio at different life stages. The amounts, and thus costs, of oxygen and bicarbonate are dependent on the feed required. Oxygen required is estimated to be 0.60 kg oxygen per kg feed, which includes an oxygen transfer efficiency of 75%. Bicarbonate required is estimated to be 0.20 kg bicarbonate per kg feed, which includes a base chemical availability of 75%. The cost of the electricity is determined by the RAS design, which identified all pumps and motors required for operation. The number, and thus cost, of eggs required is estimated by the assumed mortality rates at different life stages. Labor costs for the LBCC-RAS model farm include management (biological and maintenance), fish culture technicians, laboratory technicians, maintenance mechanics, and primary processing staff. It is assumed that insurance cost for the first year of operation is 4% of standing biomass, and then that declines to 2% of standing biomass in the following years. The ratio between interest and cash for capital cost and first year operating cost was 60/40, and an interest rate of 6% was used. Depreciation of each item was estimated using a straight line approach, meaning depreciation cost was charged evenly throughout the useful life of each capital item. Maintenance cost was estimated to be 10% of the total variable cost. To capture unknown costs, a contingency cost is also included which was assumed to be 10%

of the total cost. The increase with 2% for the first 5 years and 3% for the rest are also applied for each cost item due to unforeseen future changes, same as the ONP system.

Sales and Income

It takes approximately one year for salmon to grow to market size, therefore, there is no harvest for Year 1 and a proportionally smaller harvest for Year 2. In Year 3 and onwards, a constant harvest of 3300 MT is assumed for the ONP and LBCC-RAS systems. The price used here is the export market price of fresh gutted salmon in the US market, which is approximately 5.97 US$/kg or 41.8 NOK/kg averaged weekly price for the year 2014 (Statistics Norway, 2015). It is also assumed that the price for salmon in the future would increase in a similar way as the cost items, i.e., increased by 2% for the first five years and 3% for the rest. However, preliminary sales of Atlantic salmon produced by a LBCC-RAS farm have commanded a significant price premium (Guy Dean, Albion Fisheries (Vancouver, BC), personal communication, September 4, 2014), here a 30% price premium is assumed which is approximately 7.76 US$/kg. The total sales revenue is calculated based on export price and annual harvest.

Carbon Footprint

The carbon footprint is the sum of potential climate impacts that a product causes from a defined part of its life cycle. The carbon footprint was calculated using life cycle assessment (LCA) methodology that is a tool for environmental assessment (ISO, 2006a,b). It assesses the inputs of energy and material to the system and from that calculates potential environmental impacts caused by the resource use and outputs to nature in the form of emissions, waste and products. This LCA includes both direct emissions from the feed and salmon production and indirect emissions caused by production and distribution of the commodities and infrastructure that underpin the salmon life cycle.

The potential climate impact, the global warming potential, is calculated by characterizing all emission and impacts into CO_2 equivalents (CO_2eq) according to their radiative properties based on IPCC guidelines (IPCC, 2007).

The goal of the carbon footprint was to compare the potential climate impacts from different ways of providing a retailer in Seattle, WA (US) with Atlantic salmon:

1a) Salmon from a LBCC-RAS system in the US running on electricity generated from a source that uses a typical mix of coal, gas, nuclear, wind and hydropower. Salmon is assumed to be transported fresh to the retailer 250 km by truck.

1b) Salmon from a LBCC-RAS System in the US running on electricity generated from a source that uses 90% hydropower and 10% coal. Salmon is assumed to be transported fresh to the retailer 250 km by truck.

2a) Salmon from a Norwegian ONP system. Salmon is assumed to be transported fresh, first with truck in Norway to Oslo, 520 km, and then with airfreight to Seattle, 7328 km.

2b) Salmon from a Norwegian ONP system. Salmon is assumed to be transported frozen, first with truck in Norway to Oslo, 520 km, and then with ship from Ålesund, Norway, to Seattle through the Panama Canal, 16,473 km.

The functional unit for the assessment, the basis for comparison, was 1 kg of gutted salmon with head on, at the retailer gate. For each case, the assessment included the complete production system, from production of feed ingredients, smolt production and construction of facilities, equipment and transports.

It was assumed that the salmon was gutted close to the production facility and that all byproducts, such as guts, skin and trimmings were utilized mainly for feed production. Mass allocation was applied meaning that the carbon footprint up to slaughter was allocated between the head-on-and-gutted salmon and the byproducts based on their mass. Thus, per unit of mass live salmon and head on and gutted salmon have the same carbon footprint. Important cut offs, processes that are not included in the assessment include: slaughtering process, treatment of the biosolids from the LBCC-RAS system, and transport infrastructure.

Carbon Footprint Data

Table 5 presents important activity data for the carbon footprint of the two systems. Data for the LBCC-RAS system was derived from the concept-level design. Data for the Norwegian ONP system is gathered from industry actors and industry statistics (Winther et al., 2009; Hognes et al., 2011, 2014). Data on the climate impacts from capital and operational inputs were modeled with data from the LCA inventory database Ecoinvent v3.1 (2013). Since many of the operations performed at the ONP farm are performed by sub-contractors, and the extent of the activities, e.g., cleaning and priming of nets, are dependent of exact location, these data are based on the assumption of a representative production model.

Table 5. Inventory data for carbon footprint for two production models (LBCC-RAS system and ONP system) for a 3300 MT HOG Atlantic salmon farm. All numbers are per ton of salmon produced or transported.

	Unit	LBCC-RAS System	ONP system
Feed, economic FCR	ton	1.09	1.27
Concrete	kg	82.5	–
Steel, reinforcing	kg	14.40	0.63
Steel, chromium 18/8 steel	kg	–	0.70
Glass fiber	kg	8.93	–
Nylon	kg	–	1.01
Polypropylene	kg	–	1.79
Polyethylene	kg	–	0.28
Fuel	l	–	10.50
Electricity	kWh	5460	–
Oxygen (liquid)	kg	656	–
Lime (calcium carbonate)	kg	219	–
EPS for transport packaging	kg	25	25
Ice	kg	300	300

Both the LBCC-RAS and ONP systems are modeled using the same feed. Based on LCAs of the average Norwegian salmon feed in 2012, the feed is associated with a carbon footprint of 2.5 kg CO_2eq/kg feed at the feed factory gate. This is a feed with the following composition: 12% marine oil; 19% marine protein; 19% oil from crops; 39% protein from crops; 8% starch from crops and 3% micro ingredients (minerals, vitamins, pigments and other). This carbon footprint reflects a feed where 50% of the soy in the feed is equal to the average Brazilian soy, as modeled by the Agrifootprint database (Centre for Design and Society of the RMIT University, 2014), and the remaining coming from old farms where climate impacts from land use change is not included (Hognes et al., 2014).

Electricity for the LBCC-RAS system in case 1b is modeled as being generated from 90% hydropower and 10% coal power with data from Ecoinvent v3.1 (2013). This case is included as an illustrative case for what is possible if this type of electricity is available. Electricity loss of 3.5% was included for the transmission of the power and transformation from high to medium voltage. This associated the electricity with a carbon footprint of 0.04 kg CO_2eq/kWh. For comparison, the Ecoinvent v3.1 database also provides a dataset that describes the electricity available in the regional entity of the North American Electric Reliability Corporation (NERC), that gives a carbon footprint of 0.64 kg CO_2eq/kWh. This was the electricity data used for the LBCC-RAS system in case 1a.

Road transport was modeled with a truck carrying 20 tons of fish, consuming 3.7 L of diesel per 10 km and has a carbon footprint of 0.09 kg CO_2eq/tkm; this also includes fuel used for the refrigeration system and emission of refrigerants (Winther et al., 2009). The fuel consumption reflects a modern truck. For the ONP system in case 2a, airfreight was modeled using data for a Boeing 747–400 from the Agrifootprint database, with an emission factor of 1.18 kg CO_2eq/tkm (Centre for Design and Society of the RMIT University, 2014). This plane is assumed to use 100% of its load capacity (3600 tons) and the emissions include landing and takeoff for a flight of approximately 10,000 km. For the ONP system in case 2b, ship transport was modeled with data for a ship of 120,000 tons (dry weight) utilizing 80% of its capacity, with an emission factor of 0.004 kg CO_2eq/tkm. Emissions from preparing for the return of the ship and re-loading is included in this data. Fuel for running refrigeration systems and emissions of refrigerants were also included with an emission factor of 0.1 kg CO_2eq/h (Winther et al., 2009).

RESULTS

Financial Analysis

Capital Cost

Tables 3 reports the capital cost of ONP and LBCC-RAS systems. In the ONP system, the largest cost is license fees, which are almost 80% of the total capital cost, while the physical structure cost only accounts for 20%. For LBCC-RAS, the largest cost is the RAS system which is half of the total cost; 18% of the LBCC-RAS capital cost is for building structures. The capital cost of LBCC-RAS is 80% higher than that of the ONP system given the same production capacity. It is important to note that the replacement costs of some cost items are not included in this table, but incorporated into the cash flow analysis.

Operating Cost

The operating cost breakdowns for the two systems are presented in Table 4 and Fig. 2. The total operating costs for the two systems are relatively similar, with LBCC-RAS only 10% higher than the ONP system. Without interest and depreciation, the two production systems have an almost equal operating cost, 4.30 US$/kg for ONP and 4.37 US$/kg for LBCC-RAS. Feed is the single biggest cost item accounting for 41% and 34% of the total operating cost for the ONP and the LBCC-RAS systems, respectively. It is worthwhile to note that these operating costs are subject to change with site selection due to differences in power costs, feed shipping costs and other factors. For example, operating costs presented here do not include the cost of heating or cooling that may or may not be required based on the geographic location of the LBCC-RAS facility.

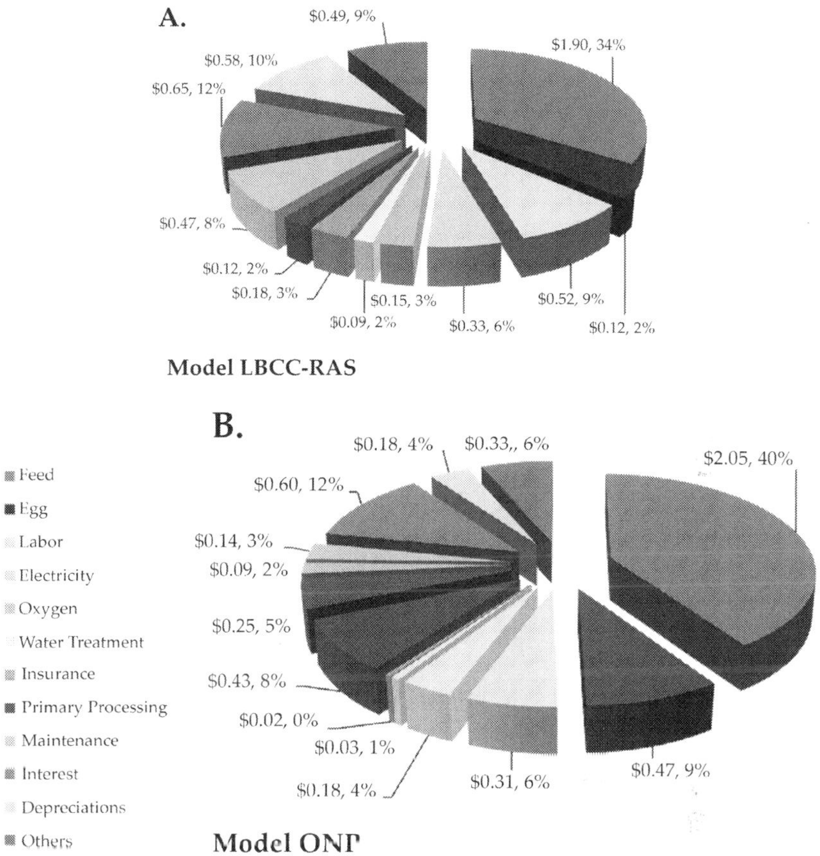

Figure 2. Estimated production costs (US$/kg HOG) according to the investments, product price estimates and the biological production plans for a model 3300 MT HOG Atlantic salmon LBCC-RAS farm (A) and ONP farm (B).

Financial Indicators

The financial analysis is conducted for a period of 15 years; the discount rate is set to seven percent. The summary of the financial analysis is presented in Table 6. Overall, the ONP model system is financially better than the LBCC-RAS model system, even when the LBCC-RAS is selling product with a price premium. All three cases generate positive operating margins, indicating that from a production operating perspective, all are financially viable. The LBCC-RAS system selling salmon at a price premium is comparatively as profitable as the ONP system, even though its NPV is negative (−20,340,000 US$) and its return on investment (9.01%) is lower than the ONP system's ROI (17.77%). However, when selling salmon at the same price as the ONP system, the LBCC-RAS system is barely financially profitable and not an attractive investment. To be comparable with an ONP system, the LBCC-RAS system must command higher market price to breakeven or be profitable.

Table 6. Economic indicators for a 3,300 MT HOG LBCC-RAS and ONP Atlantic salmon farm. Also presented are indicators for the LBCC-RAS farm selling salmon with a 30% price premium.

Economic indicator	ONP system	LBCC-RAS system	LBCC-RAS system premium price
Operating (gross) margin	38.39%	17.56%	40.64%
Profit margin	23.62%	(–)	18.18%
NPV (million US$)	3.54	−120.20	−20.34
IRR before EBIT	15.96%	(–)	13.28%
IRR	7.94%	(–)	2.67%
ROI	17.77%	(–)	9.01%
Break-even production (MT)	1251	3307	2387
Pay-back period (year)	5.63	(–)	11.10
Break-even price (US$)	5.33	(–)	6.44

The IRR can be considered as the true expected yield from an investment. The IRR before EBIT for the LBCC-RAS with price premium is calculated to be 13.28%. The real IRR for the LBCC-RAS with price premium is 2.67%. The discount rate of 7% used here is below the IRR before EBIT and thus the LBCC-RAS would be an investment that results in a positive NPV. However, the discount rate of 7% used here is also above real IRR, and that investment in LBCC-RAS results in a negative NPV. Investors must make investment decisions based on her expectation(s) on return, whether using the IRR of 13.28% or 2.67%.

Sensitivity Analysis

The financial results are very sensitive to some factors. For instance, prices have substantial influence on the results, and are subject to short- and long-term fluctuations due to dynamics in supply and demand. Feed is the largest cost item, so any changes in feed price and feed utilization have large impacts on the economic performance of the operations. Recent figures have suggested the cost of feed has increased gradually. The assumption for feed conversion ratio during growout is one of the most critical values in the estimation because it drives the largest component of the cost of production—feed cost during growout. Performance data from repeated Freshwater Institute trials indicate a feed conversion ratio less than 1.1 during the final growout phase (Summerfelt et al., 2013); utilizing lower FCR values during final growout instead of 1.1 would reduce the cost of production, by potentially up to 6%. Feed is also the major factor influencing the carbon footprint. Other factors such as mortality rates, power cost and mortality also have impacts on financial performance.

Carbon Footprint Results

If the alternative is intercontinental export of fresh salmon by air, then a modern and efficient LBCC-RAS system close to the market can be a more climate friendly alternative, even when running on electric power that mainly originates from fossil fuels (7.4 versus 15.2 kg CO_2eq per kg HOG salmon at retailer gate in Seattle). If the LBCC-RAS system is running on 90% hydropower the carbon footprint of the LBCC-RAS salmon is further reduced to 4.1 kg CO_2eq per kg HOG salmon at the retailer gate. The most climate friendly alternative of all is to ship frozen salmon from Norway with a modern container ship, 3.8 kg CO_2eq per kg HOG salmon at the retailer gate. A frozen product is not directly comparable with a fresh, but with modern freezing technologies, the quality of frozen products is not necessarily inferior to fresh.

At the producer gate, before transport to the retailer in Seattle, the production systems have climate impacts per unit produced of 3.4 versus 3.7 and 7.0 kg CO_2eq/kg salmon live-weight for the ONP and the LBCC-RAS using hydropower or average fossil fuel based electricity, respectively (Table 7 and Fig. 3).

Table 7. Estimated carbon footprint with component contributions at the producer gate and the retailer gate for the following scenarios: (1a) Salmon from a LBCC-RAS system in the US running on a typical electricity mix; (1b) Salmon from a LBCC-RAS system in the US running on electricity generated predominantly from hydropower; (2a) Salmon from a Norwegian ONP system transported by airfreight to Seattle; (2b) Salmon from a Norwegian ONP system transported by ship to Seattle.

	1a)	1b)	2a)	2b)
Feed production	2.69	2.69	3.21	3.21
Construction of facility and equipment	0.39	0.39	0.02	0.02
Grow out and smolt (fuel and electricity)	3.48	0.21	0.16	0.16
Oxygen and lime	0.44	0.44	–	–
At producer gate (live weight)	7.01	3.73	3.39	3.39
Transport, road	0.03	0.03	0.06	0.062
Transport, air or water	–	–	11.40	0.09
Packaging and ice	0.37	0.37	0.37	0.11
Refrigeration during transport	0.00	0.00	0.00	0.10
At retailer gate (HOG)	7.41	4.14	15.22	3.75

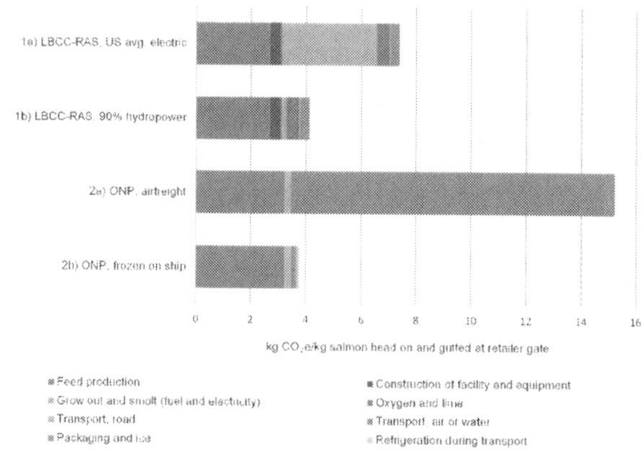

Figure 3. Estimated carbon footprint with component contributions at the producer gate and the retailer gate for the following scenarios: (1a) Salmon from a LBCC-RAS system in the US running on a typical electricity mix; (1b) Salmon from a LBCC-RAS system in the US running on electricity generated predominantly from hydropower; (2a) Salmon from a Norwegian ONP system transported by airfreight to Seattle; (2b) Salmon from a Norwegian ONP system transported by ship to Seattle.

The more general findings confirmed what previous LCAs have found that fish feed is the dominant climate aspect for the selected salmon products, but that energy used in growout and emissions from transports are also important. Production and maintenance of equipment and production facilities are not important climate aspects compared to feed production, transport and water treatment.

DISCUSSION

Given current technology development and possible increases in market price for salmon and production input factors, the ONP system still remains the most profitable, even at this relatively small scale. To achieve comparative financial performance, the LBCC-RAS system requires a price premium, at least 25% higher than current market prices. This is mainly due to considerably higher capital cost for the LBCC-RAS system. However, the difference in operating costs between both systems is relatively small. If the feed conversion ratio can be further improved from 1.1 to 1.0 for LBCC-RAS systems, the gap will be even smaller since feed is the most important cost item. However, improvements in feed conversion ratio are also likely to happen in ONP systems, so the difference in the future for optimized systems is hard to predict. It is important to note that ONP systems are just for the growout phase in Norway, and that salmon now spend more of their lifecycle in LBCC-RAS smolt production facilities (Dalsgaard et al., 2013). Additionally, other costs such as managing sea lice and loss due to disease could further increase the operating cost of ONP systems significantly (Liu and Bjelland, 2014). The largest limiting factor for using LBCC-RAS system appears to be the capital cost. Thus, there are economic incentives for advancing technological innovations of LBCC-RAS systems that can reduce capital cost to become more competitive with ONP systems.

LBCC-RAS systems are not a new technology, and have been used for the last twenty years for growing out both freshwater species, such as eel and catfish, and marine species like trout and sea bass (Martins et al., 2010; Badiola et al., 2012). There is increasing interest in applying LBCC-RAS for the salmon smolt stage in Nordic countries and Europe (Dalsgaard et al., 2013). However, due to low returns on investment and and a history of failures when the technology was not well advanced, LBCC-RAS have not been used widely.

Economic incentives have been proven to be more effective than traditional command and control policy (Bailly and Willmann, 2001; Liu et al., 2013). Market-based economic instruments such as taxes, subsidies, fees/charges and eco-labeling can create incentives for the industry to foster cost-effective technology innovation and adaptation such as LBCC-RAS systems or other closed containment systems (Rosten et al., 2013). However, such incentive-based approaches have to be executed with the vectors of market and social forces such as environmental policy and consumers. Eco-labeling farmed products would be a market-driving power to change consumers' purchasing behavior. Concerned consumers are likely willing to pay more for the products which are produced in an environmental sustainable way. Subsidies and taxes can be used to stimulate cost-effective technology innovation and adaptation, e.g., rewarding improved environmental performance from capturing and controlling waste streams in closed-containment systems or eliminating sea lice infestation. While environmental policies may also have a role, in Norway, "green" concessions for salmon farming require the aquaculture industry to employ technological and operational innovations and solutions to reduce the incidence of salmon lice and escapes. These technologies require upfront investment which can be significant, but over the long run, such technological innovation would increase social license to operate through improved environmental performance and reduced conflict with other resource users, perceived market payoffs through reduced costs to obtain and maintain a license to operate, and monitor and mitigate negative impacts, e.g., costs of recapturing escapes. Captured nutrient laden waste streams associated with LBCC-RAS may also result in ancillary revenue streams, e.g., aquaponics.

The carbon footprint analysis showed that, with respect to climate impact, producing close to the market is preferable by a good margin, especially when the LBCC-RAS system utilized electricity generated from 90% hydropower and the alternative is to export fish fresh, fast and a long distance. Even if salmon is LBCC-RAS produced with electricity based on fossil fuels, intercontinental export of fresh fish on airplanes is not a preferable option. However, environmental considerations involving high inputs of electricity should be followed up with a discussion of what is the environmentally optimum way of using available electricity. Electricity is of the highest energy quality available, and many industrial and infrastructure processes do not have an alternative to electricity. Export of frozen salmon was the best option of all, but cannot be directly compared with fresh salmon. Still, this result points to a future option, with product development, improvement of logistic chain management, to maintain quality through the transport, and market acceptance, frozen intercontinental export has the potential to compete with local LBCC-RAS products. Another important assumption regarding transport is that most intercontinental export of fresh Norwegian salmon is done with flights that also carry passengers. Thus a more precise comparison should include details and insight into how it is reasonable to allocate the fuel used and corresponding emissions between goods and passengers. In addition to this, the LCA data that is available on flight transport is highly variable. This indicates that more precision on the exact age/technology and size of the aircrafts being used should be included.

The carbon footprint contained several cut-offs and assumptions that limits the conclusions that can be drawn, e.g., the same data on feed was used for salmon production in the US and Norway. There are likely to be differences in the carbon footprint of the feeds that would actually be used. A potentially important cut off is that treatment of the biosolids was not included. Biosolids could be seen as both waste and a resource, but either way handling it will involve the use of both energy and transports together with emissions from the biosolids itself. Still, this aspect was left out because it would be difficult to compare to the ONP system, where there is no biosolids capture and waste feed and feces is discharged directly in the ocean.

Most often, the concentrated effluent of LBCC-RAS systems now in operation in North America and Europe are treated in order to meet stringent wastewater discharge permits. Thus a flow-through system will have a higher eutrophication potential. However, if the concentrated effluent of a LBCC-RAS is not treated there is no such advantage to be obtained. Rosten et al. (2013) suggests a classification system for closed containment systems from 1 to 4, where category 4 is the most closed system towards the external environment applying treatment of both inlet and outlet of a LBCC-RAS system. Acidification and toxic potentials are strongly connected to energy consumption and thus similar to climate impacts with regards to where and why they occur.

Aquaculture technologies have been compared with LCA previously; our assessment was compared with a selection of peer reviewed literature (Table 8). This selection of literature points to the same main conclusions: feed production is a dominating factor for carbon footprint in salmon aquaculture, and for LBCC-RAS, the use of energy for water treatment can be equally important and equipment and infrastructure is of minor importance. The importance of energy used for water treatment depends on how this energy is produced. The literature also shows that important parameters for the LCA, such as the FCR and energy used for water treatment varies considerably. This study has not gone into the details to explain these differences, but important reasons are probably that the studies rely on different assumptions, experimental data and site specific properties. These differences make it difficult to compare the final carbon footprint among studies. In addition to differences in the aquaculture systems that are compared, it is also not possible to be sure that the data on feed that are used are comparable. Finally, there are also methodical differences, e.g., Ayer and Tyedmers (2009) used allocation based on the energy content in the different outputs rather than their mass and Samuel-Fitwi et al. (2013) used system expansion.

Table 8. Data from published studies on LCAs of LBCC-RAS for rearing salmonids.

System analyzed and method	Electricity consumption (kWh/kg)	Feed efficiency	Carbon footprint of product (kg CO_2eq/kg)	Reference
Salmon production with marine net-pen, marine floating bag, land-based saltwater flow through and a land-based freshwater RAS. Assessment from feed and smolt production to farm gate	Net-pen: 1.49	Net-pen: 1.30	Net-pen: 2.07	Ayer and Tyedmers (2009)
	Land, flow through: 13.4	Land, flow through: 1.17	Bag: 1.90	
	Land, recirculating: 22.6	Land, recirculating: 1.45	Land, flow through: 2.77	
	Electricity mix 80% fossil fuels		Land, recirculating: 28.20	
Trout production with a flow through system and a hypothetical recirculating system. From feed production to fish ready for slaughter	Flow through low pumping: 2.36	Flow through: 1.10	Flow through: 2.02	d'Orbcastel et al. (2009)
	Recirculation: 10.7	Recirculation: 0.80	Recirculation: 1.60–2.04	
	86.6% nuclear energy			

Rainbow trout production in flow through systems (extensive and intensive) and recirculating system. From feed production to fish ready for slaughter	Intensive flow through: 2.55	Flow through: 0.91–1.2	Flow through extensive: 2.24	Samuel-Fitwi et al. (2013)
	Recirculating: 19.6	Recirculating: 0.86	Flow through intensive: 3.56	
	Electricity based on fossil fuels		Recirculating: 13.60	
Rainbow trout production with flow-through, recirculating and semi-closed system. From feed production to fish ready for slaughter	Flow-through: 0.65	Flow through: 1.15	Flow through: 1.16	Dekamin et al. (2015)
	Recirculating: 8.1	Recirculating: 1.47	Recirculating: 6.10	
	Semi-closed: 7.6	Semi-closed: 1.57	Semi-closed: 6.38	
Salmon production within a floating tank, flow-through, solid-walled aquaculture system. From feed production to fish ready for slaughter	Actual production cycle: 7.3	Hatchery: 1.5	Actual: 3.87	McGrath et al. (2015)
	Intended production cycle: 4.6	Grow out actual: 1.46	Intended: 3.03	
		Grow out intended 1.37		

The conclusion with regards to the hypothesis that a LBCC-RAS produced salmon will have a higher carbon footprint than one from an ONP system is solely dependent on what carbon dioxide emission the electricity production is attributed with and the method and form that the product is transported to market with. If the electricity for the LBCC-RAS is considered to be primarily hydropower then the carbon footprint for the two systems at the producer gate are relatively close (3.39 and 3.73 kg CO_2eq/kg salmon live-weight). If the electricity for the LBCC-RAS is considered to be the average US mix dominated by fossil fuels, then the LBCC-RAS has a higher carbon footprint at the producer gate (7.01 versus 3.39 kg CO_2eq/kg salmon live-weight). The carbon footprint demonstrates the importance of the emissions associated with electricity generation for LBCC-RAS systems.

In a market where electric power is a commodity in short supply, and where power markets are connected through economy and/or the grid, it is challenging to argue that power is supplied from one specific source. On top of this, renewable energy, such as hydropower, is often sold to clients that pay extra for a certificate to claim that their electricity is produced from renewable sources. For this system to work, as well as for carbon footprint, it would require a mechanism that ensures that the sum of certificates that are sold do not exceed the renewable power that is actually available and that everybody who does not buy certificates uses a carbon footprint of their electricity that does not include the renewables that are sold with certificates. This is what is then called the residue mix. As far as these authors know, no such system exists today and it is recognized to be "good practice" to use the average production mix in the grid where the electricity use takes place. The grid here being what is physically and/or economically connected.

Extending the carbon footprint to include transport to market for the most likely production systems, fresh salmon produced in LBCC-RAS systems close to a US market that use an average US electricity mix and fresh salmon produced in Norway in ONP systems shipped to the same market by airfreight, yields the result that LBCC-RAS has a much smaller carbon footprint, 7.41 versus 15.22 CO_2eq/kg salmon HOG, respectively. In this case the carbon footprint associated with transport is the dominant factor for ONP-produced salmon, accounting for more carbon footprint than the entire production on a kg salmon HOG basis (Fig. 3).

CONCLUSIONS

In this paper, we compare the economic and environmental performance of the Norwegian open net pen system in the sea and the US land-based, closed containment water recirculating aquaculture system for the same production capacity targeting the same US market. The scale used for the open net pen system is smaller than the average operation scale in Norway, so both systems could be scaled up to higher production capacity. This will result in reduction in cost due to scale of economy. However, the main findings are drawn:

- Capital cost for land-based closed containment water recirculating salmon farming systems is significantly greater than capital cost for traditional open net pen salmon farming systems, but increasing net pen site license costs in Norway are bringing the capital costs closer.
- Production cost for land-based closed containment water recirculating salmon farming systems is approximately the same as production cost for traditional open net pen salmon

farming systems at this scale, when excluding interest and depreciation.

- Return on investment for traditional open net pen salmon farming at this scale is twice that of land-based closed containment water recirculating salmon farming, when land-based produced salmon are sold at a price premium.
- Internal rate of return for earnings before interest and tax for traditional open net pen salmon farming at this scale is only slightly greater than that of land-based closed containment water recirculating salmon farming, when land-based produced salmon are sold at a price premium.
- The carbon footprint of salmon produced in land-based closed containment water recirculating aquaculture systems that are using a typical US electricity mix based on fossil fuels is twice that of salmon produced in traditional open net pen systems, when delivery to the market is not included.
- The carbon footprint of salmon produced in land-based closed containment water recirculating aquaculture systems delivered to market in the US is less than half of that for salmon produced in traditional open net pen systems in Norway that is delivered to the US by air freight.

ACKNOWLEDGEMENTS

The authors thank the Agriculture Research Service of the United States Department of Agriculture (Agreement No. 59-1930-1-130) and the Norwegian Research Councilfor funding this work.

REFERENCES

1. Aardal, J., 2014. Grønn Konsesion til Bjørøya: Bjørøya Fiskeoppdrett er tildelt grønn konsesjon for oppdrett av laks. Namdalsavisa, Available at: http://www. namdalsavisa.no/Nyhet/article7264369. ece.
2. Ayer, N.W., Tyedmers, P.H., 2009. Assessing alternative aquaculture technologies: life cycle assessment of salmonid culture systems in Canada. J. Clean. Prod. 17, 362–373.

3. Badiola, M., Mendiola, D., Bostock, J., 2012. Recirculating aquaculture systems (RAS) analysis: main issues on management and future challenges. Aquacult. Eng. 51, 26–35.
4. Bailly, D., Willmann, R., 2001. Promoting sustainable aquaculture through economic and other incentives. In: Subasinghe, R.P., Phillips, M.J., Bueno, P., Hough, C., McGladdery, S.E., Arthur, J.R. (Eds.), Aquaculture in the Third Millennium. FAO, Rome, pp. 95–101.
5. Bergheim, A., Drengstig, A., Ulgenes, Y., Fivelstad, S., 2009. Production of Atlantic salmon smolts in Europe—current characteristics and future trends. Aquacult. Eng. 41, 46–52.
6. Centre for Design and Society of the RMIT University, 2014. Agrifootprint [Data File]. PRé Consultants, Amersfoort, The Netherlands.
7. Dalsgaard, J., Lund, I., Thorarinsdottir, R., Drengstig, A., Arvonen, K., Pedersen, P.B., 2013. Farming different species in RAS in Nordic countries: current status and future perspectives. Aquacult. Eng. 53, 2–13.
8. Dekamin, M., Veisi, H., Safari, E., Liaghati, H., Khoshbakht, K., Dekamin, M.G., 2015. Life cycle assessment for rainbow trout (Oncorhynchus mykiss) production systems: a case study for Iran. J. Clean. Prod. 91, 43–55.
9. d'Orbcastel, E.R., Blancheton, J.-P., Aubin, J., 2009. Towards environmentally sustainable aquaculture: comparison between two trout farming systems using life cycle assessment. Aquacult. Eng. 40, 113–119.
10. De Ionno, P.N., Wines, G.L., Jones, P.L., Collins, R.O., 2006. A bioeconomic evaluation of a commercial scale recirculating finfish growout system—an Australian perspective. Aquaculture 259, 315–327.
11. Directorate of Fisheries, 2014. Profitability Survey on the Production of Atlantic Salmon and Rainbow Trout. Statistics–Norwegian Aquaculture, Available from Directorate of Fisheries website, http://www.fiskeridir.no/english/statistics.
12. Ecoinvent v3.1., 2013. LCA background database. Access can be purchased at http://www.ecoinvent.org/database/ecoinvent-

version-3/ecoinvent-v30/.

13. Färe, R., Grosskopf, S., Roland, B.-E., Weber, W.L., 2005. License fees: the case of Norwegian salmon farming. In: Working Paper Series in Economics and Management No.07/05 October 2005. University of Tromsø.
14. Gempesaw, C.M., Supitaningsih, I., Bacon, J.R., Heinen, J., Hankins, J., Wang, J., 1993. Economic analysis of an intensive aquaculture recirculating system for trout production. In: Wang, J. (Ed.), Techniques for Modern Aquaculture. American Society of Agricultural Engineers, St. Joseph, MI, pp. 263–277.
15. Heinsbroek, L.T.N., Kamstra, A., 1990. Design and performance of water recirculation systems for eel culture. Aquacult. Eng. 9, 187–207.
16. Hognes, E.S., Ziegler, F., Sund, V., 2011. Carbon Footprint and Area Use of Farmed Norwegian Salmon (SINTEF Fisheries and Aquaculture Report: A22673). SINTEF Fisheries and Aquaculture, Trondheim, Norway.
17. Hognes, E.S., Angus, G., Ziegler, F., 2014. Handbook for Greenhouse Gas Assessment of Seafood Products (SINTEF Fisheries and Aquaculture Report: A26745). SINTEF Fisheries and Aquaculture, Trondheim, Norway.
18. International Organization for Standardization [ISO], 2006a. ISO 14040:2006(E). Environmental Management–Life Cycle Assessment–Principles and Framework. ISO, Geneva, Switzerland.
19. International Organization for Standardization [ISO], 2006b. ISO 14044:2006(E). Environmental Management–Life Cycle Assessment–Requirements and Guidelines. ISO, Geneva, Switzerland.
20. Intergovernmental Panel on Climate Change [IPCC], 2007. Contribution of Working Group I. In: Solomon, S., Qin, D., Manning, M., Chen, Z., Marquis, M., Averyt, K.B., Tignor, M., Miller, H.L. (Eds.), Fourth Assessment Report of the Intergovernmental Panel on Climate Change. Cambridge University Press, Cambridge, UK.
21. Kolarevic, J., Baeverfjord, G., Takle, H., Ytteborg, E., Reiten, B.K.M., Nergård, S., Terjesen, B.F., 2014. Performance and welfare of Atlantic salmon smolt reared in recirculating or flow through

aquaculture systems. Aquaculture 432, 15–25.

22. Kumar, G., Engle, C., 2011. The effect of hybrid catfish fingerling prices on the relative profitability of hybrid channel catfish. J. World Aquacult. Soc. 42, 469–483.
23. Liu, Y., Bjelland, H.V., 2014. Estimating costs of sea lice control strategy in Norway. Prev. Vet. Med. 117, 469–477.
24. Liu, Y., Chuenpagdee, R., Sumaila, R., 2013. How governable is salmon aquaculture? In: Bavinck, M., Chuenpagdee, R., Jentoft, S., Kooiman, J. (Eds.), Governability of Fisheries and Aquaculture: Theory and Applications. Springer, New York, pp. 201–218.
25. Liu, Y., Sumaila, U.R., 2007. Economic analysis of netcage versus sea-bag production systems for salmon aquaculture in British Columbia. Aquacult. Econ. Manag. 11, 371–395.
26. Losordo, T.M., Westerman, P.W., 1994. An analysis of biological, economic, and engineering factors affecting the cost of fish production in recirculating aquaculture systems. J. World Aquacult. Soc. 25 (2), 193–203.
27. Marine Harvest ASA., 2014. Salmon Farming Industry Handbook 2014. Available from Marine Harvest web site, (June) http://www.marineharvest.com/globalassets/investors/handbook/handbook-2014.pdf.
28. Martins, C.I.M., Eding, E.H., Verdegem, M.C., Heinsbroek, L.T., Schneider, O., Blancheton, J.-P., Verreth, J.A.J., 2010. New developments in recirculating aquaculture systems in Europe: a perspective on environmental sustainability. Aquacult. Eng. 43, 83–93.
29. Mattilsynet, 2011. Per Generation Mortality for Atlantic salmon. Available from Mattilsynet website, http://www.mattilsynet.no/fisk og akvakultur/.
30. McGrath, K.P., Pelletier, N.L., Tyedmers, P.H., 2015. Life cycle assessment of a novel closed-containment salmon aquaculture technology. Environ. Sci. Technol. 49, 5628–5636.
31. Muir, J., 1981. Management and cost implications in recirculating water systems. In: Allen, L.J., Kinney, E.C. (Eds.), Proceedings of the Bio-Engineering Symposium for Fish Culture (FCS Publ. 1). American Fisheries Society, Bethesda, MD, pp. 116–127.

32. Naylor, R., Hindar, K., Fleming, I., Goldburg, R., Williams, S., Volpe, J., Whoriskey, F., Eagle, J., Kelso, D., Mangel, M., 2005. Fugitive salmon: assessing the risks of escaped fish from net-pen aquaculture. Bioscience 55, 427–437.
33. Norway, 2008. Regulations on the Operation of Aquaculture Facilities (FOR 2008-06-17 No. 822). Available from LOVDATA website, https://lovdata.no/dokument/SF/forskrift/2008-06-17-822?q=akvakulturloven.
34. Rosten, T., Terjesen, B.F., Ulgenes, Y., Henriksen, K., Biering, E., Winther, U., 2013. lukkede oppdrettsanlegg i sjø–økt kunnskap er nødvendig. VANN 01 2013, pp. 5–13.
35. RS Means, 2010. Facilities construction cost data, 2010. Access can be purchased at http://www.rsmeans.com/65205.aspx.
36. Samuel-Fitwi, B., Nagel, F., Meyer, S., Schroeder, J.P., Schulz, C., 2013. Comparative life cycle assessment (LCA) of raising rainbow trout (Oncorhynchus mykiss) in different production systems. Aquacult. Eng. 54, 85–92.
37. Statistics Norway, 2015. Export of Salmon. Available from Statistics Norway website, https://www.ssb.no/en/utenriksokonomi/statistikker/laks.
38. Summerfelt, S., Christianson, L., 2014. Fish farming in land-based closed-containment systems. World Aquacult. (March), 18–22.
39. Summerfelt, S., Waldrop, T., Good, C., Davidson, J., Backover, P., Vinci, B., Carr, J., 2013. Freshwater Growout Trial of St. John River Strain Atlantic Salmon in a Commercial-Scale, Land-Based, Closed-containment System, January. The Conservation Fund, Shepherdstown, WV.
40. Summerfelt, S.T., Wilton, G., Roberts, D., Rimmer, T., Fonkalsrud, K., 2004. Developments in recirculating aquaculture systems for Arctic char culture in North America. Aquacult. Eng. 30, 31–71.
41. Timmons, M.B., Ebeling, J.M., 2010. Recirculating Aquaculture, 2nd ed. Cayuga Aqua Ventures, LLC, Ithaca, NY.
42. United States Department of Agriculture Economic Research Service [USDA ERS], 2015. U.S. Atlantic Salmon Imports, Volume by Selected Sources—All Years [data file]. Available from USDA

ERS web site, http://www.ers.usda.gov/dataproducts/aquaculture-data.aspx.

43. United States National Oceanic and Atmospheric Administration National Marine Fisheries Service [NOAA], 2013. FishWatch—Atlantic Salmon. Retrieved from http://www.fishwatch.gov/seafood profiles/species/salmon/species pages/atlantic salmon farmed.htm (26.03.15.).
44. Winther, U., Ziegler, F., Skontorp Hognes, E., Emanuelsson, A., Sund, V., Ellingsen, H., 2009. Carbon Footprint and Energy use of Norwegian Seafood Products (SINTEF Fisheries and Aquaculture Report: SFH80 A096068). SINTEF Fisheries and Aquaculture, Trondheim, Norway.
45. Ziegler, F., Winther, U., Hognes, E.S., Emanuelsson, A., Sund, V., Ellingsen, H., 2013. The carbon footprint of Norwegian seafood products on the global seafood market. J. Ind. Ecol. 17, 103–116

5 PURPOSES AND DEGREES OF COMMODIFICATION: ECONOMIC INSTRUMENTS FOR BIODIVERSITY AND ECOSYSTEM SERVICES NEED NOT RELY ON MARKETS OR MONETARY VALUATION

Thomas Hahn[a], Constance McDermott[b], Claudia Ituarte-Lima[a], Maria Schultz[a], Tom Green[c], Magnus Tuvendal[a]

[a] Stockholm University, Stockholm Resilience Centre, SE-10691 Stockholm, Sweden

[b] University of Oxford, Environmental Change Institute, Oxford OX1 3QY, United Kingdom

[c] Simon Fraser University, Centre for Dialogue, Vancouver, British Columbia, Canada V6B 1L6

ABSTRACT

Commodification of nature refers to the expansion of market trade to previously non-marketed spheres. This is a contested issue both in the scientific literature and in policy deliberations. The aim of this paper is to analytically clarify and distinguish between different purposes and degrees of commodification and to focus attention to the safeguards: the detailed institutional design. We identify six degrees of commodification and find that all ecosystem services policies are associated with some degree of commodification but only the two highest degrees can properly be associated with neoliberalisation of nature. For example, most payments for ecosystem services (PES) are subsidy-like government compensations not based on monetary valuation of nature. Biodiversity offsets can be designed as market schemes or non-market regulations; the cost-effectiveness of markets cannot be assumed. To avoid the confusion around the concept 'market-based instrument' we suggest replacing it with 'economic instruments' since relying on the price signal is not the same thing as relying on the market. We provide a comprehensive framework emphasising the diversity in institutional design, valuation approaches and role of markets. This provides flexibility and options for policy integration of biodiversity and ecosystem services in different countries according to their political and cultural context.

Keywords: Biodiversity financing mechanisms, Innovative financial mechanisms, Financialization, Non-monetary valuation, Adaptive governance

INTRODUCTION

Valuation of ecosystem services (ES) and the use of economic instruments are increasingly becoming part of the international discussions on scaling-up biodiversity financing. The Convention of Biological Diversity (CBD) states that "biodiversity values" should be integrated into development strategies, planning processes, national accounts, and reporting systems (Aichi Biodiversity Target 2) and calls for the elimination of harmful subsidies as well as the development of "positive incentives for the conservation and sustainable use of biodiversity" (Target 3).[1]

The focus on biodiversity values and 'Innovative Financial Mechanisms' (IFMs) has for some actors become extremely controversial within the CBD process, especially the use of economic instruments like payments for ecosystem services (PES) and biodiversity offsets.[2] Without appropriate institutional arrangements that safeguard (ensure) biodiversity and equity, there is a risk that economic instruments, as well as other types of policy instruments, will not contribute towards the three CBD objectives (Ituarte-Lima et al., 2014). These are (i) conservation of biological diversity, (ii) the sustainable use of its components, and iii) the fair and equitable sharing of the benefits arising out of the utilisation of genetic resources. The CBD calls for a broader governance approach to valuation and financing so that the IFMs do not "undermine achievement of the Convention's three objectives" (CBD, 2010). This motivates a focus on safeguards, which we define as the specific factors in the institutional design and implementation procedures aimed to ensure good environmental and social outcomes.

The risks of using market-based instruments for financing biodiversity and ES range from ethical arguments about transforming human-nature relations by "commodity fetishism" (Kosoy and Corbera, 2010) and crowding-out moral obligations as motives for nature protection e.g. (Luck et al., 2012), to instrumentalist arguments focused on the efficiency and equity of the processes and outcomes of such schemes e.g. (Corbera et al., 2007). This is often associated to neoliberalism. Based on McAfee and Shapiro (2010), we define neoliberal ES policies as instruments designed on the premise that the market allocates scarce ecological resources more efficiently than 'command-and-control' regulations and treaties.

Normative framings of ES and commodification are important but sometimes become an obstacle to addressing the empirical instrumental question of how different economic incentive schemes actually perform (Gómez-Baggethun and Ruiz-Pérez, 2011: 622; Dempsey and Robertson, 2012). In this paper we emphasise an instrumentalist approach. The aim is to analytically clarify and distinguish between different purposes and degrees of commodification and to refocus attention on the detailed institutional design of policy instruments, in particular controversial economic instruments.

This paper interrogates the concepts of commodification, valuation, and markets in order to build a framework for policy integration of ES, i.e. addressing and integrating ES concerns in sector policy making (Nilsson and Persson, 2003). Based on this framework, we further analyse the foundations for payments for ecosystem services (PES) and biodiversity offsets, to explore a menu of options for tailoring these instruments to accommodate country-specific concerns. Since biodiversity and most ES are much more difficult to measure than carbon dioxide and other emissions, our framework of commodification is not directly transferable to pollution quotas, carbon markets or emission-trading systems.

SIX DEGREES OF COMMODIFICATION

Commodification of biodiversity and ES means, broadly speaking, the expansion of market trade to previously non-marketed areas of the environment (Luck et al., 2012). This is often described as a process related to idea of commensurability underlying monetary valuation (Aldred, 2002; Vatn, 2009). Kosoy and Corbera (2010) identify three necessary stages in commodification: first defining an ecosystem service, second assigning an exchange-value, and third create a market. Adding property rights, Gómez-Baggethun and Ruiz-Pérez (2011) identify four main stages of commodification: (i) economic utilitarian framing, (ii) monetary valuation, (iii) appropriating the value of ES through formalisation of property rights, and (iv) commercialisation, i.e. market trade by PES or offsets.

However, these stages need not be consecutive and the process is not necessarily unidirectional or irreversible (Gómez-Baggethun and Ruiz-Pérez, 2011), hence we use degrees rather than stages. Based on Muradian et al. (2010: 1206), we define the degree of commodification as the extent to which the value of biodiversity or an ecosystem service has become a tradable commodity.

We find it useful to analyse commodification in terms of policy integration and in this more empirical context at least two more degrees need to be added, as well as the zero degree (policy integration without commodification). The justification for these degrees is discussed below. Empirically, the degree of commodification is a matter of the institutional design of a particular policy instrument. The stated purpose of introducing this policy instrument is tailored to the specific ideological orientation of the government and can be observed e.g. in national legislation and underlying government bills. Together, the degree (institutional design) and stated purpose (justification) influence the legitimacy of the instrument. When the detailed institutional arrangements are analysed we find that what are generally described as PES and biodiversity offsets may involve different degrees of commercialisation and hence commodification. Hence, we find it useful to distinguish between six degrees (plus the zero degree) of commodification:

0. "No commodification" (zero degree) includes intrinsic or relational appreciation of ecosystems, in which the rationale for protecting nature is nature itself. 'Relational' include indigenous cosmologies emphasising reciprocity and cyclical processes (MA, 2005a: 86–87) as well as interaction for reasons of spirituality and even subsistence farming (Turnhout et al., 2013). The policy options include long-established policy instruments with no commodification such as national parks and nature reserves as well as the more recent notion of 'environmental flow' in water governance[3] when justified by deontological ethics (moral duties) or nature's intrinsic values. This also includes approaches linked to the rights of nature or the inalienable rights of indigenous peoples to sustain their cultural and sacred practices (the viability of biocultural sites for cultural and sacred practices are often linked with protecting intrinsic values of biodiversity).

International legal and policy options involving no commodification include new legal paradigms recognising rights of nature which have been characterised as "Earth Jurisprudence" (Burdon, 2011). As part of this, Higgins et al. (2013) question the belief that the market will provide effective and efficient remedies and have proposed "Ecocide" as part of international criminal law which would aim to pre-empt, prevent and prohibit mass damage, destruction or loss of ecosystems whether committed during or outside of war-time as well as impose an associated legal duty of care upon persons in positions of superior responsibility. Other international initiatives include the World People's Conference on Climate Change and the Rights of Mother Earth,[4] Global Alliance for the Rights of Nature, and the Community Environmental Legal Defence Fund (Daly, 2012). At the national level, examples of intrinsic values include the recent Constitutional recognition of the rights of nature in Ecuador (Burdon, 2011)[5] and rights of Mother Earth and good living in Bolivia's Law 071 and Law 300.[6]

1. The first degree of commodification arises under the instrumental (or even economic) framing of nature, though without explicit efforts at valuation. The separation of humans and nature and hence an instrumental

view of nature can be found already in the works of Francis Bacon (1561–1626) (Merchant, 1980). The expansion of this instrumental framing to include ecosystem processes was popularised by e.g. Ehrlich and Mooney (1983) and Daily (1997). Since The Millennium Ecosystem Assessment (MA, 2005) ecosystem services and instrumental framing have become mainstream in environmental policy although this may not involve monetary (economic) valuation (Dempsey and Robertson, 2012).

Even if instrumental framing has often been put forward as the first step of commodification (e.g. Kosoy and Corbera (2010) and Gómez-Baggethun and Ruiz-Pérez (2011)), there is little information on how this can be assessed. We suggest that the difference between an intrinsic (relational) and instrumental framing (zero vs. first degree) can be assessed by analysing the institutional design of the policy instrument, including the detailed management prescriptions (see Tables 1 and 2). There is no sharp demarcation line and many countries may use both framings for the same policy instrument.

Table 1. Degrees of commodification in terms of policy instruments for biodiversity and ecosystem services.

Degree of commodif.	Main category	Examples
0	Moral suasion and regulations justified by intrinsic value	• Information appealing to moral responsibility. • Recognising social equity and nature's intrinsic value, e.g. endangered species acts and nature reserves
1	Non-monetary regulations based on instrumental arguments	• Nature reserves and other land-use plans focusing on nature's instrumental value to human wellbeing
2	Non-monetary regulations based on physical metrics (units of nature)	• Ecological compensation with no role for price signals or market transactions
3	Non-monetary regulations designed to maximise economic efficiency	• City park designed and managed to maximise calculated recreation values

4	Economic instruments (not traded)	• Taxes and subsidies • Subsidy-like PES paid by governments
5	Economic instruments (voluntary market trade)	• Market-like PES • Markets for ecosystem services (MES), e.g. biodiversity offsets trading conservation credits
6	Financial instruments	• Forest bonds • Biodiversity derivatives

2. The second degree of commodification occurs when policy makers introduce "new" property rights and liabilities which involve measurements of biodiversity or ecosystem service units but without monetary valuation or price signals. An example of this is biodiversity offsets or ecological compensation and the idea is that a company that develops land and water resources for e.g. mining, housing, industries, infrastructure, etc. should compensate for the degradation of biodiversity and ES by investing in these values elsewhere (Conway et al., 2013). This could be voluntary or mandatory. Degradation (i.e. negative impacts) should, according to the mitigation hierarchy, first be avoided by choosing a site with less ecological values for the development project. Once a site has been approved for exploitation, degradation should be minimised. The third step of the mitigation hierarchy (sometimes included in the second step) is that the developer takes rehabilitation or restoration measures on the ecosystems impacted and the final step consists of offset measures outside the developed area to compensate for significant adverse residual impacts (Dickie et al., 2010).

Biodiversity offsets are often used synonymously with ecological compensation but while biodiversity offsets specifically seek measurable units to deliver no net loss (or net gain) of conservation outcomes, "ecological compensation is a broader term" (Conway et al., 2013: 2). In the scope of this paper, we focus on mandatory programs and let the use of market transactions define the difference between these instruments, so that biodiversity offsets use the market while ecological compensation does not (we will elaborate on the meaning of 'markets' later).

With this demarcation, ecological compensation belongs to the 2nd degree of commodification while biodiversity offsets belong to the 5th level. Ecological compensation is common in the European Union, e.g. development within Natura 2000 sites needs to be compensated for and this is done case-by-case by government agencies. On ordinary (non-protected) land, the German compensation pools is an established policy instrument where the weighing of degradation and compensation as well as the decision of how much and which land should be used for compensation are done by municipal or private agencies, appointed by the state. Having a pool of compensation land to choose from facilitates the process of matching sites and enables the agency to consider habitat connectivity like green corridors. Even if the German compensation pools sometimes are referred to as "habitat banks" (e.g. Conway et al., 2013: 106), it is clear that there is very limited role for the market in determining price or quality (Conway et al., 2013:113–115). The Western Cape Biodiversity Offset program in South Africa (Dickie et al., 2010: 51) would also be an ecological compensation program with our definition (Koh et al., 2014).

3. The third degree involves deliberate efforts to express or 'demonstrate' (TEEB, 2010: 11) the value of nature in monetary terms (monetary valuation). As mentioned already, this has been referred to as step towards commodification. Still, we are hesitant to include it as a degree since monetary valuation is 'only' an analytical calculation which is not necessary as decision support to justify economic instruments. We have chosen to include it because it introduces commensurability (expressing ES in monetary terms) (Aldred, 2002). Furthermore, Neuteleers and Engelen (2015) argue that monetary valuation contributes to real commodification and therefore "paves the way (discursively and sometimes technically) for commodification to happen" (Gómez-Baggethun and Ruiz-Pérez, 2011: 624).

For example, a nature reserve targeted at biodiversity conservation has probably different management prescriptions compared to a nature reserve designed and managed to maximise recreation values based on cost-benefit calculus (Table 1). Furthermore, when governments demand a cost-benefit analysis of nature reserves, although in the short run institutional design may remain unaltered, in the long run alteration seems likely if nature reserves are provided by private and public landowners just like subsidy-like PES program.

Sometimes 'valuation' is used as shorthand for 'monetary valuation' although 'valuation' is much broader and may involve anything between two extreme approaches; (i) a qualitative understanding and appreciation of the importance of the underlying ecological processes (Daily, 1997) and (ii) a monetary expression of the final ES "directly enjoyed" by people (Boyd and Banzhaf, 2007: 619). The first approach is central for ecological economics. It emphasises multi-functionality and the role of biodiversity (in genes, species, and ecosystems) to support and sustain livelihoods, especially the role of resilience. Such valuation is mainly expressed in non-monetary terms, focusing on understanding the link between ecosystem processes and human wellbeing. It belongs to the first degree of commodification. The second approach could be appropriate when the purpose of valuing ES is to integrate these values in national accounting and reporting systems (Fisher et al., 2009), e.g. in accordance to Aichi Target 2.

In between these extremes are two versions of monetary valuation that belong to the same box in Table 2 but still are ontologically different. The first version can be exemplified by replacement cost and avoided cost methods which are the most common methods for estimating the value of regulating services in monetary terms (Pascual et al., 2010: 206). These methods focus on the *consequences* of particular choices rather than preferences. The second version include methods analysing stated or revealed preferences. These are rooted in the neoclassical idea that the value of nature depends on human preferences. As a contrast, ecological economics is rooted in the idea that nature has value to human wellbeing regardless whether we understand it or not (Daily et al., 2000).

Table 2. Framework for ES valuation and policy integration

	Information in		
	Qualitative terms	**Quantitative terms**	**Monetary terms**
Stated purpose (as observed in national legislation)	Concern for non-measurable objectives like social equity, precautionary principle and safeguarding the insurance value of biodiversity.	Concern for reaching quantitative targets in cost-effective ways without expressing targets in monetary terms.	Concern for economic efficiency expressed in monetary terms and justified as a means to internalise externalities.

Methods for describing values. Decision-support.	SWOT analysis, identification, historical assessment, narratives, stakeholder consultation, Delphi methods, multicriteria analysis.	Technical/scientific mapping and assessment of trends e.g. water flows, pollination, and species abundance, multicriteria analysis.	Cost-benefit analysis is the frame, methods for monetary valuation include replacement cost, contingent valuation, and hedonic pricing.
Policy integration by non-monetary regulation and without market trade (DC 0, 1, 2, 3)	Land use planning, protected areas, and species acts targeted to intrinsic values (DC0) or instrumental values (DC1).	Land use planning, protected areas, and species acts designed to reach measurable targets (DC0 or DC1 depending on institutional design).	Land use planning and protected areas designed to maximise economic efficiency. (DC1 or DC3 depending on institutional design).
	N/A	Liability for ecological compensation, using physical metrics, e.g. German compensation pools (CD2).	N/A
Economic instruments (no market trade). (DC 4)	Subsidy-like PES paid by governments targeted at high biodiversity, multiple ES or poverty areas, e.g. EU agri-env. and PES in Ecuador and Costa Rica.	Subsidy-like PES paid by governments targeted at well-defined measurable units, e.g. PES for carnivores in Sweden.	PES when payment is informed by the stated value of the targeted ecosystem service, not by opportunity costs, e.g. Wimmera, Australia.
Economic instruments (market trade) (DC 5)	N/A	Biodiversity offsets trading conservation credits, e.g. US habitat banking, and market-like PES financed by users, e.g. Vittel watershed in France.	
		Monetary value set by market actors.	

Financial instruments (DC 6)	N/A	N/A	Monetary value set by markets, e.g. biodiversity derivatives.

DC=Degree of Commodification. N/A=Not applicable

The examples provided are described in text.

An important objection to using stated or revealed preferences to inform policy is that the "outcome of economic valuation is in this respect not more informed than the people whose values are being assessed" (Daily et al., 2000: 396). The ideal, to simulate markets, is also inappropriate because market prices do not only reflect present knowledge but also present institutions (Bromley, 1990; Vatn, 2010). Finding 4 of the Millennium Ecosystem Assessment states that the present ecosystem challenges can be met but this requires "significant changes in policies, institutions and practices that are not currently under way."[7] These significant changes would dramatically alter all relative prices in the economy and make "economic advice derived within the neoclassical paradigm... especially inappropriate" (Rammel and van den Bergh, 2003: 122).

In other words, it is questionable to use price tags, derived from existing undesirable institutions, preferences, and wealth distribution, to inform us on what institutions would be 'efficient.' To do so would imply a strong bias in favour of status quo (Schmid, 1987: 213). This approach to valuation reflects the consumer tastes and institutional arrangements that together have undermined sustainability. We know from game theory that it is impossible to escape the suboptimal non-cooperative Nash equilibrium of the prisoners' dilemma as long as the market behaviour and institutions, which results in the Nash equilibrium, are used as a norm for valuation and collective action.

4. In the fourth degree taxes and subsidies are used to enhance ecosystem values. These monetary incentives employ the price signal in the Pigouvian sense to internalise externalities and evoke behavioural change but do not create markets: taxes and subsidies are not traded. The level of the tax or subsidy may or may not be informed by a monetary valuation, hence the third degree of commodification is not needed for

the fourth degree. Often the government seeks to determine a level of taxation or subsidy that results in the desired effect. For example, the level of a subsidy/payment for farming organically is set to compensate a sufficient number of farmers for the forgone net benefits (opportunity costs) compared to conventional farming. In other words, it is the harvest decrease of wheat and potatoes that is commodified, not the expected increase in biodiversity and ES.

Indeed, the large public PES (agri-environment) programs in Europe are "Pigouvian-type PES" (Sattler and Matzdorf, 2013: 3) and hard to distinguish from more traditional subsidy programs. Sometimes this is called compensation for ecosystem services (CES) rather than PES (McAfee and Shapiro, 2010). Globally, CES or government-financed subsidy-like PES constitutes 97–99 percent of all biodiversity-PES, the lower figure for developing countries (Milder et al., 2010; Vatn, 2015), if we exclude biodiversity offsets and eco-certification like organic food in our definition of PES. Subsidy-like PES belong to the 4th degree of commodification whereas the more market-oriented Coasean-type PES (Sattler and Matzdorf, 2013), what Wunder et al. (2008) refer to as pure PES, belong to the 5th degree (Table 1). We explain more fully market and market-based below.

The well-known Costa Rican PES program *Pago por Servicios Ambimentales*(PSA) was launched in 1996 as a "neoliberal market mechanism" but should more properly be labelled a government "subsidy in disguise" (Fletcher and Breitling, 2012): 402). This is because it was largely financed through a carbon tax and water tariffs, enabled by a new Forest Law that banned land-use change but not sustainable use, and targeted high poverty areas and "biological corridors" (Matulis, 2013: 256). If the institutions of PSA would be consistent to the stated purpose, this program would fit in the 5th degree of commodification, not the 4th degree (Table 2). Another national PES program in Ecuador, Programa Socio Bosque (PSB) has avoided connotations to payments and market (and may therefore be called CES). Instead the stated purpose of the government in Ecuador has been combining ecosystem conservation with poverty alleviation (De Koning et al., 2011) by connecting it to the rights of Nature (Madre Tierra) as well as people's rights to a good way of living (buen vivir).[8] Such payments can be seen as rewards for good stewardship rather than economic incentive (Muradian et al., 2013). The degree of commodification is the same for the Costa Rican PES and the Ecuadorian PES/CES, but the governments used different stated purposes to adapt to different cultural or political contexts (Farooqui and Schultz, 2012).

Government-financed biodiversity-PES schemes typically target a commodity with fuzzy characteristics, based on assumptions about the relationship between a certain land use and the provision of desired but often non-explicit outcomes for biodiversity and ES (Wunder et al., 2008:839). For example, European agri-environment schemes pay farmers for grazing land and organic farming although it is not clear exactly which species or ES benefit from such land use. The same is true for Costa Rica where forest cover is regarded as a proxy for biodiversity conservation (Porras et al., 2013). As a result, it is rather one hectare of forest or agricultural land under a specific land-use that is commodified, not the desired species or ES themselves (Muradian et al., 2010). Hence, unlike carbon-PES, government-financed subsidy-like biodiversity-PES should not be interpreted as price tags on specific units of nature although it constitutes a high degree (4) of commodification.

Wunder et al. (2008):843) identified only one PES program where providers are paid according to measured units actually delivered and that is payments by the Swedish government to indigenous Sami communities for documented presence and reproductions of wolves, wolverine and lynx on reindeer grazing land. A system where the monetary value is determined by a government auction, e.g. the Wimmera catchment pilot program for salinity control in Australia (Wunder et al., 2008 :838), can be regarded an 'incomplete market with intermediary' (Vatn, 2015 :227) where the government is acting on behalf of the real users (Table 2).

5. In environmental and ecological economics, distinctions are usually made between taxes, subsidies and cap-and-trade; these are considered as three separate economic instruments. These are founded in legislation but distinguished from conventional (non-monetary) regulations (Table 1). Economic instruments and regulations are in turn distinguished from policy instruments based on information or moral suasion (Common and Stagl, 2005: 409).

The fifth degree includes market-traded biodiversity offsets and other markets for ecosystem services (MES) resembling cap-and-trade systems.

Most are mandatory and the largest programs are in the USA: wetland mitigation, stream mitigation, and conservation banking (OECD, 2013). Conservation banking is legally mandated biodiversity offsets, modelled after wetland banking (McKenney and Kiesecker, 2010). The general idea, according to guidance by the US agency Fish and Wildlife Service (FWS, 2003), is that market actors (e.g. landowners willing to restore ecological values to compensate for other actors' degradation) join a habitat bank and are issued conservation credits by a government agency and subsequently allowed to sell these credits to a developer (e.g. public transportation agency or private developer). In this way ownership of credits is transferred from the government to the bank, after which the credits can be sold and re-sold on market conditions. The government however controls this market by creating it and by determining the number of credits (the cap), the initial distribution and commands the demand for the credits, i.e. the rules for who needs to buy credits and how many they need to buy.

Non-mandatory biodiversity offsets with other buyers than governmental bodies also belong to this level, as well as non-mandatory PES. These user-financed PES schemes are more market-like since they are fully voluntary (not only on the provider side). Besides, they are usually focused on a single ecosystem service like the Vittel watershed protection in France (Wunder et al., 2008:838–839) which also increases the degree of commodification. Globally, these non-mandatory offsets/banking and PES programs embrace USD 10-17 million/year, which is only 0.5–1 percent of the payments within the mandatory programs (Milder et al., 2010).[9]

6. Financialisation is the sixth degree of 'complete commodification' and describes how the traded commodity is re-packaged and re-sold as financial instruments (e.g. bonds or derivatives). The financial flows of PES or MES schemes are the basis (underlying value) for this. Financialisation uses the ecosystems, which are commodified by PES or MES schemes, as collateral for investments and thereby significantly enhances the degree of commodification. Based on Sullivan (2012) and Dempsey and Robertson (2012) we define financialisation as a process in which financial actors invest in units of conserved nature and turn these investments into financial instruments which are traded on financial markets.

The forest bonds proposed by WWF and Global Canopy Programme offer an example of financialisiation. They enable the forest owners, who issue forest bonds, "to raise large-scale finance now that will be repaid by existing and anticipated future income" (Cranford et al., 2011: 6). The issuer of the bonds "will need to convince investors that the cash flows they plan to pay the bond back with are sufficiently secure and predictable" (ibid.). A reliable global REDD+ program[10] would provide such predictability but "support from the public sector through regulations or other commitments will be needed to ensure that these cash flows materialise" (ibid.). Goldman Sachs is one of the financial partners in this proposal and a potential investor in such bonds.

Biodiversity derivatives is another form of financialisation which has been discussed by Mandel et al. (2009). The government might buy a ten-year biodiversity derivative for a species of concern wherein a predefined amount of funds would be released by the seller if a species' population falls below a threshold. Basically, the seller (issuer or writer of the derivative) is making the bet that the species population will not fall below the threshold, or that it can take measures to prevent such an occurrence whose costs are within the range wherein they can make a profit given their proceeds from issuing the derivative.

FRAMEWORK FOR VALUATION AND POLICY INTEGRATION

In a broad economic sense valuation is about expressing a willingness to make sacrifices to gain or protect something. In the context of policy integration, the role of valuation is to raise awareness and provide decision-support expressed in qualitative, quantitative, and/or monetary terms (Table 2). The monetary valuation could be based on analytical calculations or the price from hypothetical or real markets. Table 2provides a comprehensive and pluralistic framework of the relation between valuation and policy integration: there is no unique relation between the instrument chosen for policy integration (horizontal axis) and the political justification for these instruments or the way nature's values are described in the columns. Table 2 is an attempt to address both purposes and degrees of commodification.

Much of the confusion in the discussion of commodification can be attributed the misunderstanding that economic instruments assumes monetary valuation. Table 2decouples the technical calculation of monetary valuation from policy instruments; monetary valuation is not needed for the lower degrees (1–2 and 4) of commodification and in the higher degrees (5–6) the price is a result of market trade. Table 2 also suggests that both proponents and critics to neoliberalism are making mistakes when they associate economic instruments with neoliberal frameworks, rather than assessing the actual institutions and performance (Dempsey and Robertson, 2012).

In reality, a PES scheme for grazing land can be justified by a qualitative analysis of vulnerable species depending on grazing land or targeted to forest communities as a tool for poverty alleviation. Reversely, a nature reserve can be justified by a cost-benefit analysis emphasising its extraordinary value for recreation or public health. Again, it is the institutional design of a particular policy instrument that determines the degree of commodification. However, public perceptions of commodification and the political legitimacy of the policy instrument will also be influenced by the stated purpose and the valuation framework used to justify the instrument.

We do not suggest that any method (column) of policy integration is better than the other. Qualitative information has many similarities to SWOT (Strength, Weakness, Opportunity, Threat) analysis in business (Hanson et al., 2008) and is common in business thinking on resilience (Hamel and Vlikangas, 2003). The right column represents the ideal in neoclassical economics, wherein all values are demonstrated in monetary terms in support of efficiency calculations.

'MARKET-BASED INSTRUMENTS' IS A CONFUSING TERM

'Market-based instruments' are generally assumed to include taxes, subsidies and various cap-and-trade systems (e.g. Pearce (2013)) but often include a "wide a range of hybrid instruments" (Muradian and Gómez-Baggethun, 2013: 1118). Here we want to emphasise the confusion between the role of prices and the role of markets. A tax definitely uses the price signal but it is not a market since the tax is imposed, not traded.

The price *mechanism* (or market mechanism) is something else: it can be described as the autonomous mechanism that determines the price in a market economy, as an equilibrium between supply and demand. The price *signal* is the result of the price mechanism in competitive markets and also in less competitive markets where a large producer or consumer can influence the price. However, for taxes and subsidies, the price signal is a result of a political decision to internalise externalities in the Pigouvian sense. 'Market-based instruments' could properly be used for the instruments under the 5th and 6th degrees of commodification but we prefer the more general term 'economic instruments' to describe all policy instruments relying on the price signal (monetary incentives), i.e. degree 4–6 (Tables 1 and 2).[11]

Voluntary market-like Coasean-type PES and biodiversity offsets (mandatory or voluntary) trading conservation credits constitute the fifth degree of commodification because the price is the result of supply and demand of the market. The price signal is determined by the government in the 4th degree but by the market in the 5th and 6th degrees of commodification. Markets for ecosystem services (MES) indeed require more, not less, regulations than the previous degrees if MES are to contribute to environmental benefits (Glicksman and Kaime, 2013). Principles of international law and safeguards can contribute to designing and implementing MES including ensuring accountability in environmental governance (Ituarte-Lima et al., 2012).[12] Hence, 'markets' (reliance on market transactions) should not be understood ideologically as opposite to regulations even though they are often presented as such (Penca, 2013; Fletcher and Breitling, 2012). Moreover, the transaction costs associated with MES are often high compared to taxes, subsidies, and regulations, which suggests that *a priori* assumptions about cost-effectiveness should be avoided (Gómez-Baggethun and Muradian, 2015).

DISCUSSION

The Diversity of PES Designs

Table 2 suggests there is no direct link between the methodological approach to valuation (non-monetary or monetary) and the choice of policy instrument. Concerning PES, Table 2 clarifies three issues. First, subsidy-like government-financed PES, also called Pigouvian-type PES or Compensation for ecosystem services (CES), represent a lower degree of commodification compared to the much smaller market-like Coasean-type PES.

Second, subsidy-like PES can be further differentiated based on stated purpose and use of valuation framework, as suggested by the three columns. If society wishes to use monetary incentives (or rewards) but avoid that monetary valuation reduces "dignities" (intrinsic values) to "commodities" (instrumental values), CES may be an option.

Third, the stated purpose may sometimes appear to be inconsistent to the actual institutional design of the program. The national PES program of Costa Rica and CES program of Ecuador are similar in terms of the degree of commodification, but are different in the particular design. In both countries, these differences have been exaggerated by the political justifications for the programs. Rather than criticising this, we suggest it illustrates how flexible the design of PES/CES programs can be and how they can be implemented and justified to accommodate country-specific concerns to gain legitimacy which is of key importance within the CBD (Farooqui and Schultz, 2012; Ogwal and Schultz, 2014).

Sometimes the definition of PES includes legally mandated private payments which are required to offset environmental impacts, e.g. wetland mitigation or habitat banking, as well as the price premium for eco-certified products that reaches the producers of ES (Scherr and McNeely, 2008; Milder et al., 2010). However, we believe it makes sense analytically i) to distinguish between PES and liabilities to compensate for damage (offsets or ecological compensation) regardless if they are mandatory or not; and ii) to distinguish between economic instruments on the one hand and eco-certification on the other. Although an eco-label often functions as a price premium for producers (monetary incentive), it can also be regarded, from a policy perspective, as an instrument based on information and moral suasion (Common and Stagl, 2005: 409).

Biodiversity Offsets and Ecological Compensation

Ecological compensation schemes, including biodiversity offsets, are often controversial. Proponents emphasise that it is fair (analogous to the

Polluter Pays Principle) that developers are required to pay for restoration activities so as to offset the degradation they cause and this may also steer development away from areas with high values for biodiversity and ES because these are expensive to compensate for (Conway et al., 2013): vii). Landowners who invest in biodiversity and ES can enter contracts and receive payments from the developers rather than from tax-payers. Interviews with private landowners in Germany and Australia have noted the economic attractiveness of entering a long-term maintenance contract with a compensation agency, as it ensured a stable source of income (Tucker et al., 2013:193; O'Connor, 2009:15).

Critics of biodiversity offsets such as Forest Peoples Programme (2011) are concerned that ecosystems, as well as their functions for livelihood opportunities, are not fully replaceable. Local people in one region normally depend on the biodiversity and ES in that area for their livelihoods and other benefits. There is a risk that such schemes will offer incentives for developers and administrators "to downplay or ignore the requirement to first avoid and reduce their impacts under the false impression that any impact can be compensated for" (Quétier and Lavorel, 2011: 2991. This is sometimes referred to as the restoration myth or a 'license to trash' (Dickie et al., 2010:237). If biodiversity offset policies allow for the approval of land exploitation that would otherwise not have been accepted, biodiversity objectives as well as livelihood opportunities for local communities can be compromised.

This can be avoided only if the avoidance step of the mitigations hierarchy is respected, which means that 'no-go' areas are stipulated (Conway et al., 2013: viii) and "current rules to decide whether developments should go ahead do not change" (Dickie et al., 2010: 237). Hence, the proponents emphasise that introduction of a biodiversity offset scheme must not result in less stringent regulations.

Despite this agreement, the largest offset program, US Wetland mitigation, has focused too much on the compensation part and neglected the earlier stages of the mitigation hierarchy (Hough and Robertson, 2009). The result is poor performance (Kihslinger, 2008; Turner et al., 2001). For example, an evaluation of 391 wetland offset projects in Massachusetts showed that 54% were not in compliance with the wetland regulations (Brown and Veneman, 2001) and Ambrose and Lee

(2004) found that 46% of the 250 sites surveyed in California failed to replace key wetland ES which had been determined by the state.

When the whole spectrum of institutional design from non-market liabilities (ecological compensation) to market-like trade (biodiversity offsets) is acknowledged, considerable flexibility is shown. This is promising since it allows for the emergence of country-specific programs with a cultural-political adaptation to gain legitimacy (Hahn, 2011). This is also a risk because the devil is in the detail; small changes in the institutional design may have large consequences for the performance. We give four examples:

First, weighing the development interest against the conservation interest at the site of proposed development must not be compromised by the (false) promise that the damage can and will be entirely compensated for elsewhere. A successful program should result in a *reduction* of permits obtained by developers because the liability to finance restoration at the compensation site increases costs, thereby deterring development projects with marginal profitability.

Second, and related to the first, even an ideally designed program may fail for ecological reasons, if restoration turns out to be much more complicated than anticipated (Hilderbrand et al., 2005; Maron et al., 2012).

Third, a successful program would align the interest of (city) planners with the interest of developers. Planners often try to steer development away from remaining green areas and developers would also avoid green land with high biological values since these would be expensive to compensate for.

Fourth, simple metrics (one acre equals one credit) and use of market trade might have lowered transaction costs of the large US programs but this cannot be regarded as cost-effective if the achieved effect, the actual performance of the program, is low. The evaluations referred to above suggest low performance. Briggs et al. (2009) warn that "[w]ithout careful regulation, habitat banks could offer low-cost compensation as a result of cutting corners on conservation, and the market would reward poorly managed banks and thus harm conservation efforts" (p. 117). The dichotomy government regulations vs. market is false (Vatn, 2015) and this is particularly obvious for ecological compensation/offsets since good performance seem to be a function of a comprehensive government regulation (Dickie et al., 2010: 237), i.e. safeguards.

CONCLUSION

The issue of monetary valuation of biodiversity and ES is a never-ending debate within ecological economics because it runs counter to one of the central beliefs within this community, namely that natural capital cannot be substituted for. However, there is an emerging scientific consensus that the price signals used in PES are in general set to compensate for opportunity costs and hence cannot be regarded as a monetary valuation of the desired ES. We also find that biodiversity offsets, designed with strong safeguards, would properly be named 'ecological compensation' and have very little to do with market trade or valuation of ES. Hence, these economic instruments should not by default be associated to neither monetary valuation nor neoliberalism.

Dempsey and Robertson (2012) argue that many social scientists "have largely positioned themselves as critical" to PES and other ES policies and associated these to commodification and "the neoliberalization of nature" (pp. 762, 772). Instead Dempsey and Robertson suggest "a useful solidarity with people engaged, through ES policy, in opposition to business-as-usual resource development" (p. 773). Our results support that ES policies in general can be useful to counter some of the market failures of business-as-usual and hence should not be understood as 'market-based instruments' or 'neoliberalisation of nature.' At the same time, we have used 'commodification' as an analytical, not normative, term and found that all ES policies, by using an instrumental (human wellbeing) perspective, do involve at least low degrees of commodification.

The strength of markets, searching for lowest cost for provision, may result in low quality for complex 'products' such as conservation of biodiversity and ES, unless combined with strong safeguards including enforcement. Markets may entail low transaction costs and be cost-effective regarding simple products but this should not be assumed for complex products.

The institutional diversity and plurality in understanding and implementing various biodiversity financing mechanisms, especially economic instruments, gives flexibility for the 194 Parties of the CBDto adapt these instruments to their national political and cultural context. This flexibility in turn enhances prospects for reaching international agreement on biodiversity financing mechanisms and enhances policy options in each country.

ACKNOWLEDGEMENTS

Research for this paper was funded by Formas, Sweden, through the projects SAPES (no. 2009-1876-15428-58) and Safeguarding Biodiversity (no. 2013-11763-24670-36), as well as by Mistra through a core grant to Stockholm Resilience Centre.

REFERENCES

1. Aldred, J., 2002. Cost-benefit analysis, incommensurability and rough equality. Environ. Values 11, 27–47.
2. Ambrose, R., Lee, S., 2004. Guidance Document for Compensatory Mitigation Projects Permitted Under Clean Water Act Section 401 by the Los Angeles Regional Quality Control Board. California Environmental Protection Agency, Los Angeles.
3. Boyd, J., Banzhaf, S., 2007. What are ecosystem services? The need for standardized environmental accounting units. Ecol. Econ. 63, 616–626.
4. Briggs, B.D.J., Hill, D.A., Gillespie, R., 2009. Habitat banking—how it could work in the UK. J. Nat. Conserv. 17, 112–122.
5. Bromley, D.W., 1990. The ideology of efficiency: Searching for a theory of policy analysis. J. Environ. Econ. Manag. 19, 86–107.
6. Brown, S., Veneman, P., 2001. Effectiveness of compensatory wetland mitigation in Massachusetts, USA. Wetlands 21, 508–518.
7. Burdon, P., 2011. Exploring Wild Law: the Philosophy of Earth Jurisprudence. Wakefield Press, Adelaide.
8. CBD, 2010. Convention of Biological Diversity, point 8(c) of CBD COP10 Decision X/3, ⟨www.cbd.int/decision/cop/default.shtml?id¼12269⟩ accessed 10.12.13.
9. Common, M., Stagl, S., 2005. Ecological Economics: An Introduction. Cambridge University Press.

10. Conway, M., Rayment, M., White, A., Berman, S., 2013. Exploring potential demand for and supply of habitat banking in the EU and appropriate design elements for a habitat banking scheme. Final Report submitted to DG Environment by ICF GHK Consulting Ltd in association with BIO Intelligence Service.
11. Corbera, E., Brown, K., Adger, W.N., 2007. The equity and legitimacy of markets for ecosystem services. Dev. chang. 38, 587–613.
12. Cranford, M., Henderson, I., Mitchell, A., Kidney, S., Kanak, D., 2011. Unlocking Forest Bonds – A High-Level Workshop on Innovative Finance for Tropical Forests. Workshop Report. WWF Forest & Climate Initiative, Global Canopy Programme and Climate Bonds Initiative, ⟨www.theredddesk.org/fr/node/5627⟩.
13. Daily, G.C., 1997. Nature's services: societal dependence on natural ecosystems. Island Pr.
14. Daily, G., Söderqvist, T., Aniyar, S., Arrow, K., Dasgupta, P., Ehrlich, P., Folke, C., Jansson, A., Jansson, B.O., Kautsky, N., Levin, S., Lubchenco, J., Mäler, K.G., Simpson, D., Starrett, D., Tilman, D., Walker, B., 2000. The value of nature and the nature of value. Science 289, 395–396.
15. Daly, E., 2012. The Ecuadorian example: the first ever vindications of constitutional rights of nature. Rev. Eur. Commun. Int. Environ. Law 21, 63–66. http://dx.doi.org/10.1111/j.1467-9388.2012.00744.x.
16. De Koning, F., Aguiñaga, M., Bravo, M., Chiu, M., Lascano, M., Lozada, T., Suarez, L., 2011. Bridging the gap between forest conservation and poverty alleviation: the Ecuadorian Socio Bosque program. Environ. Sci. Policy 14, 531–542.
17. Dempsey, J., Robertson, M.M., 2012. Ecosystem services Tensions, impurities, and points of engagement within neoliberalism. Prog. Hum. Geogr. 36, 758–779.
18. Dickie, I., Tucker, G., Ozdemiroglu, E., Cranford, M., Ohlenburg, H., Wende, W., Chapman, D., Donovan, C., Bishop, J., Brans, E., Reinders, R., Ekstrom, J., ten_-Kate, K., Treweek, J., 2010. The use of market-based instruments for biodiversity protection: The case of habitat banking. Technical Report by eftec IEEP and others. ⟨http://ec.europa.eu/environment/enveco/pdf/eftec_habitat_technical_

report.pdf⟩.

19. Ehrlich, P.R., Mooney, H.A., 1983. Extinction, substitution, and ecosystem services. BioScience 33, 248–254.
20. Farooqui, M.F., Schultz, M., 2012. Co-chairs' Summary of Dialogue Seminar on Scaling up Biodiversity Finance. Quito 6–9 March 2012, pp. 1-49. ⟨www.cbd.int/doc/meetings/fin/ds-fb-01/official/ds-fb-01-02-en.pdf⟩.
21. Fisher, B., Turner, R.K., Morling, P., 2009. Defining and classifying ecosystem services for decision making. Ecol. Econ. 68, 643–653.
22. Fletcher, R., Breitling, J., 2012. Market mechanism or subsidy in disguise? Governing payment for environmental services in Costa Rica. Geoforum 43 (3), 402–411.
23. Forest Peoples Programme, 2011. Submission to the Convention on Biological Diversity relating to innovative financial mechanisms and the rights of indigenous peoples and local communities.
24. FWS, 2003. Fish and Wildlife Service. Guidance for the Establishment, Use, and Operation of Conservation Banks. United States Department of the Interior.
25. Glicksman, R.L., Kaime, T., 2013. A comparative analysis of accountability mechanisms for ecosystem services markets in the United States and the European Union. Transnatl. Environ. Law 2, 259–283.
26. Gómez-Baggethun, E., Ruiz-Pérez, M., 2011. Economic valuation and the commodification of ecosystem services. Prog. Phys. Geogr. 35, 613–628.
27. Gómez-Baggethun, E., Muradian, R., 2015. In markets we trust? Setting the boundaries of market-based instruments in ecosystem services governance. Ecol. Econ. 117, 217–224.
28. Groenfeldt, D., Schmidt, J.J., 2013. Ethics and water governance. Ecol. Soc. 18. http: //dx.doi.org/10.5751/ES-04629-180114.
29. Hahn, T., 2011. Self-organized governance networks for ecosystem management: who is accountable? Ecology and Society 16 (2), 18 ⟨http://www.ecolo gyandsociety.org/vol16/iss2/art18/⟩.
30. Hamel, G., Vlikangas, L., 2003. The quest for resilience. Harv. Bus. Rev., 52–63.

31. Hanson, C., Ranganathan, J., Iceland, C., Finisdore, J., 2008. The corporate ecosystem services review, Guidelines for Identifying Business Risks and Opportunities Arising From Ecosystem Change. WRI, WBCSD and Meridian Institute.
32. Higgins, P., Short, D., South, N., 2013. Protecting the planet: a proposal for a law of ecocide. Crime, Law Soc. Chang. 59, 251–266.
33. Hilderbrand, R.H., Watts, A.C., Randle, A.M., 2005. The myths of restoration ecology. Ecol. Soc. 10, 19.
34. Hough, P., Robertson, M., 2009. Mitigation under section 404 of the Clean Water Act: where it comes from, what it means. Wetl. Ecol. Manag. 17, 15–33.
35. Ituarte-Lima, C., Schultz, M., Hahn, T. and Cornell, S., 2012. 'Safeguards in scaling-up biodiversity financing and possible guiding principles', Stockholm Resilience Centre at the Stockholm University, Information document for the 11th Conference of the Parties of the Convention on Biological Diversity, (UNEP/CBD/COP/11/INF/7). ⟨http://www.cbd.int/cop11/doc/⟩ and Notification for Parties of CBD to submit comments ⟨https://www.cbd.int/doc/notifications/2013/ntf-2013-025-financial-en.pdf⟩.
36. Ituarte-Lima, C., Schultz, M., Hahn, T., McDermott, C., and Cornell, S., 2014, Biodiversity financing and safeguards: lessons learned and proposed guidelines, Stockholm: SwedBio/Stockholm Resilience Centre at Stockholm University, Information Document UNEP/CBD/COP/12/INF/27 for the 12th Conference of the Parties of the Convention on Biological Diversity in Pyeongchang Korea. ⟨http://www.cbd.int/doc/?meeting¼cop-12⟩.
37. Kihslinger, R., 2008. Success of wetland mitigation projects. Natl. Wetl. Newsl. 30 (2), 14–16.
38. Koh, N.S., Hahn, T., Ituarte-Lima, C., 2014. A comparative analysis of ecological compensation programs: The effect of program design on the social and ecological outcomes. Working Paper, Uppsala University. ⟨http://uu.diva-portal. org/smash/record.jsf?pid¼diva2%3A772933&dswid¼1454⟩.
39. Kosoy, N., Corbera, E., 2010. Payments for ecosystem services as commodity fetishism. Ecol. Econ. 69, 1228–1236.
40. Luck, G.W., Chan, K.M., Eser, U., Gómez-Baggethun, E., Matzdorf,

B., Norton, B., Potschin, M.B., 2012. Ethical considerations in on-ground applications of the ecosystem services concept. BioScience 62, 1020–1029.

41. MA, 2005a. Ecosystems and Human Well-being: Synthesis Report, The Millennium Ecosystem Assessment. Island Press.
42. Mandel, J.T., Donlan, C.J., Armstrong, J., 2009. A derivative approach to endangered species conservation. Front. Ecol. Environ. 8, 44–49.
43. Maron, M., Hobbs, R.J., Moilanen, A., Matthews, J.W., Christie, K., Gardner, T.A., Keith, D.A., Lindenmayer, D.B., McAlpine, C.A., 2012. Faustian bargains? Restoration realities in the context of biodiversity offset policies. Biol. Conserv. 155, 141–148.
44. Matulis, B.S., 2013. The narrowing gap between vision and execution: neoliberalization of PES in Costa Rica. Geoforum 44, 253–260.
45. McAfee, K., Shapiro, E.N., 2010. Payments for ecosystem services in Mexico: nature, neoliberalism, social movements, and the state. Ann. Assoc. Am. Geogr. 100, 579–599.
46. McKenney, B.A., Kiesecker, J.M., 2010. Policy development for biodiversity offsets: a review of offset frameworks. Environ. Manag. 45, 165–176. Merchant, C., 1980. Death Nat.
47. Milder, J.C., Scherr, S.J., Bracer, C., 2010. Trends and future potential of payment for ecosystem services to alleviate rural poverty in developing countries. Ecol. Soc. 15, 2.
48. Muradian, R., Corbera, E., Pascual, U., Kosoy, N., May, P.H., 2010. Reconciling theory and practice: an alternative conceptual framework for understanding payments for environmental services. Ecol. Econ. 69, 1202–1208.
49. Muradian, R., Arsel, M., Pellegrini, L., Adaman, F., Aguilar, B., Agarwal, B., Corbera, E., Ezzine de Blas, D., Farley, J., Froger, G., 2013. Payments for ecosystem services and the fatal attraction of win-win solutions. conserv. Lett. 6, 274–279.
50. Muradian, R., Gómez-Baggethun, E., 2013. The institutional dimension of "marketbased instruments" for governing ecosystem services: introduction to the special issue. Soc. Nat. Resour. 26, 1113–1121.

51. Neuteleers, S., Engelen, B., 2015. Talking money: how market-based valuation can undermine environmental protection. Ecol. Econ. 117, 253–260.
52. Nilsson, M., Persson, A., 2003. Framework for analysing environmental policy integration. J. Environ. Policy Plan. 5, 333–359.
53. O'Connor, N.R.M., 2009. Evaluation after two years of operations. Maylands Department of Sustainability and Environment, Victoria. Retrieved from ⟨http://www.depi.vic.gov.au/__data/assets/pdf_file/0017/205019/BushBrokerEvaluationReportFinal2009.pdf⟩.
54. OECD, 2013. Scaling-up Finance Mechanisms for Biodiversity. OECD Publishing http://dx.doi.org/10.1787/9789264193833-en.
55. Ogwal, S.F., Schultz, M., 2014. Co-Chairs′ Summary of Second Dialogue Seminar on Scaling up Finance for Biodiversity, Quito. Secretariat of the Convention on Biological Diversity, Montreal 9–12 April 2014. ⟨www.cbd.int/doc/meetings/fin/ds-fb-02/official/ds-fb-02-report-en.pdf⟩.
56. Pascual, U., Muradian, R., Brander, L., Gómez-Baggethun, E., Martín-López, B., Verma, M., et al., 2010. The Economics of Ecosystems and Biodiversity and Ecological and Economic Foundations. In: Kumar, Pushpam (Ed.), 2010. Earthscan, London and Washington, pp. 183–240.
57. Pearce, D., 2013. Blueprint 2: Greening the World Economy. Routledge.
58. Penca, J., 2013. Marketing the market: the ideology of market mechanisms for biodiversity conservation. Transnatl. Environ. Law 2, 235–257.
59. Porras, I., Barton, D.N., Miranda, M., Chacón-Cascante, A., 2013. Learning from 20 Years of Payments for Ecosystem Services in Costa Rica. International Institute for Environment and Development, London.
60. Quétier, F., Lavorel, S., 2011. Assessing ecological equivalence in biodiversity offset schemes: key issues and solutions. Biol. Conserv. 144, 2991–2999.
61. Rammel, C., van den Bergh, J.C., 2003. Evolutionary policies for

sustainable development: adaptive flexibility and risk minimising. Ecol. Econ. 47, 121–133.

62. Sattler, C., Matzdorf, B., 2013. PES in a nutshell: From definitions and origins to PES in practice-approaches, design process and innovative aspects. Ecosyst. Serv. 6, 2–11.
63. Scherr, S.J., McNeely, J.A., 2008. Biodiversity conservation and agricultural sustainability: towards a new paradigm of 'ecoagriculture' landscapes. Philos. Trans. R. Soc. B 363 (1491), 477–494.
64. Schmid, A., 1987. Property, Power, and Public Choice: An Inquiry Into Law and Economics, Second edition. Praeger, New York.
65. Sullivan, S., 2012. Banking nature? The spectacular financialisation of environmental conservation. Antipode 45, 198–217.
66. TEEB, 2010. The Economics of Ecosystems and Biodiversity: Mainstreaming the Economics of Nature: A synthesis of the approach, conclusions and recommendations of TEEB.
67. Tucker, G., Allen, B., Conway, M., Dickie, I., Hart, K., Rayment, M., et al., 2013. Policy Options for an EU No Net Loss Initiative. Institute for European Environmental Policy, London.
68. Turner, E., Redmond, A., Zedler, J., 2001. Count it by acre or function – mitigation adds up to net loss of wetlands. Natl. Wetl. Newsl., 5–16.
69. Turnhout, E., Waterton, C., Neves, K., Buizer, M., 2013. Rethinking biodiversity: from goods and services to "living with". Conserv. Lett. 6, 154–161.
70. Vatn, A., 2009. An institutional analysis of methods for environmental appraisal. Ecol. Econ. 68, 2207–2215.
71. Vatn, A., 2010. An institutional analysis of payments for environmental services. Ecol. Econ. 69, 1245–1252.
72. Vatn, A., 2015. Markets in environmental governance. From theory to practice. Ecol. Econ. http://dx.doi.org/10.1016/j.ecolecon.2014.07.017
73. Wunder, S., Engel, S., Pagiola, S., 2008. Taking stock: a comparative analysis of payments for environmental services programs in developed and developing countries. Ecol. Econ. 65, 834–852.

6 THE SCIENCE, POLICY AND PRACTICE OF NATURE-BASED SOLUTIONS: AN INTERDISCIPLINARY PERSPECTIVE

Carsten Nesshöver[ad], Timo Assmuth[e], Katherine N. Irvine[f], Graciela M. Rusch[h], Kerry A. Waylen[f], Ben Delbaere[j], Dagmar Haase[ck], Lawrence Jones-Walters[l], Hans Keune[m], Eszter Kovacs[no], Kinga Krauze[p], Mart Külvik[q], Freddy Rey[r], Jiskavan Dijk[h], Odd Inge Vistad[i], Mark E. Wilkinson[g], Heidi Wittmer[b]

[a] UFZ-Helmholtz Centre for Environmental Research, Department of Conservation Biology & UFZ Science-Policy Expert Group, Permoser Str. 15, 04318 Leipzig, Germany

[b] UFZ-Helmholtz Centre for Environmental Research, Department of Environmental Politics & UFZ Science-Policy Expert Group, Permoser Str. 15, 04318 Leipzig, Germany

[c] UFZ-Helmholtz Centre for Environmental Research, Department of Computational Landscape Ecology, Permoser Str. 15, 04318 Leipzig, Germany

[d] iDiv - German Centre for integrative Biodiversity Research Halle-Jena-Leipzig, Deutscher Platz 5e, 04103 Leipzig, Germany

[e] Finnish Environment Institute (SYKE) Environmental Policy Centre, FI-00251 Helsinki, Finland

[f] Social, Economic and Geographical Sciences, The James Hutton Institute, Craigiebuckler, Aberdeen AB15 8QH, United Kingdom

[g] Environmental and Biochemical Sciences, The James Hutton Institute, Craigiebuckler, Aberdeen AB15 8QH, United Kingdom

[h] Norwegian Institute for Nature Research, Terrestrial Ecology Department, Sluppen, 7485 Trondheim, Norway

[I] Norwegian Institute for Nature Research, Human Dimension Department, Fakkelgarden, 2624 Lillehammer, Norway

[j] ECNC-European Centre for Nature Conservation, 5000 LG Tilburg, The Netherlands

[k] Humboldt Universität zu Berlin, Institute of Geography, Lab for Landscape Ecology, Rudower Chaussee 16, 12489 Berlin, Germany

[l] Wageningen Environmental Research (Alterra), Wageningen UR, Wageningen Campus, 6700 Wageningen, The Netherlands

[m] INBO & Belgian Biodiversity Platform & University of Antwerp, Kliniekstraat 25, 1070 Brussels, Belgium

[n] Szent István University, Institute of Nature Conservation and Landscape Management, Páter K. u. 1., H-2100 Gödöllő, Hungary

[o] Environmental Social Science Research Group (ESSRG), Rómer Flóris u. 38, Budapest H-1024, Hungary

[p] European Regional Centre for Ecohydrology of the Polish Academy of Sciences, Tylna 3, 90-364 Lodz, Poland

[q] Estonian University of Life Sciences, Institute of Agricultural and Environmental Sciences, Kreutzwaldi 1, Tartu 51014, Estonia

[r] Univ. Grenoble Alpes, Irstea, UR EMGR, 2 rue de la Papeterie, BP 76, 38402 Saint-Martin-d'Hères, France

ABSTRACT

In this paper, we reflect on the implications for science, policy and practice of the recently introduced concept of Nature-Based Solutions (NBS), with a focus on the European context. First, we analyse NBS in relation to similar concepts, and reflect on its relationship to sustainability as an overarching framework. From this, we derive a set of questions to be addressed and propose a general framework for how these might be addressed in NBS projects by funders, researchers, policy-makers and practitioners. We conclude that:

(1) NBS need to be developed and discussed in relation to existing concepts to clarify their added value;

(2) When considering and implementing NBS, the 'relabelling' of related concepts and the misuse of the concept have to be prevented in order to avoid misunderstanding, duplication and unintended consequences;

(3) NBS as currently framed by the European Commission provides an opportunity for: a) transdisciplinary research into the design and implementation of solutions based on nature; and b) overcoming a bias towards development alternatives with narrow perspectives that focus on short-term economic gains and effectiveness;

(4) The strength of the NBS concept is its integrative, systemic approach which prevents it from becoming just another "green communication tool" that provides justification for a classical model of natural resource exploitation and management measures.

To realise their full potential, NBS must be developed by including the experience of all relevant stakeholders such that 'solutions' contribute to achieving all dimensions of sustainability. As NBS are developed, we must also moderate the expectations placed on them since the precedent provided by other initiatives whose aim was to manage nature sustainably demonstrates that we should not expect NBS to be cheap and easy, at least not in the short-term.

Keywords: Sustainability, Ecosystem services, Ecosystem management, Environmental governance

INTRODUCTION: NATURE-BASED SOLUTIONS AS A NEW TERM IN SCIENCE, POLICY AND PRACTICE

Those working in science, policy and practice related to the management of the natural environment regularly encounter new ideas and terminologies. In the late 1980s the phrase 'sustainable development' was defined by the United Nations (UN) Brundtland Commission (Brundtland et al., 1987) and then the term 'biodiversity' emerged from the field of conservation biology (Takacs, 1996; Wilson, 1988). More recently, the use of 'natural capital', from the fields of applied ecology and ecological economics (Daly and Farley, 2011; Jansson, 1994; Costanza and Daly, 1992; Schumacher, 1973) and the idea of 'ecosystem services' has become widespread (Costanza et al., 1998; Daily, 1997; MA, 2005; Gómez-Baggethun et al., 2010). These concepts are reflected in policy agreements including the adoption of the Convention on Biological Diversity (CBD) in 1992 (UNEP, 1993) and its Ecosystem Approach (UNEP/CBD, 2000), the UN's Millennium Ecosystem Assessment (MA, 2005) and its follow-up activities such as the Intergovernmental Platform on Biodiversity and Ecosystem Services (IPBES) (UNEP, 2010). They also shape programmes of research. For example, until 2014 the European Union's Framework Programmes for research supported many projects focussing on ecosystem services (Admiraal et al., 2016). Over the past twenty years, an increasing number of perspectives have reflected an anthropocentric view of the management of natural resources, including biodiversity and the environment (Nesshöver et al., 2015), with a focus on the benefits that nature may provide for humans (Díaz et al., 2015; MA, 2005; TEEB, 2010a).

The most recent entry to this discourse is 'Nature-Based Solutions' (NBS), a concept introduced specifically to promote nature as a means for providing solutions to climate mitigation and adaptation challenges (Cohen-Schacham et al., 2016; IUCN, 2012). Within Europe, policy-makers have integrated the concept into their new framework programme for research and innovation, 'Horizon 2020', providing a new narrative involving biodiversity and ecosystem services aligned with goals of innovation for growth and job creation (European Commission, 2015c), and with a potential opening for transformational pathways towards sustainable societal development (Maes and Jacobs, 2015).

Despite limited research about the concept to date (Eggermont et al., 2015; Maes and Jacobs, 2015), the term has already diversified. In the United States 'nature-based infrastructure' and 'engineering with nature' are more common as descriptions for actions to support resilience and to reduce flood risk (US Army Corps of Engineers, 2013), and the IUCN and EC definitions of NBS provide alternate perspectives on its remit and purpose (Table 1). For the EC, NBS is understood as actions that 'aim to help societies address a variety of environmental, social and economic challenges in sustainable ways. They are actions which are inspired by, supported by or copied from nature' (European Commission, 2015c, p. 5).

Table 1. Definitions for the concept of nature-based solutions.

	International Union for Conservation of Nature	**European Union Directorate General on Research and Innovation**
Definition/rationale	'the potential power of nature and the solutions it can provide to global challenges in fields such as climate change, food security, social and economic development' (p. 1) 'Healthy, diverse and well managed ecosystems lay the foundation for practical, nature-based solutions to global problems.' (page 16) '[Nature-Based Solutions] offer unique and effective solutions to climate change.... The underlying approach is a pro-active application of the sustainable management and conservation of natural resources to address major global challenges (food security, disaster risk reduction, economy).' (Annex I, p. 24–25) 'The following principles are an initial attempt to provide some guidance on what type of interventions could (or should not) be considered as a nature-based solution. 1. The intervention delivers an effective solution to a major global challenge using nature 2. The intervention provides biodiversity benefits in terms of diverse, well-managed ecosystems 3. The intervention is cost effective relative to other solutions 4. The rationale behind the intervention can be easily and compellingly communicated 5. The intervention can be measured, verified and replicated 6. The intervention respects and reinforces communities' rights over natural resources 7. The intervention harnesses both public and private sources of funding' (Annex 1, p. 24–25).	'Nature-Based Solutions harness the power and sophistication of nature to turn environmental, social and economic challenges into innovation opportunities.' (p. 2) '[Nature-Based Solutions] are actions inspired by, supported by or copied from nature; both using and enhancing existing solutions to challenges, as well as exploring more novel solutions, for example, mimicking how non-human organisms and communities cope with environmental extremes. Nature-Based Solutions use the features and complex system processes of nature, such as its ability to store carbon and regulate water flows...' (p. 24)

Aim of employing concept	'Nature-Based Solutions offer multiple benefits simultaneously and therefore efficiently. This Programme Area focuses initially on nature-based solutions to climate change (including disaster risk reduction), food security, and economic and social development, but will …explore opportunities to broaden this approach to sectors such as health and access to energy' (p. 5) 'Healthy and restored ecosystems make cost-effective contributions to meeting global challenges of climate change, food security and economic and social development' (Table 1, p. 7). 'Apart from providing effective solutions to major global challenges, nature-based solutions also deliver clear biodiversity benefits in terms of diverse, well-managed and functioning ecosystems. They must be cost efficient relative to other solutions. As nature-based solutions are designed to reach beyond the conservation community they need to be easily and compellingly communicated as well as being measurable, verifiable and replicable. Finally they must be designed and implemented in such a way as to respect and reinforce communities' rights over natural resources.' (p. 16) 'helping the conservation community reach beyond its traditional constituencies, build new alliances and broaden its messaging beyond the immediate imperatives of addressing biodiversity threats' (p. 24)	'Nature-Based Solutions aim to help societies address a variety of environmental, social and economic challenges in sustainable ways.' (p. 24) '…in order to achieve desired outcomes, such as reduced disaster risk and an environment that improves human well-being and socially inclusive green growth (p. 24). 'Four principal goals have been identified that can be addressed by nature-based solutions: 1. Enhancing sustainable urbanisation… can stimulate economic growth as well as improving the environment, making cities more attractive, and enhancing human well-being. 2. Restoring degraded ecosystems… can improve the resilience of ecosystems, enabling them to deliver vital ecosystem services and also to meet other societal challenges. 3. Developing climate change adaptation and mitigation…can provide more resilient responses and enhance the storage of carbon. 4. Improving risk management and resilience…can lead to greater benefits than conventional methods and offer synergies in reducing multiple risks.' (p. 4) NBS solutions to strengthen community cohesion, involvement of society, with the aim of re-connecting people with nature. (p. 21)
Reference	IUCN, 2012. The IUCN Global Programme 2013–16, Adopted by the IUCN World Conservation Congress, September 2012.	European Commission 2015. Towards an EU Research and Innovation policy agenda for Nature-Based Solutions & *Re*-Naturing Cities. Final Report of the Horizon 2020 Expert Group on 'Nature-Based Solutions' and Re-Naturing Cities. Directorate-General for Research and Innovation,

Due to these broad framings, the meaning of NBS can appear vague, and the links to pre-existing concepts may be unclear. It has been suggested that the conceptual flexibility associated with a vague or loosely-defined term can risk missing important opportunities to improve the management of natural resources (e.g. Waylen et al., 2014). A comprehensive formulation would definitely help to stimulate discussion and innovation, and facilitate communication among the communities of science, policy and practice (Abson et al., 2014; Brand and Jax,

2007; Star and Griesemer, 1989). However, there is also a danger of oversimplification, reinventing the wheel, (non)deliberate misuse, or generating new, unforeseen trade-offs in decision-making (Bennett et al., 2009; Ring et al., 2010). It is thus important that such a concept, which has emerged at the science-policy-practice interface, is analysed and placed in the context of existing terms and terminologies, and that potential overlaps as well as differences are acknowledged.

In this paper, we reflect critically on the concept from the viewpoint of NBS contributions to sustainable development in Europe. With an aim of providing a basis for researchers and other actors involved in NBS projects, we examine how NBS relate to existing concepts and what implications can be drawn for NBS research, its applications and to inform policies. Firstly, we consider how NBS refer to and build upon existing and comparable concepts, and how NBS relate to sustainability in general. Secondly, we identify how NBS relate to cross-disciplinary research and implementation challenges and present a framework that can aid implementation of interventions intended to work with nature in order to tackle societal challenges. We end by reflecting on the added value of the NBS concept, the opportunities and challenges that it offers to natural resource and ecosystem management, and the role of research.

THE CONCEPTUAL CONTEXT OF NATURE-BASED SOLUTIONS

Differences and Similarities with other Concepts used For Ecosystem Management

The NBS concept aims to explicitly link positive outcomes for society ('solutions') with a notion of 'nature' as something helpful for these aims. It may, therefore, relate to or overlap with other concepts defined and used to inform ecosystem management for societal benefit, although these relationships are seldom acknowledged explicitly (see European Commission, 2015c). Six such concepts with associated definitions, objectives and examples are analysed in Table 2. We have focussed on those most commonly used in contemporary literature about ecosystem management for sustainability, societal benefit and human well-being. None of them has a single uncontested definition, but they are commonly in use within science, policy and practice.

Table 2. Non-exhaustive overview of different concepts related to nature-based solutions

	Problem solving techniques		**Approaches to management**			
Concept	Ecological Engineering (EE) and Catchment Systems Engineering (CSE)	Green/Blue Infrastructure (GI/BI)	Ecosystem Approach (EA)	Ecosystem-based Adaptation//Mitigation (EBA)	Ecosystem Services Approach/Framework (ES)	Natural Capital (NC)
Definition	The earliest definition of Ecological Engineering came from Odum, with 'those cases in which the energy supplied by man is small relative to the natural sources, but sufficient to produce large effects in the resulting patterns and processes' (Odum, 1962, as cited in Mitsch and Jørgensen, 2004, p. 25). It has since been redefined as 'the design of sustainable ecosystems that integrate human society with its natural environment for the benefit of both (Mitsch and Jorgensen, 1989, p. 365). It has also been defined as 'actions using and/or acting for nature' (Rey et al., 2015 p. 1336). CSE developed by Quinn et al. (2010), is defined as 'an interventionist approach to altering the catchment scale runoff regime and nutrient dynamics through the manipulation of hydrological flow pathways to manage water quality and quantity sustainably' (Wilkinson et al., 2014) p. 1247). In an even broader context, the concept and practice of ecological restoration can be linked here as well (Aronson et al., 2007).	'A strategically planned and managed, spatially interconnected network of multi-functional natural, semi-natural and man-made green and blue features including agricultural land, green corridors, urban parks, forest reserves, wetlands, rivers, coastal sand other aquatic ecosystems' (European Commission, 2013a, p. 3). Green infrastructure (land-based) can include, terrestrial protected areas, field margins in intensive agricultural land, ecoducts and tunnels for animals, parks and green roofs in cities. Blue infrastructure (water related) includes coastal areas, rivers, lakes, wetlands but also designed elements such as artificial channels, ponds, water reservoirs, retention basins and tanks as well as urban waste water networks(CEEWEB and ECNC, 2013; European Commission, 2013b; Haase, 2015; Naumann et al., 2010).	A strategy for decentralised, participatory and systemic natural resource management. 'It is based on the application of appropriate scientific methodologies focused on levels of biological organization which encompass the essential processes, functions and interactions among organisms and their environment. It recognizes that humans, with their cultural diversity, are an integral component of ecosystems. It is implemented via 12 principles that are "complementary and interlinked" (UNEP/CBD, 2000); section B, paragraph 6), that include: 1) The objectives of management of land, water and living resources are a matter of societal choice. 3) Ecosystem managers should consider the effects (actual or potential) of their activities on adjacent and other ecosystems. 5) Conservation of ecosystem structure and functioning, in order to maintain ecosystem services, should be a priority target of the ecosystem approach. 6) Ecosystems must be managed within the limits of their functioning. 7) The ecosystem approach should be undertaken at the appropriate spatial and temporal scales. 9) Management must recognize the change is inevitable.	'The adaptation policies and measures that take into account the role of ecosystem services in reducing the vulnerability of society to climate change, in a multi-sectoral and multi-scale approach. EBA involves national and regional governments, local communities, private companies and NGOs in addressing the different pressures on ecosystem services, including land use change and climate change, and managing ecosystems to increase the resilience of people and economic sectors to climate change' (Vignola et al., 2009, p. 692).	An approach to understand how natural systems can benefit humans, by 'linkages between ecosystem structures and process functioning and consequent outcomes which lead directly or indirectly to valued human welfare benefits (gains or losses)' (Turner and Daily, 2008, p. 27). These goods and services provided by ecosystems are 'Ecosystem services (ESs)'. They include provisioning services (e.g. food, water, heating and building material from ecosystems), cultural services (e.g. possibilities for recreation, tourism, education, sense of place), regulatory services (e.g. protection against flood or erosion, climate regulation) and supporting services (e.g. soil formation or nutrient cycling) (Alcamo et al., 2003; MA, 2005). If natural capital is the stock of assets, ecosystem services are the flows of benefits derived from those assets (Daily et al., 2011).	Natural capital is the stock of living and non-living parts of the natural system that directly and indirectly yield benefits to humans. Definitions vary in their scope and focus (Wackernagel and Rees, 1997): e.g. Daily et al. (2011, p. 3)define 'living natural capital' as 'Earth's lands and waters and their biodiversity' whilst other definitions include geological and biophysical components (e.g.(Natural Capital Coalition, 2016) or may explicitly encompass interactions and processes that form natural systems (Natural Capital Initiative, 2016). Definitions usually include both renewable and non-renewable resources (Daly and Farley, 2011). Costanza et al. (1998) also include the information stored in natural systems. Some scholars (e.g.Berkes and Folke, 1992) consider the services provided by the natural system as part of the stock as well, but this is normally separated.

Aim of employing concept	'The defined strategies are designed to mimic or to adapt the natural functioning of ecosystems to reach a target ecosystem, within an overall goal of sustainable development and with minimal and preferably biological and/or endogenous inputs' (Rey et al., 2015). CSE seeks first to describe catchment function (or role) as the principal driver for evaluating how it should be managed in the future (Wilkinson et al., 2014). The term 'systems' in CSE relates to both the natural and human functioning of a catchment as ultimately the stakeholders must agree with the interventions proposed (Wilkinson et al., 2014).	To provide '…ecological, economic and social benefits through natural solutions. It helps us to understand the value of the benefits that nature provides to human society and to mobilise investments to sustain and enhance them. It also helps avoid relying on infrastructure that is expensive to build when nature can often provide cheaper, more durable solutions.' (European Commission, 2013b p. 2). It aims to naturally regulate storm flows, flood risk water, air, temperatures, greenhouse gases, and ecosystem quality.	The aim of EA under the CBD is to 'promote conservation and sustainable use in an equitable way' (UNEP/CBD, 2000; Section A, Paragraph 1). It is expected that this concept can guide and enable fair management of natural resources in order to reflect and sustain different needs and values. It notes that conservation and use are to be balanced, and highlights that the objectives for management will vary between different groups and so require societal choice (Waylen et al., 2014).	It is expected that EBA will reduce society's vulnerability to climate change (Vignola et al., 2009).	It is expected that using ecosystem services to understand and describe how nature benefits humans, will help to inform and improve social and political processes so as to improve the management and governance of ecosystems (Primmer et al., 2015).	By enabling natural systems to be valued and managed equally to other forms of capital (financial, human, social, manufactured) (Natural Capital Initiative, 2016) it is expected that considering natural capital can help to improve decision-making for various sectors, including businesses - leading to more sustainable business models that are resilient, efficient, and secure (Natural Capital Coalition, 2016) – but also policy and other sectors.

Example	Vegetation has been used to mitigate hillslope instability, thus reducing several ecological and human problems (Stokes et al., 2014). Belford Catchment, UK uses a range of different natural flood risk management measures to reduce flood risk whilst delivering multiple benefits, UK (Wilkinson et al., 2014; Wilkinson et al., 2010).	Scheldt Estuary, Belgium uses a natural wetland to absorb and slow the flow of heavy rainfall, so reducing flood risks (Morris, 2007). Climate change adaptation in the city of London (Jones and Somper, 2014).	Multi-stakeholder systemic management of the Thanet Natura 2000 site, Kent, UK (Pound, 2008).	Durban's Municipal Climate Protection Programme (Roberts et al., 2012).	Planning for protected areas management under the ecosystem services framework in Doñana, Spain (Palomo et al., 2011). Planners in Vitoria-Gasteiz, Spain restored a river ecosystem along one of the city's main urban arteries, improving the city's sewage system and enhancing the city's ecological resilience. This river restoration will both slow storm water flows and prevent clean rainwater from entering the sewage system (Kopperoinen et al., 2014).	The Puma company's environmental profit and loss accounting system (PUMA SE, 2011).
Potential relation to NBS	CSE is a version of NBS: both are focusing on tackling societal challenges but CSE specifically focuses on catchment-scale working and manipulating hydrological processes in order to benefit humans.	Similar to NBS in some areas and can sometimes be synonymous though differences between "infrastructure" vs. "solution".	EBA aims to balance conservation and management for human needs. It is not equivalent to NBS, but its principles can be used in the design of NBS to improve the range of stakeholders engaged and to balance different interests.	EBA should be part of NBS, to ensure solutions are climate-adapted.	ES concepts can be an excellent way to consider solutions during NBS design and appraisal; however, their use should not be restricted to single or few ES and their beneficiaries.	The NC concept can help demonstrate the role of nature in meeting human needs, and hence the value of considering NBS versus other types of interventions.

Ecological Engineering and Catchment Systems Engineering (including ecological restoration), and Green/Blue Infrastructure (Table 2, columns 1 and 2 respectively) represent targeted approaches for solving specific activity or land-use problems. Encompassing a variety of activities and interventions, they clearly seek to apply natural alternatives to complement technology-based infrastructure and hence can be seen as NBS applications. For example, catchment systems engineering can help to reduce flood risk by attenuating runoff within a catchment (e.g. working with natural processes through the creation/restoration of ponds, wetlands, leaky barriers) thereby providing other multiple benefits (e.g. improving diffuse pollution) (Wilkinson et al., 2014). The difference between the approaches is the notion that NBS are explicitly considered as alternatives to and choices against human-made infrastructure that require large investment in materials and energy.

The concepts of Ecosystem Approach (EA) and Ecosystem-Based Adaptation (EBA) (Table 2, columns 3 and 4 respectively) are also approaches that seek to manage the natural environment in a way that balances benefits for nature and society. In contrast to the first two concepts, EA and EBA incorporate a systemic approach to understanding those relationships. Greater emphasis is thus placed on ecosystem complexity, change and resilience. The need to decentralise and involve stakeholders is also integral, highlighting the importance of considering different interests and conflicts (Waylen et al., 2014). This emphasis on an inclusive or participatory approach is not always shared by other concepts.

The final columns of Table 2 describe two interrelated concepts thought to have relevance to decision-making, particularly in the face of multiple and potentially competing priorities and aspirations. Ecosystem Services (ES, column 5) describe how society depends on nature (MA, 2005). So far, many studies of ecosystem services still tend to focus on only a single or few services (Abson et al., 2014), although disclosing multi-functionality and multiple benefits is at the core of the ES approach. The concept of Natural Capital (NC, column 6) uses the terminology of 'capital' that is sometimes associated with economy and business, and considers the stock of assets from which ecosystem services flow (Gómez-Baggethun et al., 2010; Schröter et al., 2012). ES and NC are both helpful for illustrating that the variety of goods and services generated by ecosystems (Mulder et al., 2015) depends on how natural systems are protected and managed. However, it can be challenging to actually use these concepts to support decisions (de Groot et al., 2010; Hails and Ormerod, 2013; Hauck et al., 2013; Martinez-Harms et al., 2015).

The four management concepts (columns 1–4) vary along a gradient of whether they take either a reductionist approach or a more holistic one, open to multiple objectives, issues and interactions. They also differ in focus. Furthermore, ecological engineering solutions tend to focus more on human-led physical interventions aided by technology to achieve specific targets. However, regardless of problem framing, it must be recognised that NBS will always be intervening in complex socio-ecological systems and indeed many prior studies have identified the problems of approaches that are based on overly reductionist or narrow processes (Mazzocchi, 2008). It is therefore essential to describe a problem for NBS in terms of multiple perspectives that account for the various links within and between ecological and social systems (Angelstam et al., 2013), and to consider the multiple social and environmental consequences of any intervention. Even in Europe, this is still not regularly the case. For example, the rehabilitation of wetland for flood protection might have positive impacts on multiple ecosystem services such as climate regulation, water purification, provision of habitats and ecotourism, but may at the same time involve trade-offs, with negative impacts on local farming livelihoods. Existing concepts such as ecosystem services could thereby usefully inform NBS, potentially providing a common currency for evaluating the consequences of differing solutions. Previous and current experiences of using all these concepts can assist in understanding the opportunities and challenges for NBS, whilst the NBS concept can connect useful insights from each. Experiences from the efforts to apply and implement these existing concepts, together with insights from Sections 2.2 and 2.3 therefore inform the priorities and challenges we present in Section 3.

NBS and Sustainability

Like many concepts, NBS makes an explicit link to the pillars of sustainable development, putting social, environmental and economic dimensions, at least conceptually, at the same level of importance. In Europe, there has been a strong emphasis on the role of NBS in fostering innovation in an environmental market 'to position Europe as a world leader, both in Research & Innovation on nature-based solutions and in the global market for nature-based solutions' (European Commission, 2015c, p. 6). This could be problematic if understood narrowly in terms of market driven approaches and cash flows. Enhanced capacity to understand, intervene in and manage multiple objectives in complex socio-ecological systems could, on the other hand, offer new opportunities to tackle cross-cutting societal challenges.

The NBS concept does not explicitly address whether the conservation and protection of biodiversity is a goal or simply a prerequisite or basis for NBS. Both views are possible (e.g. as mentioned in the formulation of NBS by the European Commission, 2015c). However, in general NBS appear to be focused on managing and providing multiple ecosystem services at intermediate levels of human use intensity (Schneiders et al., 2012), rather than at actions exclusively directed towards biodiversity conservation. If NBS are also intended to conserve biodiversity and ecosystems, this requires explicit articulation and stronger recognition in NBS projects. These issues relate to the notion of 'strong' versus 'weak' sustainability: whilst weak sustainability allows the substitution of different forms of capital (man-made or technological, human, social, natural), strong sustainability underlines the primary role of natural capital in sustaining human life, and does not present natural capital as totally substitutable (Neumayer, 2003).

Lastly, the terminology of 'solutions' can lead to assumptions that problems and needs are clear and agreed. However, as many issues pertaining to biodiversity and ecosystem management are complex, there may not even be an agreement about the problems to be solved, let alone the type of solutions needed (Game et al., 2014). For these reasons, there is a need to better recognize connections within and between societies and ecosystems, but also to accept uncertainty. Reflection, dialogue and democratic negotiation about the notions of sustainability and NBS is essential to safeguard socio-environmental justice (Ekins et al., 2003). To achieve this, examples and principles are available from attempts to implement other concepts such as the Ecosystem Approach. This is essential to allow social and environmental sustainability to be achieved.

It will also allow social (and social-ecological) innovation to be fostered, which is one of the desired components within most formulations of NBS (United Nations, 2016).

The Challenges of NBS as A New Umbrella Concept for Biodiversity and Ecosystem Stewardship

A central challenge for an 'umbrella concept' like NBS and other frameworks is where to draw the line as to what is considered as 'nature' or 'natural'. Many interventions may involve specific uses or manipulations of organisms and ecosystem processes; hence requiring decisions about acceptable levels of human intervention. For example, are genetically modified organisms or biomimicry developments considered as NBS? Additionally, there is potentially a multitude of solutions that could use nature, ranging from small scale land management to ecosystem restoration, the greening of artificial surfaces like green rooftops or green walls in cities, or broad-scale climate change mitigation and adaptation measures such as afforestation, natural flood control and, potentially, geoengineering. The definition provided by the European Commission (e.g. 2015) encompasses most of the above-mentioned examples, except those that "artificially alter nature, such as genetically modified organisms" (p. 24). The existence of a variety of ways to frame and define the concept is not necessarily problematic, as long as each case makes explicit its rationale and particular interpretation of NBS. Engaging in pluralistic reflection about alternative framings and conceptualizations can in itself be useful for identifying what is meant by NBS and the expectations for 'solutions' in any particular context.

There are several aspects of the NBS design process that can help to define what makes NBS sustainable, and why. Below we present these as questions. These can provide a logical approach to distinguishing unwanted or even potentially harmful aspects of the 'solutions' chosen.

- The range and nature of the problems to be 'solved' need specification on a case-by-case basis: Are the problems transient or persistent? By which communities and/or stakeholders are they mainly experienced? What are the qualities of the problems that have particular implications for finding solutions?
- Framing of nature is challenging: Is biotic nature only or at least primarily meant, or are abiotic (bio-physical, chemical)

ecosystem components and processes included? Is nature/are ecosystems taken as a whole, including for example water, matter and energy cycles as well as landscapes and urban environments with artificial structures and humans? And, is self-regulatory potential considered an aspect of naturalness and sustainability?

- Most NBS will likely include some degree of alteration and/or 'design' of nature (e.g., trade-offs by favouring/choosing one ecosystem service over another, by selecting certain assemblies of species); what is the level of integrity or 'naturalness' in order to still be considered a NBS?
- Is it possible to consider all the possible pros and cons of an intervention? Is it important to consider both benefits and costs in detail? For example, in relation to human health, NBS may bring multiple health benefits (Hartig et al., 2014), but may also contribute to health risks, such as allergies and infectious diseases (Keune et al., 2013).
- How does the 'solution' deal with the complexity of problems, as systemic problems with multiple trade-offs cannot be easily broken down to 'easy' solutions?
- Do 'solutions' rely on technical/physical types of innovation, and to what extent are intellectual and social innovations considered? The common use of terms such as 'cost-effectiveness' or 'innovation' (often perceived as technological) might indicate a preference for the former, but a combination will have more chances of contributing to sustainability (see also discussion on 'innovation' in van den Hove et al. (2012)).
- How can NBS ensure that all relevant stakeholders are considered and democratically involved, taking into account social cohesion and equity? How can conflicting goals and interests be reconciled when creating and choosing options for NBS? How can any differing outcomes of NBS be anticipated and evaluated, and how can fairness be judged?

Many aspects of these questions also arise when using other existing concepts, see for example the controversies about the role and application of the ecosystem services concept (Fitter, 2013; Gómez-Baggethun and

Muradian, 2015; Silvertown, 2015). However, the ostensible focus of NBS on the 'solutions' domain may help to tackle some challenges, by explicitly focusing attention on the actions and inputs needed.

For the practical implementation of NBS, we suggest use of these questions together with a core framework of elements discussed in the following section. These elements provide guidance to stimulate the exploration of new pathways to tackle societal challenges, yet allow the scope of NBS to be clearly bounded for different areas (Eggermont et al., 2015; Maes and Jacobs, 2015). Science, policy and practice groups will be needed to tackle these challenges. Research scientists would especially need to develop a new role to support this framework, based on transdisciplinarity and a systemic approach to problem solving and management. In the following section, we present the elements we consider relevant for enabling effective and equitable development of NBS. For example, the rehabilitation of wetland for flood protection might have positive impacts on multiple ecosystem services such as climate regulation, water purification, provision of habitats and ecotourism, but may at the same time involve trade-offs, with negative impacts on local farming livelihoods.

NATURE-BASED SOLUTIONS IN PRACTICE: KEY ELEMENTS FOR THE OPERATIONALIZATION OF THE NBS CONCEPT

As with all other concepts, NBS will need to be embedded in the existing policy mix including biodiversity protection measures, spatial planning, environmental assessment or economic incentives, as well as in practical applications and trials (Barton et al., 2014). At the same time, the institutional context will need to evolve to enable the necessary shifts to take place (Maes and Jacobs, 2015). In most cases, the critical decisions about NBS design, costs, location and scale as well as levels of management intensity will involve a wide range of stakeholders who may have different ideas and pre-existing ways of managing their problems. Further, NBS will have to become accepted alternatives to other solutions, which may not be nature-based or may be unsustainable. For this, we consider the five steps described in Sections 3.1 through 3.5 as key elements to be addressed in NBS projects, jointly by researchers and other actors (Fig. 1 – designing NBS projects, see also Cohen-Schacham et al. (2016)). We include examples from current literature for illustration and make use of the main elements identified in transdisciplinary research and successful practice in general (e.g., Jahn et al., 2012).

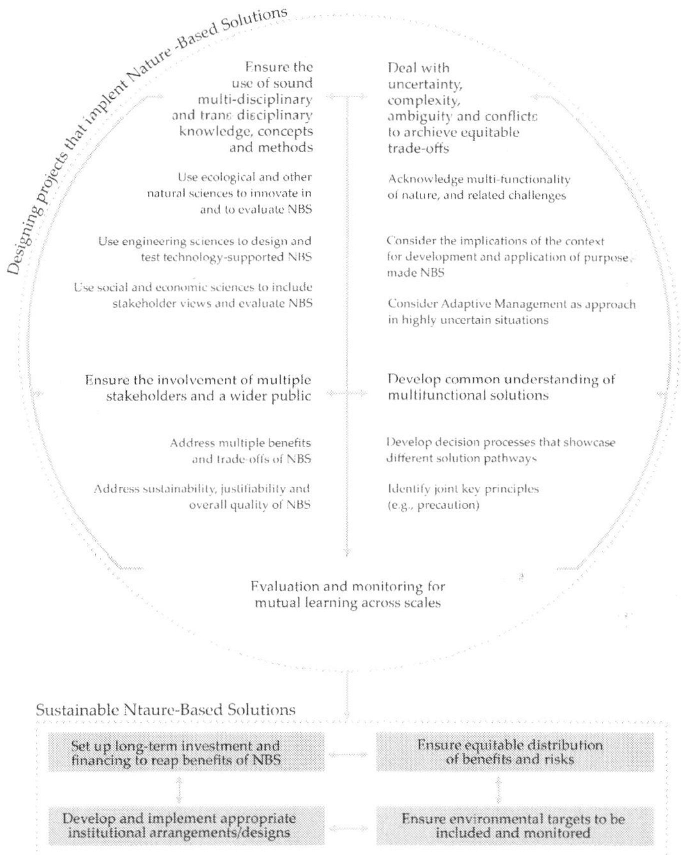

Figure 1. Processes and dynamics in designing and implementing nature-based solutions, including multidisciplinary and transdisciplinary knowledge creation and use to achieve successful solutions balancing economic, social and ecological targets.

Dealing with Uncertainty and Complexity: the Adaptive Management Approach as an Example

Uncertainty will be a prevailing characteristic when designing NBS, given that in many cases NBS deal with complex socio-ecological systems whose responses to management and natural factors are often non-linear, heterogeneous and incompletely known (Seastedt et al., 2008; Suding et al., 2004). One way of coping with uncertainty, and limiting the risks of triggering negative responses, is to manage for increased resilience (Lindenmayer et al., 2008). Some of these practices are incorporated in adaptive management, which is an approach recommended for dealing with uncertainty, complexity and dynamics. It includes devising flexible ways to maximize learning opportunities by applying different strategies, and the consideration of practices as experiments by ensuring that management treatments are replicated and responses are carefully monitored (Assmuth and Hildén, 2008; Assmuth et al., 2010; Lindenmayer et al., 2008). Although adaptive management has proven difficult in practice (Roe et al., 2005; Westgate et al., 2013); the broad experience gathered so far would help to support effective NBS implementation.

Adaptive management that includes structured monitoring of ecological responses is rarely implemented (Lindenmayer et al., 2008); which reflects the challenges and costs of such endeavours. However, there are several models developed for planning and management of, for instance, protected areas and other demarcated nature areas, such as Recreation Opportunity Spectrum, Limits of Acceptable Change, Visitor Impact Management, and Visitor Experience and Resource Protection (Newsome et al., 2012). These can all be characterized by their use of adaptive management, where goals and actions are adapted to ongoing changes (ecological and social). Such operational models, also abound in other areas of ecosystem and natural resource use, often include socio-economic perspectives, and as such could feed into the development and application of NBS. The development of an evidence-base that holds details of the circumstances in which NBS work is an essential component of effective adaptive management (see below).

Adaptive management should be coupled with some description of the critical dimensions of the solution, including societal factors. This would ensure that NBS would not only function in ecological terms but would also avoid that they are economically unfeasible or socially untenable. In some cases, social factors can be incorporated by simply placing ecological entities and processes in a societal context, in others, when feasible and relevant, through more detailed and explicit analysis of the socio-ecological links. In both cases, socio-ecological modelling can be part of a transdisciplinary learning process, where approaches to problems and solutions, including values and goals, are deliberated by experts, practitioners and stakeholders, and adapted where and when needed (cf. Sections 3.2 and 3.5). Such an approach can yield important opportunities for citizen participation and can result in the inclusion of local knowledge, environmental justice and local economic growth and an increased understanding of uncertainties and complexities (Irvine and Kaplan, 2001).

Ensuring the Involvement of Multiple Stakeholders

Any societal problem or issue is likely to be affected by or have effects on several groups of stakeholders. Their involvement is thought to bring three types of benefit to the process of planning and delivering improvements in environmental management (Blackstock and Richards, 2007): (i) 'substantive' benefits, as stakeholders' perspectives, conditions and knowledge inform and improve planning (van den Hove, 2000); (ii) 'instrumental' benefits, as the process becomes better understood and more acceptable to stakeholders, and hence better supported (Parkins and Mitchell, 2005); and (iii) 'normative' benefits, as stakeholder involvement increases the legitimacy of the process, and generally supports democracy (Schultz et al., 2010). These benefits could be translated into the design of NBS, but require that stakeholders are meaningfully involved and empowered through the NBS process (e.g., Waylen et al., 2015a; Wyborn, 2015).

The participative process may be more or less complex depending on the issues to be considered. In particular with NBS that would involve important trade-offs (see Section 3.5) a participatory process becomes critical, and can be especially pertinent, when their aim is to better connect people with their local natural resources. There is now much guidance and many ideas about how to facilitate collaborative efforts and engage relevant stakeholders, for example via analytical-deliberative approaches (Fish, 2011; Keune and Dendoncker, 2013). These approaches can be demanding in terms of time and skills, but can ensure co-design, innovation, ownership and later stewardship of NBS (Armitage et al., 2007; Jones-Walters and Çil, 2011; Reed, 2008). Finally, stakeholder engagement is also relevant for sharing of knowledge and learning across and between cases (Keune et al., 2015).

Ensuring the sound use of Multi- and Transdisciplinary knowledge

NBS projects will need to be combined with an increase of interdisciplinary work across scientific domains. We see this as an endeavour that goes beyond the collaboration that presently takes place, for example, between ecological sciences and engineering in many restoration projects, or when ecological and social sciences jointly address the use of ecosystem services. There is a strong basis for such collaboration due to the many integrative research activities that have taken place in the past, but NBS projects are likely to increase the demands for input and flexibility from different disciplines.

As society will necessarily shape the design, selection, implementation and consequences of NBS, social sciences can help us to understand the potential and pitfalls of NBS to inform the design of new NBS, and to improve our general understanding of environmental governance. Social science research can provide insights into how different choices may affect support for and implementation of NBS, about how to foster stakeholder deliberation and empowerment (e.g. Reed, 2008), and the identification of relevant, shared social values (Kenter et al., 2014) and collective actions. Some of this expertise is derived from studies of existing related approaches (Table 2). For example, the field of ecosystem services has advanced in relation to the development of methods to address the challenge of economic externalities that may occur when exploiting natural resources (e.g. via Payments for Ecosystem Services projects such as those described by Caro-Borrero et al. (2015)). However, it also embraces the assessment of multiple and non-monetary benefits from nature (Díaz et al., 2015), indicating that experience about this and other methods to elicit motivation for choices should also be considered for promoting change (Santangeli et al., 2016). Existing governance structures and institutional processes will also affect this issue, and may foster or hinder the search for alternative NBS, their acceptance and ultimate success. The need to adapt governance structures must also therefore be taken into account (Rauschmayer and Wittmer, 2006; Vignola et al., 2013; Waylen et al., 2015b).

There is also a continuing need for NBS to be knowledge-based in terms of ecological functioning. Despite the wealth of available ecological knowledge, there are still considerable gaps in translating it into actions that can help to design sustainable systems as an integral part of biodiversity and natural resource management (MA, 2005). There are presently major challenges in relation to the European MAES - Mapping and Assessing Ecosystem Services project which has set out to map and evaluate ecosystem services in terms of their condition and functionality and for the Natural Capital Accounting (NCA) project, precisely at the link between ecological research and economics (Maes et al., 2013). The gaps in knowledge are often still of a fundamental character and need to be addressed in order to inform future NBS projects.

Furthermore, there is a reverse flow of knowledge from practical action to science, guiding the formulation of new, even fundamental questions and of answers to them. Ecology and related disciplines need to provide up-to-date and broader conceptual frameworks (Suding et al., 2004) and effective practical tools (Hobbs and Harris, 2001) to support this two-way flow. Ecological disturbance, ecosystem dynamics and tipping points, population viability, vegetation structural diversity, the role of keystone species and functional groups, response diversity (Zang et al., 2014) and landscape connectivity are considered to be critical features that underpin the persistence of ecological functions and the responses of ecosystems to climate change, disturbances and environmental variability (Lavorel et al., 2015). In addition, ecological models that show the possible outcomes of human interventions would be useful tools for planning NBS (for example the European Commission's NCA).

The representation of the ecological system can be at different degrees of complexity, quantitative or conceptual, but it has to capture the essence of the system and its dynamics in order to address the problem at hand; which argues for the use of ecological knowledge from as many sources as possible, including that from practitioners and the local experience.

The collection of data on the ecological aspects before and after NBS implementation, together with information on the actual NBS measures taken as well as related socio-economic parameters, forms a fundamental component of any evidence-base on NBS (see also Section 3.5).

Developing common Understanding of Multifunctional Solutions, Trade-offs and Natural Adaptation

The involvement of relevant actors and a responsive use of natural and social sciences as outlined will allow NBS to set their goals based on a common understanding of the available options, their relative costs as well as social and ecological impacts (Fig. 1). This is challenging as the direct and indirect impacts will often be uncertain and may take time before they become evident, and the estimates of both impacts and costs will be dependent on value judgments. The understanding of them is therefore provisional, but helps to create informed and transparent processes for developing and applying NBS.

Implementing the NBS concept implies the fulfilment of multiple goals. Ideally, clear win-win options can be found, but trade-offs between alternative forms of using nature are ubiquitous (Barton et al., 2014), and cannot be solved through exact calculus and utility maximization. This is also due to incomplete knowledge (see Section 3.3) and conflicting value attributions. Instead, they require other kinds of approach, such as qualitative, multi-criteria, iterative and experimental, particularly in line with adaptive management (Prato, 2007; see also Section 3.1). In addition, NBS projects will face situations where several options can be appropriate for a particular site, catchment or region, each of them with different associated costs, benefits, impacts and risks (Eggermont et al., 2015) and involving different levels of conflict over the use of space (Schindler et al., 2014). Specifically, maintaining ecosystem and landscape-level ecological functions often implies hard trade-offs with other land use options, including provisioning ecosystem services (Schröter et al., 2014). NBS will need to address such trade-offs and set goals that maintain or restore ecological structures and functions based on best available knowledge and on an agreement regarding the impacts, securing compatibility with the level of use and disturbance.

Complexity in ecosystem responses presents considerable challenges in terms of setting objectives, selecting the appropriate solution and setting quality standards. It requires flexible and transparent models of key structures and processes that can gather the best possible knowledge from science and practice, that are flexible enough to incorporate new knowledge (for instance acquired in a process of adaptive management and which combine evidence with precaution) and that make uncertainty explicit (Rumpff et al., 2011). This however raises the important question of how to maintain natural adaptability when designing NBS, specifically how to adjust ecosystem structure, process and function in response to natural and human disturbance. This can be achieved for instance, by incorporating ranges of desired or accepted variability in the formulation of NBS goals, by ensuring that the means for planning and implementation of NBS are adaptable, by focusing on ecosystem processes in addition to composition, by prioritizing native species and local resources, and by following-up and acting on changes in ecosystems, technologies and in society.

If this is achieved, NBS will foster the transformation in the ecosystem management paradigm, which is to replace the increase-efficiency and single-objective management (e.g., separating conservation, land use and water issues), with strategies that build on maintaining ecological resilience taking into account the multi-functionality of landscapes and ecosystems (Hansen and Pauleit, 2014; Schindler et al., 2014).

Evaluate and Monitor for Mutual Learning

A final important element that is often raised in biodiversity conservation and ecological restoration practice, and which is applicable to NBS, is the development of adequate measures of progress and success towards agreed goals. This often means selection of a set of easily measurable criteria for the ecological, social and economic effectiveness of the interventions (e.g., Heink et al., 2015; Hobbs and Harris, 2001), but which may result in important but less easily measurable ones being neglected.

Many indicators have the potential to be considered as success criteria, but these will need to be clearly related to the specific 'solution' goals in terms of biophysical aspects and ecosystem services (e.g. carbon sequestration, water use efficiency, pollination). The same applies to the economic and social spheres in terms of value, capital or investment/ revenue in the system or to the effects on health and well-being. In terms of the above criteria, the outcomes of NBS projects are likely to be better understood over the long-term, especially for NBS that are applied at large scales.

NBS will often deal with problems that manifest at different spatial scales from local to regional, national and even global (e.g. NBS for addressing climate or planetary change such as carbon fixation). When evaluating and monitoring the efficiency and effectiveness of NBS a nested approach across scales is therefore required. Certain local NBS may easily be monitored in a short time-span and in connection with locally specific challenges. The feedback in terms of adapting (management of) such NBS can also occur quickly. When considering larger geographical scales, however, the temporal scale of evaluation will increase as will the specificity of reaching predefined goals. As solutions at various geographical levels are not exclusive, evaluation requires the possibility of upscaling and downscaling monitoring results and coordinated processing and communication across scales.

The social sciences can be helpful here in offering participatory evaluative and co-development approaches, as well as quantitative well-being indicators (e.g., Dallimer et al., 2014). Due to the complexity of nature and societies, dealing with quality criteria for analytical, deliberative, extended peer review or participatory processes is challenging (Keune et al., 2014). Rauschmayer et al. (2009) point out that process outcomes may be valued quite differently from different stakeholder perspectives. To respond adequately, they propose the use of participatory evaluation in order to respect the legitimacy of different views on quality, for example Delphi, group-model building and other expert or stakeholder opinion solicitation and deliberation methods (Sendzimir et al., 2010). Resulting information can sometimes be most relevant for politicians and decision-makers as it provides important contextual information about the human dimension.

SUMMARIZING DISCUSSION: SOCIETAL OPPORTUNITIES AND CHALLENGES OF THE NBS CONCEPT

Like many umbrella concepts, NBS bring new challenges and opportunities from the perspectives of science, policy and practice. Some of the likely challenges to be encountered by NBS will be similar to those encountered by other concepts for managing nature (see Table 2), whilst others may be unique to NBS. Furthermore, there are still uncertainties and knowledge gaps in terms of overlap and delineation associated with linked concepts. For example, stakeholder participation and the adoption of a truly systemic perspective is extremely difficult considering the fuzziness and uncertainties regarding the basic concept. It is therefore important to use all relevant sources of expertise in developing and applying NBS, and to support and share learning within and across NBS projects, depending on their focus. Several attempts have already been outlined that identify such areas (see for example Eggermont et al., 2015; European Commission, 2015c; Cohen-Schacham et al., 2016). For us, the following four areas will allow for a further specification of the elements outlined in Section 3:

- The maintenance of biodiversity, ecological functions and/or ecosystem services in systems and at scales where human management interacts with natural processes (ecological engineering, ecosystem approach, ecosystem services management);
- The restoration of ecosystems that have been damaged by human activities (ecological restoration, ecological engineering);
- The design of solutions as a mechanism to cope with climate change and other factors of environmental variability or hazards (ecosystem services management); and,
- The sustainable utilization of nature to contribute to alleviating challenges in human well-being, including vulnerability, social justice, economy and culture (ecosystem services management).

These four focus areas will need different kinds of NBS that complement or include existing approaches within them. For example, activities in i) will need to respect existing instruments for nature protection like protected areas, regulation of use and/or incentives for nature-friendly practices, whereas activities in ii) will often build on/include ecological engineering approaches. Developing this in a transparent manner will be essential if NBS are to have an added value compared to existing approaches.

From a scientific perspective, this calls for a transdisciplinary research approach that links practitioners, policy-makers and scientists from different disciplines, and which engages with citizens and other users and producers of knowledge. Transdisciplinary science is already encouraged in areas such as the EU Horizon 2020 Programme (European Commission, 2015a); furthermore, the EU expects city authorities to lead research on NBS for 'Smart and Sustainable Cities' (European Commission, 2015b) with a link to green growth and eco-innovation. However, transdisciplinary science will bring its own challenges, and is still far from being 'mainstream' (Angelstam et al., 2013; Brandt et al., 2013; Keune et al., 2015). Practical support and guidance will be needed in order to reflect potentials and challenges of cross-sectoral and multi-actor collaboration (Fig. 1). Tackling complex social-environmental problems will often highlight conflicts among different interests or sectors. That creates a major opportunity for the NBS concept to allow and even invite 'outside the box' thinking. However, careful facilitation will be needed for conflicts to be productive in this way (Rauschmayer and Wittmer, 2006).

NBS therefore also offer opportunities for encouraging mainstreaming of environmental targets into sectors in policy, business and practice that might not traditionally consider or value the environment, thereby strengthening the potential for strong sustainability in decision making (Fig. 1). This might be the biggest strength of the NBS concept, but this can also bring the risk of 'overselling nature' (Rodriguez-Labajos and Martinez-Alier, 2013) or of encouraging a perception of ecosystems as entirely-substitutable by other assets used by humans. Long-term investment and financing will need to be set up in order to reap the (equitably distributed) benefits of NBS (Fig. 1). In that sense, NBS can become a major contributor to the wider concept of a 'green economy', which is also critically discussed regarding its sustainability perspective (Brand, 2012; Gasparatos and Willis, 2015). The implementation of NBS projects will therefore need to be embedded in the corresponding societal debates and deliberative processes. As a minimum, communication about NBS should highlight risks from overly-simplistic or optimistic framings (see for example, the 'green revolution' of the 1960s (McIntyre, 2009) or recent discussions of 'eco-gentrification' (Curran and Hamilton, 2012)). Ideally, a diversity of actors should be involved in the deliberative processes (Parkins and Mitchell, 2005) that could take place in relation to the role, scope and appropriateness of interventions premised in relation to NBS. This will also need a careful reflection on institutional arrangements that can enable NBS with such inclusive, long-term and balanced perspectives.

To have the best chance of success, NBS projects should be based on a well-balanced, clear, widely accepted and implementable set of key principles. The considerations set out above can serve as a starting point and a foundation. The IUCN (2012) definition of NBS, (see Table 1, Cohen-Schacham et al., 2016) also provides a reference point, and any rationale should further build on the 12 Malawi principles of the Ecosystem Approach, which has also been applied in a number of European contexts (e.g. Apitz et al., 2006; Waylen et al., 2015a). The principles must allow flexibility to accommodate different types of solutions in the focus areas of NBS, which is crucial for innovation (Eggermont et al., 2015) whilst ensuring that no dimension of sustainability is overlooked. Developing them in the coming years, including an integrated 'innovation' perspective on NBS will also provide an opportunity to introduce a corresponding framework for evaluation and monitoring (Fig. 1). In order to utilize the potential of NBS for managing risks, the frameworks of the International Council for Risk Governance (IRGC, 2005; IRGC, 2009) for risk governance can be helpful (cf. Klinke and Renn, 2014; Renn and Klinke, 2013).

To conclude, the new NBS concept should be perceived as an opportunity, but also as a challenge since a good understanding of ecosystem processes is needed, a diversity of actors must be engaged and a broad set of societal facts/issues needs to be included and integrated. It is a chance for sustainability science to achieve more recognition in policy, projects and practice, and to bring together ideas from all relevant actors. Key open questions about how to implement NBS will remain as is currently the case for other similar concepts such as adaptive management and the Ecosystem Approach. Whether NBS become something that goes beyond 'just another communication tool' to promote a positive view of 'nature-based' and 'sustainable' management measures, and which avoids using old tools with diverse conceptual foundations, will depend on whether these conceptual and practical challenges can be addressed when developing projects and linking them across scales, contexts and people. Bringing together the diversity of context, societal backdrop and scale will be essential if project funders are to deliver frameworks within which researchers and other actors are to implement genuine, sustainable nature-based solutions.

Funding

This work was supported by ALTER-Net (Europe's ecosystem research network, see http://www.alter-net.info) via funding the exchange that led to this paper, including a workshop held on 29–30 January 2015, hosted by the FU Berlin. GMR contribution was funded by the OpenNESS project (EU FP7 grant agreement No 308428). EK was also supported by the Hungarian Research Centre of Excellence (9878-3/2016/FEKUT). KNI, KAW and MEW were also supported by the Scottish Government Strategic Research Programme 2011–2016.

REFERENCES

1. Abson et al., 2014 D. Abson, H. Von Wehrden, S. Baumgärtner, J. Fischer, J. Hanspach, W. Härdtle, *et al.*Ecosystem services as a boundary object for sustainability Ecol. Econ., 103 (2014), pp. 29-37
2. Admiraal et al., 2016 J. Admiraal, C. Musters, G. De SnooThe loss of biodiversity conservation in EU research programmes: thematic shifts in biodiversity wording in the environment themes of EU research programmes FP7 and Horizon 2020 J. Nat. Conserv., 30 (2016), pp. 12-18
3. Alcamo et al., 2003 J. Alcamo, E.M. Bennett, Millennium Ecosystem Assessment (Program)Ecosystems and Human Well-being: A Framework for Assessment Island Press, Washington, D.C. (2003)
4. Angelstam et al., 2013 P. Angelstam, K. Andersson, M. Annerstedt, R. Axelsson, M. Elbakidze, P.Garrido, *et al.*Solving problems in social-ecological systems: definition, practice and barriers of transdisciplinary research Ambio, 42 (2013), pp. 254-265
5. Apitz et al., 2006 S.E. Apitz, M. Elliott, M. Fountain, T.S. GallowayEuropean environmental management: moving to an ecosystem approach Integr. Environ. Assess. Manag., 2 (2006), pp. 80-85

6. Armitage et al., 2007 D. Armitage, F. Berkes, N. DoubledayAdaptive Co-management: Collaboration, Learning and Multi-level Governance UBC Press (University of British Columbia), Vancouver (2007)
7. Aronson et al., 2007 J. Aronson, S.J. Milton, J.N. BlignautRestoring Natural Capital – Science. Business, and Practice Island Press (2007), p. 400
8. Assmuth and Hildén, 2008 T. Assmuth, M. HildénThe significance of information frameworks in integrated risk assessment and management Environ. Sci. Pol., 11 (2008), pp. 71-86
9. Assmuth et al., 2010 T. Assmuth, M. Hildén, C. BenighausIntegrated risk assessment and risk governance as socio-political phenomena: a synthetic view of the challenges Sci. Total Environ., 408 (2010), pp. 3943-3953
10. Barton et al., 2014 D.N. Barton, I. Ring, G.M. RuschPolicyscapes. Natur-based policy mixes for biodiversity conservation and ecosystem services provision Policy Brief No. 2. POLICYMIX Project, POLICYMIX project, Trondheim (2014)
11. Bennctt et al., 2009 E.M. Bennett, G.D. Peterson, L.J. GordonUnderstanding relationships among multiple ccosystem services Ecol. Lett., 12 (2009), pp. 1394-1404
12. Berkes and Folke, 1992 F. Berkes, C. FolkeA systems perspective on the interrelations between natural, human-made and cultural capital Ecol. Econ., 5 (1992), pp. 1-8
13. Blackstock and Richards, 2007 K.L. Blackstock, C. RichardsEvaluating stakeholder involvement in river basin planning: a Scottish case study Water Policy, 9 (2007), pp. 493-512
14. Brand, 2012 U. BrandGreen economy–the next oxymoron? No lessons learned from failures of implementing sustainable development GAIA-Econ. Respect. Sci. Soc., 21 (2012), pp. 28-32
15. Brand and Jax, 2007 F.S. Brand, K. JaxFocusing the meaning (s) of resilience: resilience as a descriptive concept and a boundary object Ecol. Soc., 12 (2007), p. 23
16. Brandt et al., 2013 P. Brandt, A. Ernst, F. Gralla, C. Luederitz, D.J. Lang, J. Newig, *et al.*A review of transdisciplinary research in sustainability science Ecol. Econ., 92 (2013), pp. 1-15

17. Brundtland et al., 1987 G. Brundtland, M. Khalid, S. Agnelli, S. Al-Athel, B. Chidzero, L. Fadika, *et al.*Our Common Future ('Brundtland Report') (1987)
18. Caro-Borrero et al., 2015 A. Caro-Borrero, E. Corbera, K.C. Neitzel, L. Almeida-Lenero"We are the city lungs": payments for ecosystem services in the outskirts of Mexico City Land Use Policy, 43 (2015), pp. 138-148
19. CEEWEB, ECNC, 2013 CEEWEB, ECNCEnriching our Society through Natural Solutions: Why and How to Make Green Infrastructure Projects a Sustainable Answer for Ecological, Social and Economic Problems? CEEWEB, Budapest (2013)
20. Cohen-Schacham et al., 2016 E. Cohen-Schacham, G. Walters, C. Janzen, S. Maginnis Iucn (Ed.), Nature-based Solutions to Address Global Societal Challenges, IUCN, Galnd, Switzerland (2016) (pp. xiii + 97)
21. Costanza and Daly, 1992 R. Costanza, H.E. DalyNatural capital and sustainable development Conserv. Biol., 6 (1992), pp. 37-46
22. Costanza et al., 1998 R. Costanza, R. D'arge, R. De Groot, S. Farber, M. Grasso, B. Hannon, *et al.*The value of ecosystem services: putting the issues in perspective Ecol. Econ., 25 (1998), pp. 67-72
23. Curran and Hamilton, 2012 W. Curran, T. HamiltonJust green enough: contesting environmental gentrification in Greenpoint, Brooklyn Local Environ., 17 (2012), pp. 1027-1042
24. Daily, 1997 G. DailyNature's Services: Societal Dependence on Natural Ecosystems Island Press (1997)
25. Daily et al., 2011 G.C. Daily, P. Kareiva, S. Polasky, T.H. Ricketts, H. TallisMainstreaming natural capital into decisions P. Kareiva, H. Tallis, T.H. Ricketts, G.C. Daily, S. Polasky (Eds.), Natural Capital. Theory & Practice of Mapping Ecosystem Services, Oxford University Press, Oxford, UK (2011), pp. 3-14
26. Dallimer et al., 2014 M. Dallimer, D. Tinch, N. Hanley, K.N. Irvine, J.R. Rouquette, P.H. Warren, *et al.*Quantifying preferences for the natural world using monetary and nonmonetary assessments of value Conserv. Biol., 28 (2014), pp. 404-413
27. Daly and Farley, 2011 H.E. Daly, J. FarleyEcological Economics: Principles and Applications Island Press (2011)

28. de Groot et al., 2010 R.S. de Groot, R. Alkemade, L. Braat, L. Hein, L. WillemenChallenges in integrating the concept of ecosystem services and values in landscape planning, management and decision making Ecol. Complex., 7 (2010), pp. 260-272
29. Díaz et al., 2015 S. Díaz, S. Demissew, J. Carabias, C. Joly, M. Lonsdale, N. Ash, *et al.*The IPBES conceptual framework — connecting nature and people Curr. Opin. Environ. Sustain., 14 (2015), pp. 1-16
30. Eggermont et al., 2015 H. Eggermont, E. Balian, J.M.N. Azevedo, V. Beumer, T. Brodin, J. Claudet, *et al.*Nature-based solutions: new influence for environmental management and research in Europe GAIA - Ecological Perspectives for Science and Society, 24 (2015), pp. 243-248
31. Ekins et al., 2003 P. Ekins, S. Simon, L. Deutsch, C. Folke, R. De GrootA framework for the practical application of the concepts of critical natural capital and strong sustainability Ecol. Econ., 44 (2003), pp. 165-185
32. European Commission, 2013a European CommissionCommunication from the Commission to the European Parliament, the Council, the Economic and Social Committee and the Committee of the Regions: Green Infrastructure (GI) — Enhancing Europe's Natural Capital, COM (2013) 249 Final European Commission, Brussels (2013)
33. European Commission, 2013b European CommissionCommunication from the Commission to the European Parliaments, the Council, the Economic and Social Committee and the Committee of the Regions: Green Infrastructure (GI) — Enhancing Europe's Natural Capital {COM (2013) 249final}, Brussels (2013)
34. European Commission, 2015a European CommissionHorizon 2020 Work Programme 2016-017 - 17. Cross-cutting Activities (Focus Areas). Brussels (2015)
35. European Commission, 2015b European CommissionHorizon 2020 Work Programme 2016–2017-12. Climate Action, Environment, Resource Efficiency and Raw Materials (European Commission Decision C (2015) 6776 of 13 October 2015). 2015, Brussels (2015)

36. European Commission, 2015c European Commission D.-G.F.R.A. Innovation (Ed.), Towards an EU Research and Innovation Policy Agenda for Nature-based Solutions & Re-naturing Cities - Final Report of the Horizon 2020 Expert Group, European Commission, Directorate-General for Research and Innovation, Brussels (2015), p. 74
37. Fish, 2011 R.D. FishEnvironmental decision making and an ecosystems approach some challenges from the perspective of social science Prog. Phys. Geogr., 35 (2011), pp. 671-680
38. Fitter, 2013 A. FitterAre ecosystem services replaceable by technology? Environ. Resour. Econ., 55 (2013), pp. 513-524
39. Game et al., 2014 E.T. Game, E. Meijaard, D. Sheil, E. Mcdonald-MaddenConservation in a wicked complex world; challenges and solutions Conserv. Lett., 7 (2014), pp. 271-277
40. Gasparatos and Willis, 2015 A. Gasparatos, K.J. WillisBiodiversity in the Green Economy Routledge (2015)
41. Gómez-Baggethun and Muradian, 2015 E. Gómez-Baggethun, R. MuradianIn markets we trust? Setting the boundaries of market-based instruments in ecosystem services governance Ecol. Econ., 117 (2015), pp. 217-224
42. Gómez-Baggethun et al., 2010 E. Gómez-Baggethun, R. De Groot, P.L. Lomas, C. MontesThe history of ecosystem services in economic theory and practice: from early notions to markets and payment schemes Ecol. Econ., 69 (2010), pp. 1209-1218
43. Haase, 2015 D. HaaseReflections about blue ecosystem services in cities Sustainability Water Qual. Ecol., 5 (2015), pp. 77-83
44. Hails and Ormerod, 2013 R.S. Hails, S.J. OrmerodEDITORIAL: ecological science for ecosystem services and the stewardship of natural capital J. Appl. Ecol., 50 (2013), pp. 807-811
45. Hansen and Pauleit, 2014 R. Hansen, S. PauleitFrom multifunctionality to multiple ecosystem services? A conceptual framework for multifunctionality in green infrastructure planning for urban areas Ambio, 43 (2014), pp. 516-529
46. Hartig et al., 2014 T. Hartig, R. Mitchell, S. De Vries, H. FrumkinNature and health Annu. Rev. Public Health, vol. 35 (35) (2014), pp. 207-20 +

47. Hauck et al., 2013 J. Hauck, C. Görg, R. Varjopuro, O. Ratamäki, K. JaxBenefits and limitations of the ecosystem services concept in environmental policy and decision making: some stakeholder perspectives Environ. Sci. Pol., 25 (2013), pp. 13-21
48. Heink et al., 2015 U. Heink, J. Hauck, K. Jax, U. SukoppRequirements for the selection of ecosystem service indicators–the case of MAES indicators Ecol. Indic. (2015)
49. Hobbs and Harris, 2001 R.J. Hobbs, J.A. HarrisRestoration ecology: repairing the earth's ecosystems in the new millennium Restor. Ecol., 9 (2001), pp. 239-246
50. IRGC, 2005 IRGC Governance, I.C.F.R. (Ed.), White paper 1 on risk governance: towards an integrative approach, International Council for Risk Governance, Geneva (2005)
51. IRGC, 2009 IRGC Governance, I.C.F.R. (Ed.), Risk governance deficits. An analysis and illustration of the most common deficits in risk governance, International Council for Risk Governance, Geneva (2009)
52. Irvine and Kaplan, 2001 K.N. Irvine, S. KaplanCoping with change: the small experiment as a strategic approach to environmental sustainability Environ. Manag., 28 (2001), pp. 713-725
53. IUCN, 2012 IUCNThe IUCN Programme 2013–2016 IUCN, Gland (2012), p. 30
54. Jahn et al., 2012 T. Jahn, M. Bergmann, F. KeilTransdisciplinarity: between mainstreaming and marginalization Ecol. Econ., 79 (2012), pp. 1-10
55. Jansson, 1994 A. JanssonInvesting in Natural Capital: The Ecological Economics Approach to Sustainability Island Press (1994)
56. Jones and Somper, 2014 S. Jones, C. SomperThe role of green infrastructure in climate change adaptation in London Geogr. J., 180 (2014), pp. 191-196
57. Jones-Walters and Çil, 2011 L. Jones-Walters, A. ÇilBiodiversity and stakeholder participation J. Nat. Conserv., 19 (2011), pp. 327-329
58. Kenter et al., 2014 J. Kenter, M. Reed, K. Irvine, E. O'brien, E. Brady, R. Bryce, *et al.*UK National Ecosystem Assessment Follow-

on Work Package Report 6: Shared, Plural and Cultural Values of Ecosystems, UNEP-WCMC, LWEC, UK (2014)

59. Keune and Dendoncker, 2013 H. Keune, N. DendonckerNegotiated complexity in ecosystem services science and policy making Ecosystem Services. Global Issues, Local Practices, Elsevier (2013), pp. 167-180
60. Keune et al., 2013 H. Keune, C. Kretsch, G. De Blust, M. Gilbert, L. Flandroy, K. Van Den Berge, *et al.*Science–policy challenges for biodiversity, public health and urbanization: examples from Belgium Environ. Res. Lett., 8 (2013), p. 025015
61. Keune et al., 2014 H. Keune, G. Koppen, B. Morrens, A. CollesExtended peer evaluation of an analytical deliberative decision support procedure in environmental health practice Eur. J. Risk Regul., 5 (1) (2014), pp. 25-35
62. Keune et al., 2015 H. Keune, N. Dendoncker, F. Popa, J. Sander, S. Kampelmann, F. Boeraeve, *et al.*Emerging ecosystem services governance issues in the Belgium ecosystem services community of practice Ecosystem Services, 16 (2015), pp. 212-219
63. Klinke and Renn, 2014 A. Klinke, O. RennExpertise and experience: a deliberative system of a functional division of labor for post-normal risk governance Innovation: The European Journal of Social Science Research, 27 (2014), pp. 442-465
64. Kopperoinen et al., 2014 L. Kopperoinen, P. Itkonen, J. NiemelaUsing expert knowledge in combining green infrastructure and ecosystem services in land use planning: an insight into a new place-based methodology Landsc. Ecol., 29 (2014), pp. 1361-1375
65. Lavorel et al., 2015 S. Lavorel, M.J. Colloff, S. Mcintyre, M.D. Doherty, H.T. Murphy, D.J. Metcalfe, *et al.*Ecological mechanisms underpinning climate adaptation services Glob. Chang. Biol., 21 (2015), pp. 12-31
66. Lindenmayer et al., 2008 D. Lindenmayer, R.J. Hobbs, R. Montague-Drake, J. Alexandra, A. Bennett, M.Burgman, *et al.*A checklist for ecological management of landscapes for conservation Ecol. Lett., 11 (2008), pp. 78-91
67. MA, 2005 MAMillennium Ecosystem Assessment - Ecosystems and Human Well-being vol. 5, Island Press, Washington, DC (2005)

68. Maes and Jacobs, 2015 J. Maes, S. JacobsNature-based solutions for Europe's sustainable development Conserv. Lett. (2015), 10.1111/conl.12216
69. Maes et al., 2013 J. Maes, A. Teller, M. Erhard, C. Liquete, L. Braat, P. Berry, *et al.* Commission E. (Ed.), Mapping and Assessment of Ecosystems and Their Services. An Analytical Framework for Ecosystem Assessments Under Action 5 of the EU Biodiversity Strategy to 2020, European Commission, Luxembourg (2013)
70. Martinez-Harms et al., 2015 M.J. Martinez-Harms, B.A. Bryan, P. Balvanera, E.A. Law, J.R. Rhodes, H.P.Possingham, *et al.*Making decisions for managing ecosystem services Biol. Conserv., 184 (2015), pp. 229-238
71. Mazzocchi, 2008 F. MazzocchiComplexity in biology EMBO Rep., 9 (2008), pp. 10-14
72. Mcintyre, 2009 B.D. McintyreInternational Assessment of Agricultural Knowledge, Science and Technology for Development (IAASTD): Synthesis Report With Executive Summary: A Synthesis of the Global and Sub-global IAASTD Reports (2009)
73. Mitsch and Jorgensen, 1989 W.J. Mitsch, S.E. JorgensenEcological Engineering: An Introduction to Ecotechnology John Wilcy & Sons, New York (1989)
74. Mitsch and Jørgensen, 2004 W.J. Mitsch, S.E. JørgensenEcological Engineering and Ecosystem Restoration John Wiley & Sons (2004)
75. Morris, 2007 J.T. MorrisEcological engineering in intertidial saltmarshes Hydrobiologia, 577 (2007), pp. 161-168
76. Mulder et al., 2015 C. Mulder, E.M. Bennett, D.A. Bohan, M. Bonkowski, S.R. Carpenter, R. Chalmers, *et al.*Chapter one - 10 years later: revisiting priorities for science and society a decade after the millennium ecosystem assessment W. Guy, A.B. David (Eds.), Adv. Ecol. Res., 53, Academic Press (2015), pp. 1-53
77. Natural Capital Coalition, 2016 Natural Capital CoalitionWebpage: What is Natural Capital? Natural Capital Coalition, London, UK (2016)
78. Natural Capital Initiative, 2016 Natural Capital InitiativeWebpage: Glossary of Terms, London, UK (2016)

79. Naumann et al., 2010 S. Naumann, M. Davis, T. Kaphengst, M. Pieterse, M. RaymentDesign, implementation and cost elements of green infrastructure projects Final Report to the European Commission, DG Environment, Contract: 577182 (2010)
80. Nesshöver et al., 2015 C. Nesshöver, C. Prip, H. WittmerBiodiversity governance: a global perspective from the Convention on Biological Diversity A. Gaspartos, K.J. Willis (Eds.), Biodiversity in the Green Economy, Routledge, London (2015), pp. 289-308
81. Neumayer, 2003 E. NeumayerWeak Versus Strong Sustainability: Exploring the Limits of Two Opposing Paradigms Edward Elgar Publishing (2003)
82. Newsome et al., 2012 D. Newsome, S.A. Moore, R.K. DowlingNatural Area Tourism: Ecology, Impacts and Management vol 58, Channel View Publications (2012)
83. Odum, 1962 H. OdumMan in the ecosystem Conn. Storrs Agric. Exp. Stat. Bull, 652 (1962) 75–75
84. Palomo et al., 2011 I. Palomo, B. Martín-López, C. López-Santiago, C. MontesParticipatory scenario planning for protected areas management under the ecosystem services framework: the Doñana social-ecological system in Southwestern Spain Ecol. Soc., 16 (2011)
85. Parkins and Mitchell, 2005 J.R. Parkins, R.E. MitchellPublic participation as public debate: a deliberative turn in natural resource management Soc. Nat. Resour., 18 (2005), pp. 529-540
86. Pound, 2008 D. PoundFrom conflict to consensus to the ecosystem approach Governing Shared Resources: Connecting Local Experience to Global Challenges, the Twelfth Biennial Conference of the International Association for the Study of Commons, Cheltenham, UK, July 14–18, 2008 (2008)
87. Prato, 2007 T. PratoAdaptive ecosystem management Sustaining Rocky Mountain Landscapes: Science, Policy, and Management for the Crown of the Continent Ecosystem, 249 (2007)
88. Primmer et al., 2015 E. Primmer, P. Jokinen, M. Blicharska, D.N. Barton, R. Bugter, M. PotschinGovernance of ecosystem services: a framework for empirical analysis Ecosyst. Serv., 16 (2015), pp. 158-166

89. Quinn et al., 2010 P. Quinn, S. Burke, M. Wilkinson, C. Hewett, J. JonczykCatchment science and engineering British Hydrological Society International Symposium (2010)
90. Rauschmayer and Wittmer, 2006 F. Rauschmayer, H. WittmerEvaluating deliberative and analytical methods for the resolution of environmental conflicts Land Use Policy, 23 (2006), pp. 108-122
91. Rauschmayer et al., 2009 F. Rauschmayer, S. van den Hove, T. KoetzParticipation in EU biodiversity governance: how far beyond rhetoric? Eviron. Plann. C. Gov. Policy, 27 (2009), p. 42
92. Reed, 2008 M.S. ReedStakeholder participation for environmental management: a literature review Biol. Conserv., 141 (2008), pp. 2417-2431
93. Renn and Klinke, 2013 O. Renn, A. KlinkeA framework of adaptive risk governance for urban planning Sustain. For., 5 (2013), pp. 2036-2059
94. Rey et al., 2015 F. Rey, L. Cécillon, T. Cordonnier, R. Jaunatre, G. LoucougarayIntegrating ecological engineering and ecological intensification from management practices to ecosystem services into a generic framework: a review Agron. Sustain. Dev. (2015), pp. 1-11
95. Ring et al., 2010 I. Ring, B. Hansjuergens, T. Elmqvist, H. Wittmer, P. SukhdevChallenges in framing the economics of ecosystems and biodiversity: the TEEB initiative Curr. Opin. Environ. Sustain., 2 (2010), pp. 15-26
96. Roberts et al., 2012 D. Roberts, R. Boon, N. Diederichs, E. Douwes, N. Govender, A. Mcinnes, *et al.*Exploring ecosystem-based adaptation in Durban, South Africa: “learning-by-doing” at the local government coal face Environ. Urban., 24 (2012), pp. 167-195
97. Rodriguez-Labajos and Martinez-Alier, 2013 B. Rodriguez-Labajos, J. Martinez-AlierThe economics of ecosystems and biodiversity: recent instances for debate Conserv. Soc., 11 (2013), pp. 326-342
98. Roe et al., 2005 E. Roe, P. Schulman, M. Van Eeten, M. De BruijneHigh-reliability bandwidth management in large technical systems: findings and implications of two case studies J. Public Adm. Res. Theory, 15 (2005), pp. 263-280

99. Rumpff et al., 2011 L. Rumpff, D. Duncan, P. Vesk, D. Keith, B. WintleState-and-transition modelling for adaptive management of native woodlands Biol. Conserv., 144 (2011), pp. 1224-1236
100. Santangeli et al., 2016 A. Santangeli, B. Arroyo, L.V. Dicks, I. Herzon, A.S. Kukkala, W.J. Sutherland, *et al.*Voluntary non-monetary approaches for implementing conservation Biol. Conserv., 197 (2016), pp. 209-214
101. Schindler et al., 2014 S. Schindler, Z. Sebesvari, C. Damm, K. Euller, V. Mauerhofer, A.Schneidergruber, *et al.*Multifunctionality of floodplain landscapes: relating management options to ecosystem services Landsc. Ecol., 29 (2014), pp. 229-244
102. Schneiders et al., 2012 A. Schneiders, T. Van Daele, W. Van Landuyt, W. Van ReethBiodiversity and ecosystem services: complementary approaches for ecosystem management? Ecol. Indic., 21 (2012), pp. 123-133
103. Schröter et al., 2012 M. Schröter, R.P. Remme, L. HeinHow and where to map supply and demand of ecosystem services for policy-relevant outcomes? Ecol. Indic., 23 (2012), pp. 220-221
104. Schröter et al., 2014 M. Schröter, G.M. Rusch, D.N. Barton, S. Blumentrath, B. NordénEcosystem Services and Opportunity Costs Shift Spatial Priorities for Conserving Forest Biodiversity (2014)
105. Schultz et al., 2010 L. Schultz, A. Duit, C. FolkeParticipation, adaptive co-management, and management performance in the world network of biosphere reserves World Dev., 39 (2010), pp. 662-671
106. Schumacher, 1973 E.F. SchumacherSmall is Beautiful: Economics as if People Mattered Harper et Row, New York (1973)
107. Se, 2011 P. SePUMA's Environmental Profit and Loss Account for the Year Ended 31 December 2010 Herzogenaurach, Germany (2011)
108. Seastedt et al., 2008 T.R. Seastedt, R.J. Hobbs, K.N. SudingManagement of novel ecosystems: are novel approaches required? Front. Ecol. Environ., 6 (2008), pp. 547-553
109. Sendzimir et al., 2010 J. Sendzimir, P. Magnuszewski, O. Barreteau, N. Ferrand, K. Daniell, D. HaaseParticipatory modeling J.E.A. Mysiak (Ed.), Guidebook for the Adaptive Water Resource

Management, Earthscan, London (2010), pp. 39-42

110. Silvertown, 2015 J. SilvertownHave ecosystem services been oversold? Trends Ecol. Evol., 30 (2015), pp. 641-648
111. Star and Griesemer, 1989 S.L. Star, J.R. GriesemerInstitutional ecology, translations' and boundary objects: amateurs and professionals in Berkeley's Museum of Vertebrate Zoology, 1907–39 Soc. Stud. Sci., 19 (1989), pp. 387-420
112. Stokes et al., 2014 A. Stokes, G.B. Douglas, T. Fourcaud, F. Giadrossich, C. Gillies, T. Hubble, *et al.*Ecological mitigation of hillslope instability: ten key issues facing researchers and practitioners Plant Soil, 377 (2014), pp. 1-23
113. Suding et al., 2004 K.N. Suding, K.L. Gross, G.R. HousemanAlternative states and positive feedbacks in restoration ecology Trends Ecol. Evol., 19 (2004), pp. 46-53
114. Takacs, 1996 D. TakacsThe Idea of Biodiversity: Philosophies of Paradise (1996)
115. TEEB, 2010a TEEBThe Economics of Ecosystems and Biodiversity: Mainstreaming the Economics of Nature - A Synthesis of the Approach, Conclusions and Recommendations of TEEB UNEP, Bonn (2010)
116. Turner and Daily, 2008 R. Turner, G. DailyThe ecosystem services framework and natural capital conservation Environ. Resour. Econ., 39 (2008), pp. 25-35
117. UNEP, 1993 UNEPConvention on Biological Diversity. 1760 UNTS 79; 31 ILM 818 (1992) UNEP, Nairobi (1993)
118. UNEP, 2010 UNEPReport of the Third ad hoc Intergovernmental and Multi-stakeholder Meeting on an Intergovernmental Science-policy Platform on Biodiversity and Ecosystem Services ("Busan Outcome") - UNEP/IPBES/3/3 UNEP, Nairobi (2010)
119. UNEP/CBD, 2000 UNEP/CBDEcosystem approach Decisions adopted by the Conference of the Parties to the Convention on Biological Diversity at its Fifth Meeting, Nairobi, 15–26 May 2000 (2000), pp. 103-109
120. United Nations, 2016 United NationsSustainable Development Goals. 2016 UN (2016)

121. US Army Corps of Engineers, 2013 US Army Corps of EngineersCoastal Risk Reduction and Resilience. CWTS 2013-3 Directorate of Civil Workds, US Army Corps of Engineers, Washington (2013)
122. van den Hove, 2000 S. van den HoveParticipatory approaches to environmental policy-making: the European Commission Climate Policy Process as a case study Ecol. Econ., 33 (2000), pp. 457-472
123. van den Hove et al., 2012 S. van den Hove, J. Mcglade, P. Mottet, M.H. DepledgeThe Innovation Union: a perfect means to confused ends? Environ. Sci. Pol., 16 (2012), pp. 73-80
124. Vignola et al., 2009 R. Vignola, B. Locatelli, C. Martinez, P. ImbachEcosystem-based adaptation to climate change: what role for policy-makers, society and scientists? Mitig. Adapt. Strateg. Glob. Chang., 14 (2009), pp. 691-696
125. Vignola et al., 2013 R. Vignola, T.L. Mcdaniels, R.W. ScholzGovernance structures for ecosystem-based adaptation: using policy-network analysis to identify key organizations for bridging information across scales and policy areas Environ. Sci. Pol., 31 (2013), pp. 71-84
126. Wackernagel and Rees, 1997 M. Wackernagel, W.E. ReesPerceptual and structural barriers to investing in natural capital: economics from an ecological footprint perspective Ecol. Econ., 20 (1997), pp. 3-24
127. Waylen et al., 2014 K. Waylen, E. Hastings, E. Banks, K. Holstead, R. Irvine, K. BlackstockThe need to disentangle key concepts from ecosystem-approach jargon Conserv. Biol., 28 (2014), pp. 1215-1224
128. Waylen et al., 2015a K.A. Waylen, K.L. Blackstock, K.L. HolsteadHow does legacy create sticking points for environmental management? Insights from challenges to implementation of the ecosystem approach Ecol. Soc., 20 (2015)
129. Waylen et al., 2015b K.A. Waylen, K.L. Blackstock, K. Marshall, J. DunglinsonThe participation-prescription tension in natural resource management: the case of diffuse pollution in Scottish water management Environmental Policy and Governance, 25 (2015), pp. 111-124

130. Westgate et al., 2013 M.J. Westgate, G.E. Likens, D.B. LindenmayerAdaptive management of biological systems: a review Biol. Conserv., 158 (2013), pp. 128-139
131. Wilkinson et al., 2010 M.E. Wilkinson, P.F. Quinn, P. WeltonRunoff management during the September 2008 floods in the Belford catchment, Northumberland J. Flood Risk Manage., 3 (2010), pp. 285-295
132. Wilkinson et al., 2014 M. Wilkinson, P. Quinn, N. Barber, J. JonczykA framework for managing runoff and pollution in the rural landscape using a catchment systems engineering approach Sci. Total Environ., 468 (2014), pp. 1245-1254
133. Wilson, 1988 E.O. WilsonBiodiversity National Academy Press, New York (1988)
134. Wyborn, 2015 C. WybornConnectivity conservation: boundary objects, science narratives and the co-production of science and practice Environ. Sci. Pol., 51 (2015), pp. 292-303
135. Zang et al., 2014 C. Zang, C. Hartl-Meier, C. Dittmar, A. Rothe, A. MenzelPatterns of drought tolerance in major European temperate forest trees: climatic drivers and levels of variability Glob. Chang. Biol., 20 (2014), pp. 3767-3779

7 THE WATER ENERGY NEXUS, AN ISO50001 WATER CASE STUDY AND THE NEED FOR A WATER VALUE SYSTEM

Brendan P.Walsh, Sean N. Murray, D.T.J.O' Sullivan

Intelligent Efficiency Research Group (IERG), School of Engineering, University College Cork, Cork, Ireland

ABSTRACT

The world's current utilisation of water, allied to the forecasted increase in our dependence on it, has led to the realisation that water as a resource needs to be managed. The scarcity and cost of water worldwide, along with water management practices within Europe, are highlighted in this paper. The heavy dependence of energy generation on water and the similar dependence of water treatment and distribution on energy, collectively termed the water–energy nexus, is detailed. A summary of the recently launched ISO14046 Water Footprint Standard along with other benchmarking measures is outlined and a case history of managing water using the Energy Management Standard ISO50001 is discussed in detail. From this, the requirement for a methodology for improvement of water management has been identified, involving a value system for water streams, which, once optimised will improve water management including efficiency and total utilisation.

Keywords: Water–energy nexus, Water management, ISO50001, ISO14046

INTRODUCTION

Background

The volume of water used in the world and our dependence on its supply is considerably understated. Freshwater is available in abundance in some regions of the world, but is becoming increasingly scarce in others. The United Nations (UN) recently stated that water links the local to the regional and brings together global questions of food security, public health, urbanisation and energy. Addressing how we use and manage water resources is central to setting the world on a more sustainable and equitable path [1].

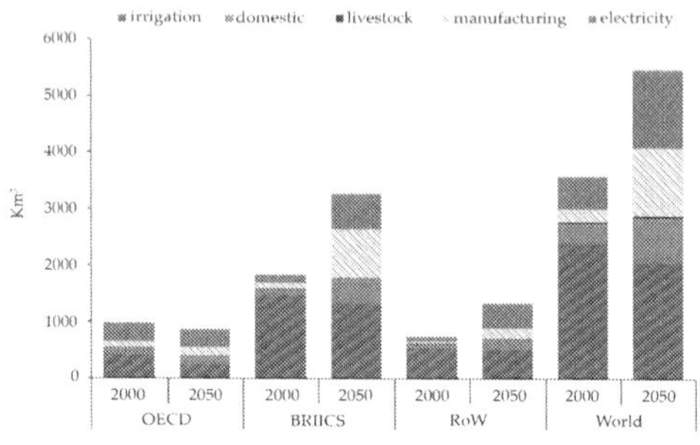

Figure 1. Water demand [3]. (*Note*: OECD: Organisation for Economic Co-Operation and Development. BRIICS: Brazil, Russia, India, Indonesia, China, South Africa. ROW: Rest of the world).

Worldwide, an estimated 768 million people remain without access to an improved source of water [2]. Global water demand (in terms of water withdrawals), illustrated in Fig. 1, are projected to increase by 55% by 2050, mainly because of the growing demands from manufacturing (400%), thermal electricity generation (140%) and domestic use (130%).

As a result, freshwater availability will be increasingly strained over this time period, and more than 40% of the global population is projected to be living in areas of severe water stress by 2050. There is clear evidence that groundwater supplies are diminishing, with an estimated 20% of the world's aquifers being over-exploited, some critically so [2].

WATER MANAGEMENT AND UTILISATION

Water Management within Europe

The level of management of water varies considerably around the world. The countries with more advanced systems have extensive programs for conservation and recycling, whilst others tend to lose substantial quantities of water in an almost regardless manner. Every year in the European Union (EU), 247,000 million m^3 of water is extracted from ground and surface water sources (streams, lakes and rivers). The largest proportion of abstracted water (44%) goes to the energy-production sector for cooling processes, with most being returned to rivers. Agriculture and food production use 24% of abstracted supplies, with 17% used for public water supply and 15% for industry. Half of the water used for manufacturing goes to the chemicals sector and petrol refineries [4].

Meanwhile, freshwater resources, which are defined as either stocks which are held in the ground (groundwater) or are available from rivers, lakes, reservoirs, etc. (surface water), show the largest availability per capita in the EU in Finland, Sweden and Slovenia. As a consequence of differences in climate and population density, the freshwater resources available are unevenly distributed, ranging from 100 m^3 per capita in Malta to 20,700 m^3 per capita in Finland. A method of analysing water stocks is the water exploitation index, which represents the total volume of water abstracted in a given year as a share of total freshwater resources. This index depends on the fresh water resources naturally available as well as the level of use of water by households, industry, energy suppliers and agriculture. The index varies widely among Member States, with the lowest shares observed in Latvia, Slovakia and Sweden (all 1%) and Ireland (2%), and the highest in Cyprus (64%), Belgium (32%), Spain (30%) and Malta (21%) [5].

Overall the abstraction and use of water resources can be considered to be sustainable in the long-term in most of Europe. However, specific regions may face problems associated with water scarcity; this is especially the case in parts of southern Europe, where it is likely that efficiency gains in relation to agricultural water use will need to be achieved in order to prevent seasonal water shortages [6].

Indeed the significance of water to the EU is illustrated through the adoption of the Directive 2000/60/EC of the European Parliament and of the Council establishing a framework for the Community action in the field of water policy, or in short, the Water Framework Directive (WFD) in 2000, with more specific directives subsequently complementing this. The WFD outlined a plan to improve the condition of all European waters by 2015. This directive highlighted that approximately 25% of groundwaters had poor chemical status due to human activities. It also identified that water scarcity, which occurs when demand for water exceeds the available natural resources, affected 11% of the European population and 17% of the EU territory [4].

Alternative sources of water are being developed and whilst there are several water purification technologies available, their deployment tends to be impaired by practicality and cost. Extensive research is ongoing in many areas, including solar-driven desalination, which ultimately form part of the solution particularly on a localised basis. Alternative approaches, including aquifer storage and recovery, offer limited but beneficial short term solutions. There are many opportunities for the joint development and management of water and energy infrastructure and technologies that maximise co-benefits and minimise negative trade-offs. An array of opportunities exists to co-produce energy and water services and to exploit the benefits of synergies, such as combined power and desalination plants, combined heat and power plants, using alternative water sources for thermal power plant cooling, and even energy recovery from sewerage water [2].

The water and energy requirements for cities of the future could be kept to a minimum, through effective planning and investment in systems designed with an integrated approach. Conservation would be the primary concern for water, along with the segregation for re-use and the treatment of the water for the intended use rather than complete treatment of all water to a high level.

In parallel with addressing the sources of water supply, wastage through leakage needs to be minimised. The EU have stated that up to 50% of water resources are being lost through leakage in water infrastructures and that the industry itself must play a major role in setting Sustainable Economic Leakage Levels (SELL) [4]. Whilst there may be an argument that the water is never lost, the lack of availability along with the wastage incurred in its sourcing and distribution needs to be considered.

The supply cost of water in countries around the world, outlined in Table 1, varies considerably and also varies depending on the use, however in industrial expenditure terms, it is consistently low. The market price is not dependant on the value of the item, but on the amount that the market permits, as with other commodities. This cost does not however incorporate any non-monetary consideration, such as drought, famine and loss of life but is simply a cost and not a value.

Table 1. OECD estimates of prices for water by broad sector usage [7].

OECD Nation	Household water supply	Industrial and commercial	Irrigation and agriculture	Average price of water supply
Netherlands	3.16	1.08	1.44	1.89
Austria	1.05	1.05	1.01	1.04
France	3.11	0.95	0.95	1.38
Greece	1.14	1.14	0.05	0.78
Spain	1.07	1.08	0.05	0.73
United States	1.25	0.51	0.05	0.60
Hungary	0.45	1.54	0.03	0.67
United Kingdom	2.28	1.68	0.02	1.33
Australia	1.64	1.64	0.02	1.10
Portugal	1.00	1.26	0.02	0.76
Turkey	1.51	1.68	0.01	1.07
Canada	0.70	1.59	0.01	0.77

Note: Data not available for all OECD member nations. Prices are in $US/ m^3 of water. Includes water supply only and excludes wastewater charges and taxes.

Local Water Management

Local water management, aside from water sourcing and storage, may be considered in terms of distribution and utilisation. Utilisation may further be divided into sectors such as domestic, commercial/retail and industrial. The patterns of utilisation and the motivating factors influencing them vary considerably between the different sectors.

Within Ireland, the management of water utilisation within the commercial/retail and industrial sectors is motivated by the costs associated with the metered supply and the metered discharge. Irish Water, Ireland's national water utility authority, has recently been established within the energy company Bord Gais and is being regulated by the Commission for Energy Regulation. As a direct result, domestic water meters are now being installed nationwide. Whilst their impending existence may be considered a financial burden on households, they shall serve to assist with the conservation of water through the reduction of domestic utilisation. Furthermore, once simple water balances of quantity pumped versus quantity metered have been performed in the distribution systems, the volumes of water and energy wasted shall become apparent. Water conservation programs within commercial/retail and domestic sectors have been successfully completed. The per capita water utilisation in New York City declined from 806 l per capita per day (lpcd) in 1980 to 481 lpcd in 2010, a drop of more than 40% [8].

Industrial Water Management

Water has many functions industrially, ranging from being used as a raw material in the pharmaceutical industry, to cleaning in manufacturing facilities and cooling in power generation stations. Typically, in a sterile manufacturing facility for example, mains water is taken in and treated by equipment and chemicals to provide several different water streams, such as de-ionised water, purified water, water for injection, boiler feed water and potable water etc. Although it is true that the supply cost is low, the additional value-added costs involving labour, materials, energy and equipment required for the operations such as purification, chemical treatment and ultimate disposal, increase the financial value of water considerably.

Whereas the financial penalty for excessive utilisation in the commercial and retail sectors is of significance, in the industrial situation it is not as onerous. Whilst an industrial facility does purchase its raw water supply and also contributes towards its treatment on discharge, the scale of the impact on water conservation is present but limited. Consequentially, water conservation investments provide a relatively low payback, which leads to difficulty with their justification. In certain industries, such as those in the food and beverage sector, the investments are justified and there have been considerable improvements made in many installations by a combination of strategic investments along with focused low-cost operational and maintenance improvements. However, in other large industries the justification is based on the commitment to improvement of the corporate and social responsibility. Smaller entities typically do not individually have adequate capital to invest in this area, however as a group, they are significant and have potential to make an impact if harnessed.

The focus of any industry is on its primary business, not on energy or water efficiency, although these may be incorporated into its larger objectives in order to achieve cost control and satisfy corporate and social responsibilities. Undeniably several large corporations successfully spearhead the debate and actively promote discussion relating to the change required to facilitate effective management of water and energy. Furthermore, consideration also needs to be given to the non-financial value, where continuity of supply is a significant concern. Quantification of this characteristic is more difficult; however it is of substantial significance. In order to assist and encourage attention in these areas, a system of assigning real or added value to water is required.

WATER–ENERGY NEXUS

On a worldwide basis, water, energy and food are considered as one nexus, with the water–energy link being a subsidiary of that. Aspects of the nexus include the strong interdependencies between water and energy generation. The increased utilisation of both water and energy has had effects on climate change and subsequently on the environment. Both are undergoing a rapidly growing demand worldwide, whilst also serving as resource constraints. Both are subject to regional quality variability, with fluctuations in supply and demand. Internationally, energy has been included in regulated markets for a long time, while water has been added more recently.

However, dissimilarities also exist, the most prominent being the relative cost of each, with the cost of exploration and treatment or generation, along with distribution and environmental taxes being allocated to energy, whereas water typically only incurs the significantly smaller abstraction, treatment and distribution costs. The business of energy is much greater than that of water and involves a correspondingly larger financial industry and consequentially, those involved in policy making are influenced by this fact. Energy data is available for most countries worldwide, usually in a variety of formats, however there is considerably less water data available due to its non-existence and this is a significant factor in it not being prioritised at all levels: locally, nationally and globally.

The nexus that exists between water and energy has several facets of even greater interest. The energy expended in the sourcing, treatment and distribution of water effectively assigns an energy footprint to the water. Thus, any wastage of water has direct implications for energy wastage and improvements made in water systems have a direct impact on energy utilisation.

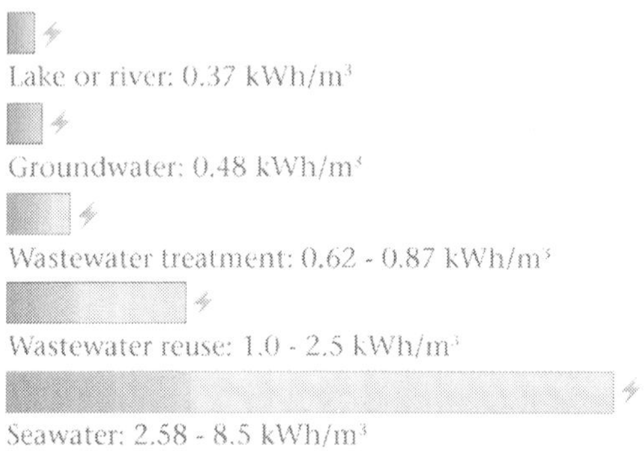

Figure 2. Energy requirement to deliver 1 m^3 water safe for human ingestion from various water sources [9]. (Excludes critical elements such as the distance the water is transported or the level of efficiency).

The energy required to deliver 1 m^3 of water safe for human ingestion from various water sources is illustrated in Fig. 2 and demonstrates several important facts. First, the dependency on energy for purification is demonstrated. Second, the significant quantity of energy required for drinking water is apparent. Third, the difference in the relative quantities of energy required is dependent on the source of water and highlights the importance of the efficient utilisation of freshwater resources.

The utilisation of energy may also be employed as an indicator of water utilisation through the application of data analytics software. The information made available from electricity meters, installed on the motors driving pumps on a farm's irrigation system for instance, represents a set of data, which when analysed correctly, can be turned into a profile of when and where water is being used and indeed lost in the irrigation system. Thus, without the addition of extra water meters, the existing information can be analysed to assist with water conservation.

Another facet of interest is the water used by energy generation systems. Cooling within power generation stations is critical and invariably involves water, either directly or indirectly. Inefficiencies within cooling systems lead to water wastage. Approximately 90% of global power generation is water intensive [2]. Thermal power plants are responsible for approximately 80% of global electricity production, and as a sector they are a large user of water. Power plant cooling is responsible for 43% of total freshwater withdrawals in Europe (more than 50% in several countries), nearly 50% in the United States of America, and more than 10% of the national water cap in China [2].

Global energy demand is expected to grow by more than one-third over the period to 2035 [2].

The International Energy Agency (IEA) estimated global water withdrawals for energy production in 2010 at 583 billion m^3, representing some 15% of the world's total withdrawals. Of this, 66 billion m^3 (bcm) was consumed (withdrawn but not returned to its source). By 2035, withdrawals could increase by 20% and consumption by 85%, driven via a shift towards higher efficiency power plants with more advanced cooling systems (that reduce water withdrawals but increase consumption) and increased production of biofuel. Local and regional impacts of biofuels could be substantial, as their production is among the most water intensive types of fuel production [10]. The current focus on the shift towards electrically powered transport systems also has a significant impact due to the intensive water demands of the generation stations. Similarly, increases in the employment of hydro-power and pumped storage are not necessarily beneficial to the water situation in their locality. African countries tend to possess significant potential for these technologies and hence an integrated approach needs to be adopted in simultaneously addressing both energy and water management.

A third facet of the water–energy nexus is the similarity in the necessity to manage water in a manner analogous to that which has been applied to energy. Historically, there have been several crises associated with energy, the most significant of which being those associated with oil. This has resulted in the evolution of measures to control the usage of oil, with one measure being a system to manage energy. Usually, standards are developed to assist with these management systems, most recently a worldwide standard ISO50001 for Energy Management was released.

The EU Directive, 2012/27/EU of the European Parliament and of the Council, The Energy Efficiency Directive [11] launched in 2012, outlines the actions deemed necessary to address the requirements of the energy situation. It specifically outlines the objective to reduce the dependence on energy imports and scarce energy resources, to limit climate change and to overcome the economic challenge. This has resulted in several measures, including the establishment of EU wide and national energy utilisation targets, the obligation to carry out energy audits periodically and the requirement for energy efficient upgrading and procurement in certain areas. The introduction of similar measures in response to the water predicament would be required in order to satisfactorily resolve the water situation.

Several countries are planning for an environment-friendly energy future; however they also now need to make provision for water scarcity and the two programs need to be incorporated. Energy and water utilisation are inextricably linked and the water–energy nexus needs to be understood and managed with any potential synergies harnessed. The neglect of either the energy supply area or water supply area, both internationally and indeed industrially, can have a significant impact on the other and lead to shortages with consequential effects.

In order to assist with this, the UN has revised its publications protocol and the inaugural annual World Water Development Report, published in March 2014, is themed upon "Water and Energy" [2]. This report outlines the current status of water abstraction, availability and utilisation world-wide, paying particular attention to the water–energy nexus. It identifies the requirement to progress with a coherent integrated approach, considerate of the state of affairs of both water and energy. Looking at the world as a range of ecosystems (from pristine nature to intensive agriculture) and recognising that ecosystems provide a variety of services to the water–energy nexus can help the management of trade-offs and ensure that short-term gains (e.g. in providing energy) do not undermine services critical for resilience and long-term environmental sustainability [12].

WATER METRICS

Water Footprint Standard ISO14046

The carbon footprint of an organisation is defined as being the total of all the green-house gases caused by the organisation. Presently, companies calculate their carbon footprint and strive to improve it by reducing their emissions. In the not-so-distant future, companies are also expected to declare a water-footprint in a similar manner to the declaration of a carbon footprint presently. This footprint will be applicable to products, processes and organisations and will be based on life cycle assessments (LCA's). It will be necessary to accurately quantify the water used, in a verifiable and consistent manner. In order to facilitate this, a new international standard entitled ISO14046 Environmental Management: Water Footprint, was launched in July 2014.

This standard specifies principles, requirements and guidelines related to water footprint assessment of products, processes and organisations based on life cycle assessment. It provides principles, requirements and guidelines for conducting and reporting a water footprint assessment as a stand-alone assessment, or as part of a more comprehensive environmental assessment. The result of a water footprint assessment is a single value or a profile of impact indicator results [13].

The expected effects of this standard are that water utilisation shall be more transparent and hence reduced, thus performing a similar function to the application of ISO50001 to water management, but in a different manner.

The Water Footprint shall serve as a measure which shall allow comparison of facilities and products. Indeed certain measures already exist, albeit in a localised or specific capacity.

Water Usage Effectiveness^{tm} and Other Benchmarks

Many multi-national companies now require measures of both the quality and quantity of supply of both energy and water. Quantitative metrics are being developed in order to allow comparison between proposed locations. They are being used by multi-national corporations as an aid in the selection of the location for future developments and also to assist with the determination of the business continuity risk to existing facilities. One such metric is water usage effectiveness (WUE^{TM}) [14], which is indicative of the quantity of water required for operating and cooling equipment. This measure has significance in the cloud computing industry and in particular to server farms, which constitute data centres. Some multi-national companies have even commenced highlighting the WUE^{TM} measure by publishing real-time dashboards on the internet that display how its data centres are performing against these metrics.

Methods also exist such as Embedded Resource Accounting (ERA), proposed by Ruddell et al. [15], which is a generalised process-orientated, input–output and network-based framework for complex systems analysis. This method may be developed as a generalised footprint methodology and used to derive footprints including the water footprint.

An essential requirement in the effective employment of metrics for benchmarking or demonstrating performance improvement, is the suitability of the metric and its ability to exclude external interferences. Water benchmarks also need to be appropriate and accurate. Certain energy benchmarks simplistically calculate energy utilised per item manufactured without consideration of volume reducing factors and thus can be misleading.

There are benefits and disadvantages to all systems, both in isolation and in combinations. For instance, intensive focus on energy efficiency alone can lead to measures being implemented which benefit energy efficiency, but lead to increased water usage. However, this detrimental effect is usually not fully considered, as the true value or cost of water utilised is not determined.

MANAGING WATER AS ENERGY: A CASE HISTORY USING ISO50001 ENERGY MANAGEMENT SYSTEMS

Background

Adoption of a standard implies that the organisation undertakes to perform its activities in a responsible, structured and efficient manner and as described in its relevant certification documentation. On implementation of a standard approach, a certification audit is executed in order to grant certification, with follow-up audits carried out routinely. Any non-compliance is identified and corrected. Typically, this results in a well-managed, sustainable and improved performance in the area covered by the standard.

Particular organisations have already adopted water as a form of energy and managed it using standards, including ISO50001 the Energy Management standard; University College Cork (UCC) being one such example. UCC, which was established in 1845 comprises over 20,000 students, was awarded certification to ISO50001 in 2011. It was the first third level institution worldwide to achieve certification, along with being the first public sector body in Ireland to be certified.

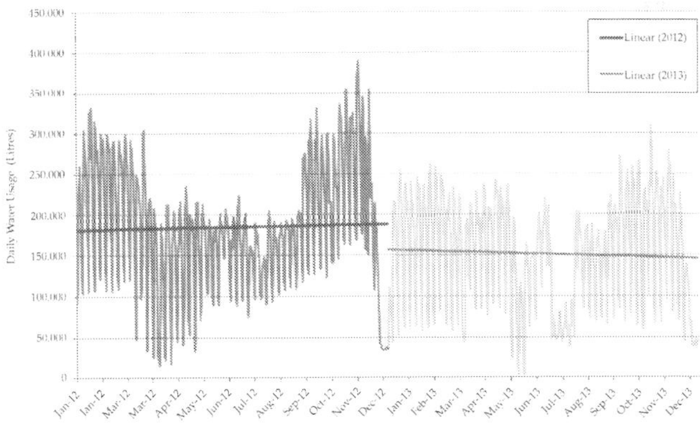

Figure 3. Water utilisation in University College Cork [16].

The trend in Fig. 3 illustrates the daily readings and linear average water usage by the main campus of UCC, since certification to the ISO standard. The data shown is for the calendar years of 2012 and 2013 and includes the significant users such as the heating boilers and sanitary facilities. The individual peaks and troughs visible represent the weekly pattern, with the monthly pattern demonstrating the student occupation characteristics of the academic year, supported by an underlying reading resulting from the presence of non-academic staff.

Implementation of the ISO Standard

The putting into practice of the standard followed the typical ISO pattern; the required variables were established and the ISO approach of Plan-Do-Check-Act was adopted in order to determine objectives, targets and action plans. A base load was determined along with a baseline utilisation. Significant energy users, or in this case, significant water users were identified and performance indicators were established.

The temporal variation in water flowrates due to student occupation levels, dictated by weekends and seasonal holidays, in conjunction with the increasing student population complicated the analysis. The variation in weather from year to year also influenced the readings.

Targeting the baseload, which is realistically essential utilisation along with wastage, was one particular area of success over the years of implementation, which saw the night time flowrate decrease from 2.3 l/s to 1.9 l/s, which is a specific reduction from leak repair and use-analysis. An early warning system was installed to alert high usage patterns. Successful monitoring identified anomalies which required investigation and ultimately leads to increased efficiency through the identification of causes of the increases. Tracking of the remediation works executed along with quantification of the losses allowed identification of the savings and also facilitated the provision of feedback to all participants.

ISO50001 requires the demonstration of continual improvement through the reduction in utilisation of energy, or in this case water. This is different to other standards which permit the improvement to be demonstrated through additional training for example. Hence, the necessity to correctly identify usage is essential and it is not permitted to use metrics such as m^3/student, for example, as an indicator of improvement when the number of students increases.

Without doubt, all facilities have their individual characteristics which increase the complexity of implementation of a standard or program. In this case, the small quantity of meters installed and the absence of a definitive quantification of occupants in a particular building at any particular time, proved to be the largest challenges. The use of virtual water meters and the use of occupant levels in the access controlled library as a substitute for the student population on campus overall, proved to be successful measures in elimination of these obstacles.

The areas requiring remediation were typical of a university campus comprising several buildings of different ages and included underground pipework leaks, faulty meters, operation of sanitary facilities during times of non-occupation, operation of laboratory facilities when not required and inefficient cleaning practices.

Inevitably all causes of anomalies in readings could not be explained; the installation of additional metering would increase the functionality of the system and potentially facilitate automated fingerprinting and fault diagnostics along with more advanced statistical analysis, however justification of such expenditure was not possible at this stage. In the absence of such systems, participation by operational personnel was of critical importance to the implementation of the UCC system and this effort allied to the commitment of senior management, which are requirements of the ISO standard, ensured the successful functioning of the program.

Results of the ISO Standard

In summary, the benefits of following the program have been realised through the 18% reduction in total annual usage from 67,434 m^3/year to 54,966 m^3/year [16]. This was achieved despite an increased level of student activity within the college. The trend in Fig. 3 also illustrates the progress being made within the 2013 year. The average utilisation at similar times of the year is lower and the annualised trend is decreasing within the second year.

THE SOLUTION TO WATER MANAGEMENT—THE TRUE VALUE OF WATER

Understanding the Requirements

Simply increasing the unit costs for metered water is not the correct solution. The financial balances within many organisations are well established and significant modification of this may jeopardise their viability with extreme consequences. The application of human intervention and management are also key components in any system. Standards and directives including ISO50001 and WFD specifically mention and require public consultation and involvement.

Novotny [17] presents a three-phase model for the relationship between water demand and energy use in Fig. 4. The model suggests that in the first phase, the energy and greenhouse gas (GHG) emission reduction is proportional to the reduction of the high water use. Separately, it has been estimated that up to 65% of water used could be saved in some US cities just by implementing simple water conservation measures, such as efficient water appliances, reduction in leaks and dry landscaping (i.e. xeriscaping) [17]. Part of this reduction in water utilisation would yield a proportional reduction in energy use, concurring with the hypothesis described in phase one. In the second phase, cities can augment their water supply through additional sources and treating and reusing stormwater. This will conserve additional water, although this will not result in as significant a decrease in energy demand as in phase one, due to the energy requirement of the treatment processes. The third phase involves advanced water treatment options such as reverse osmosis water recycling systems and desalination plants. Although these advanced systems are energy intensive, they can offer a reliable source of water and their additional energy input requirements may be countered by the use of efficient technology and renewable energy sources [17].

Thus, the focus on future sources of energy generation has a significant role to play. Table 2 details the differing water withdrawal and consumption rates for power generation technologies. The variance in the values between the technologies highlights the importance of the selection and the impact of the decisions made.

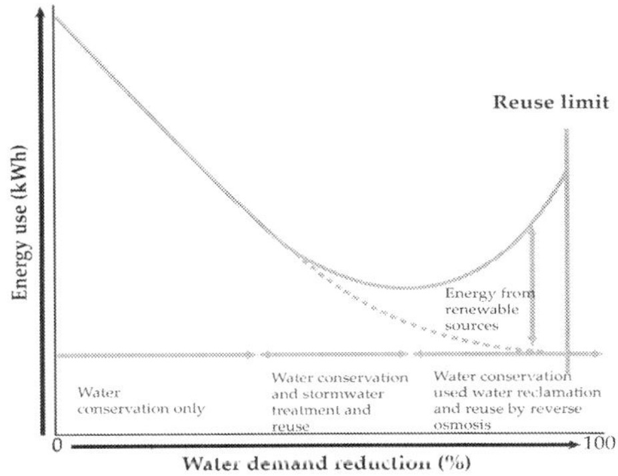

Figure 4. Three phase model for energy–water demand reduction [17].

Table 2. Median water withdrawal and consumption values by fuel type and cooling technology [18]

Fuel	**Cooling type**	**Median withdrawal**	**Median consumption**
		(m³/MW h)	**m³/MW h)**
Nuclear	Tower (recirculating)	4.2	2.5
	Once through	167.9	1
Natural gas (combined cycle)	Tower (recirculating)	1	0.7
	Once-through	43.1	0.4
Coal (supercritical/advanced)	Tower	2.3	1.9
	Once-through	85.5	0.4
Coal (with carbon capture and sequestration)	Tower	4.3	3.2
Solar photovoltaic	n/a	0.1	0.1
Wind	n/a	0	0

Furthermore, conflicting policy objectives also need to be considered, particularly allowing for environmental commitments and projections; however improving water and energy efficiency has benefits in multiple areas. As Faeth, Sovacool et al. identify, synergies exist for some options in the power sector to meet growing electricity demand in cost-effective ways that conserve water, reduce conventional air pollutants and cut greenhouse gas emissions. The least expensive option is to slow demand growth through end-use energy efficiency improvements. Not only is efficiency the cheapest approach because it avoids the need for new capacity altogether, but it also eliminates cooling water needs and emissions. The least expensive option for new generation capacity is natural gas, which has significant environmental benefits over coal, which is the dominant fuel for power production globally. Water withdrawals and consumption are less than half that of coal for the same cooling technology and there are no emissions of particulate matter and sulphur dioxide (SO_2), there is 90% lower nitrous oxide emitted (NO_x), and the carbon dioxide emissions are also less than half that of coal [19].

Ultimately, in order to achieve more effective water management, a change is needed from the traditional reactive approach, where large storage capacity is incorporated into systems. More pro-active management is required, with the installation of the facilities which allow the capture and subsequent analysis of data. The smart use of the data will then allow the identification and exploitation of more opportunities for water efficiency.

A Value Added Water System

In order to achieve this objective, an understanding is ultimately required of the true value of water. With water, as with many other substances, the more we use, the less valuable we understand it to be.

There is no standardised methodology or framework for the assignment of values to water streams at present. Such a system would facilitate analysis through simulation and eventual optimisation of the situation regarding utilisation. This would lead in due course, to more efficient systems involving employment of water of an appropriate grade with reduced overall utilisation, reduced cost and improved environmental performance. An example of a value system is shown in Table 3, which represents the water utilisation typical of a manufacturing facility, where the incoming mains water is treated and used for different purposes. The necessary treatments and associated costs add value to each stream. Additional alternative sources of water, originating from rainwater harvesting or recycling, would necessitate different treatments.

Table 3. Typical manufacturing facility value added water system.

Utility	Value added									Cumulative value volume consumed=V
	Mains supply cost	**Chemical treatment**	**Processing treatment**	**Energy consumed**	**Sampling cost**	**Equipment depreciation**	**Waste neutralisation**	**Biological waste treatment**	**Offsite dilute waste disposal**	
Drinking water	M			E_d	S_d	D_d		B	O	$V_d \times (M+E_d+S_d+D_d+B+O)$
De-ionised	M		P_{di}	E_{di}	S_{di}	D_{di}		B	O	$V_{di} \times (M+P_{di}+E_{di}+S_{di}+D_{di}+B+O)$
Water for injection	M	T_i	P_i	E_i	S_i	D_i		B	O	$V_i \times (M+T_i+P_i+E_i+S_i+D_i+B+O)$
Cooling tower water	M	T_c		E_c	S_c	D_c	N_c	B	O	$V_c \times (M+T_c+E_c+S_c+D_c+N_c+B+O)$
Chilled water	M	T_{ch}	P_{ch}	E_{ch}	S_{ch}	D_{ch}	N_{ch}	B	O	$V_{ch} \times (M+T_{ch}+P_{ch}+E_{ch}+S_{ch}+D_{ch}+N_{ch}+B+O)$
Low pressure steam	M	T_s	P_{sl}	E_{sl}	S_{sl}	D_{sl}	N_s	B	O	$V_{sl} \times (M+T_s+P_{sl}+E_{sl}+S_{sl}+D_{sl}+N_s+B+O)$
High pressure steam	M	T_s	P_{sh}	E_{sh}	S_{sh}	D_{sh}	N_s	B	O	$V_{sh} \times (M+T_s+P_{sh}+E_{sh}+S_{sh}+D_{sh}+N_s+B+O)$
Firewater	M	T_f		E_f		D_f		B	O	$V_f \times (M+T_f+E_f+D_f+B+O)$
Toilet facilities water	M			E_t		D_t		B	O	$V_t \times (M+E_t+D_t+B+O)$
Equipment washing water	M			E_w		D_w		B	O	$V_w \times (M+E_w+D_w+B+O)$

The value added system facilitates simulation modelling, with the requirements of the facility formulating the constraints for optimisation. Chew et al. [20] successfully employed a model-based approach for simultaneous water and energy reduction in a pulp and paper mill and demonstrated significant savings.

An added value water system has applicability across all spectra of water utilisation and can be adapted and weighted according to the priorities of the domain. Eventually, any successful overall water management scheme will involve a combination of sub-systems, depending on the sector. Every system will be comprised of an applied value system with components comprising legislative, financial, corporate and personal input.

CONCLUSIONS

The management of water is of increasing importance globally. Water is essential for our existence and its availability is not assured. It is anticipated that in future years, temperatures worldwide shall be higher, rainfall events shall be more concentrated and sea-levels shall be higher, all of which influence water abstraction, consumption and distribution.

The UN estimates that 1.2 billion people, or almost one-fifth of the world's population, live in areas of physical water scarcity, while another 1.6 billion people, or almost one quarter of the world's population, face economic water shortage. Furthermore, projections indicate that by 2025, eight hundred million people will be living in countries or regions with absolute water scarcity and two-thirds of the world's population could be under severe water stress conditions [21].

Water availability will become a major consideration in residential occupation and business operation around the world. The metered cost of water is also expected to increase in all parts of the world.

Concerns regarding the future availability and cost of water have stimulated interest in the management of this valuable resource. The EU Framework Programme for Research and Innovation research Horizon 2020, which commenced in 2014, predicts €500 million per year EU public investment in Water research in Europe [22].

The demand side management of water utilisation in a manner analogous to energy utilisation, using ISO50001 Energy Management

Systems, has been shown to be successful. The recently published standard ISO14046 Environmental Management: Water Footprint, shall further advance progress in this area. As well as efficiency, other factors such as wastage through ageing distribution networks, conservation measures and waste water treatment developments to facilitate recycling, also need to be considered.

Currently, 90% of energy production relies on intensive and non-reusable water models that are not sustainable [23]. Support for other energy sources, such as biofuels, needs to be targeted in its approach and considerate of the water footprint of the fuel, a case in point being biofuels from irrigated sources. Support for the development of renewable energy, which remains far below that for fossil fuels, will need to increase dramatically before it makes a significant change in the global energy mix, and by association, in water demand [2].

Increasing energy demands are restricting future water planning measures, again emphasising the requirement for a comprehensive integrated approach from policy makers. In order to achieve a system closer to a functioning circular economy, there needs to be increased focus on water. Water and energy efficiency need to be addressed simultaneously in a cohesive manner. However, in order to progress effectively, a standardised method of assignment of a true value of water needs to be developed and adopted, particularly within industrial environments.

ACKNOWLEDGEMENTS

The authors would like to acknowledge the financial support of the Irish Research Council under the Enterprise Partnership Scheme, in association with ENMS Ltd.(Grant no. EPSPG/2011/275).

REFERENCES

1. Managing water under uncertainty and risk, The United Nations World Water Development Report 4, vol. 1, 2012, UNESCO, Paris, France.
2. United Nations World Water Assessment Programme (WWAP), The United Nations World Water Development Report 2014: Water and

Energy, UNESCO, Paris, France, 2014.

3. OECD Publishing, OECD Environmental Outlook to 2050: The Consequences of Inaction. OECD Publishing, Paris, France, March 2012.
4. Directive 2000/60/EC of the European Parliament and of the Council of 23 October 2000 Establishing a Framework for Community Action in the Field of Water Policy, European Commission, The EU Water Framework Directive Information and Overview, http://dx.doi.org/10.2779/75229.
5. European Commission, Eurostat, Environmental Statistics in Europe: Facts and Figures on the Environment, News release 189/2010, 10 December 2010.
6. European Commission, Eurostat, Key Figures on Europe, 2013 digest, 11.5 Water, ISSN 1830-7892, 2013.
7. Water, Water Use and Water Pricing Around the World. Canadian Water Policy Backgrounder No. 8, Calgary, Alberta, CWF, September 2011.
8. NYC (New York City), Water Conservation Report Annual Update July 2012, New York, New York City Department of Environmental Protection, 2012.
9. World Business Council on Sustainable Development (WBCSD), Water, Energy and Climate Change: A Contribution from the Business Community, World Business Council on Sustainable Development (WBCSD), Chemin de Conches 4, 1231 ConchesGeneva, Switzerland, March 2009.
10. International Energy Agency, Water for Energy, Is Energy Becoming a Thirstier Resource?, World Energy Outlook, International Energy Agency, 9 rue de la Fédération 75739 Paris Cedex 15, France, 2012.
11. Directive 2012/27/EU of the European Parliament and of the Council of 25 October 2012, The Energy Efficiency Directive, European Commission, Official Journal of the European Union L315/1, 2012.
12. An Ecosystem Services Approach to Water and Food Security, in: E. Boelee, T. Chiramba, E. Khaka (Eds.), UNEP/ International Water Management Institute (IWMI), Nairobi/Colombo, 2011.
13. ISO 14046:2014, Environmental Management: Water Footprint:

Principles, Requirements and Guidelines. International Organisation for Standardisation ISO Central Secretariat 1, ch. de la Voie-Creuse CP 56—CH-1211 Geneva 20 Switzerland, 2014.

14. Water Usage Effectiveness (WUETM): A Green Grid Data Center Sustainability Metric, White Paper 35, The Green Grid Administration, 3855 SW 153rd Drive, Beaverton, Oregon 97003 USA.
15. Richard R. Rushford, Elizabeth A. Adams, Benjamin L. Ruddell, Generalising ecological, water and carbon footprint methods and their worldview assumptions using Embedded Resource Accounting, Water Resourc. Ind. 1–2 (2013) 77–90.
16. Data from Building Management System, Buildings and Estates Office, University College Cork, Ireland.
17. V. Novotny, Water and energy link in the cities of the future: achieving net zero carbon and pollution emissions footprint, in: V. Lazarova, K.H. Choo, P. Cornel (Eds.), 2012: Water–Energy Interactions in Water Reuse, IWA Publishing, London, 2012.
18. Data from J. Macknick, R. Newmark, G. Heath, K.C. Hallett, A Review of Operational Water Consumption and Withdrawal Factors for Electricity Generating Technologies. Technical Report NREL/TP-6A20-50900 (Golden, CO: National Renewable Energy Laboratory, March 2011).
19. Faeth, P., Sovacool, B.K. et al., A Clash of Competing Necessities: Water Adequacy and Electric Reliability in China, India, France, and Texas, CNA Corporation, 3003 Washington Boulevard, Arlington, VA 22201, IRM-2014-U-007191, July 2014.
20. I.M.L. Chew, D.C. Foo, J.-C. Bonhivers, P. Stuart, A. Alva-Argaez, L.E. Savulsecu, A model-based approach for simultaneous water and energy reduction in a pulp and paper mill, Appl. Therm. Eng. 51 (2013) 393–400.
21. Water Scarcity and Humanitarian Action: Key Emerging Trends and Challenges, OCHA Occasional Policy Briefing Series, Policy Development and Studies Branch, UN Office for the Coordination of Humanitarian Affairs (OCHA), September 2010.

22. Fragakis, Christos, Horizon 2020, Water and Marine and Maritime Research and Innovation info-day, Paris, January 2014, European Commission DG Research & innovation.
23. Bokova, Irina, Director General of UNESCO, Foreword, The United Nations World Water Development Report 2014: Water and Energy, UNESCO, Paris, France, 2014.

8 INDICATIVE ENERGY TECHNOLOGY ASSESSMENT OF UK SHALE GAS EXTRACTION

Geoffrey P. Hammond[ab], Áine O'Grady[a]

[a] Department of Mechanical Engineering, University of Bath, Bath BA2 7AY, United Kingdom

[b] Institute for Sustainable Energy and the Environment (I•SEE), University of Bath, Bath BA2 7AY, United Kingdom

ABSTRACT

There is at present much interest in unconventional sources of natural gas, especially in shale gas which is obtained by hydraulic fracturing, or 'fracking'. Boreholes are drilled and then lined with steel tubes so that a mixture of water and sand with small quantities of chemicals – the fracking fluid – can be pumped into them at very high pressure. The sand grains that wedge into the cracks induced in the shale rock by a 'perforating gun' then releases gas which returns up the tubes. In the United Kingdom (UK) exploratory drilling is at an early stage, with licences being issued to drill a limited number of test boreholes around the country. However, such activities are already meeting community resistance and controversy. Like all energy technologies it exhibits unwanted 'side-effects'; these simply differ in their level of severity between the various options. Shale gas may make, for example, a contribution to attaining the UK's statutory 'greenhouse gas' emissions targets, but only if appropriate and robust regulations are enforced. The benefits and disadvantages of shale gas fracking are therefore discussed in order to illustrate a 'balance sheet' approach. It is also argued that it is desirable to bring together experts from a range of disciplines in order to carry out energy technology assessments. That should draw on and interact with national and local stakeholders: 'actors' both large and small. Community engagement in a genuinely participative process – where the government is prepared to change course in response to the evidence and public opinion – will consequently be critically important for the adoption of any new energy option that might meet the needs of a low carbon future.

Keywords: Shale gas, Hydraulic fracturing, Resources, Economic and environmental impacts, Induced seismicity, Public acceptance, Regulation

INTRODUCTION

Background

Human development is underpinned by energy sources of various kinds that heat, power and transport its citizens in their everyday life. But all energy technologies have unwanted 'side-effects'; they simply differ in their level of severity. Hydraulic fracturing, or 'fracking', for shale gas is a particularly controversial energy option that is receiving significant development support from government in the United Kingdom (UK). Licenses have been issued by the Department of Energy and Climate Change (DECC) to drill a limited number of test boreholes around the country (see, for the case of England, [1]). These boreholes are then lined with steel tubes, and a mixture of water and sand with small quantities of chemicals – the fracking fluid – is pumped into them at very high pressure. The sand grains that wedge into the cracks induced in the shale rock by a 'perforating gun' then releases gas which returns up the tubes (see Fig. 1). The UK Government is attracted by the possible benefits of securing large quantities of shale gas for the UK as an energy 'game changer': leading to a potential 'Golden Age of Gas', according to the *International Energy Agency* (IEA) [2]. The IEA sees shale gas as contributing about 14% to global gas production by 2035. However, the exploitation of fracking will involve a range of advantages and disadvantages ('credits and debits') that will fall disproportionately on different sections of British society. So it is necessary to identify the components of a shale gas fracking 'balance sheet' of the sort employed in technology assessment [3–5] in order to evaluate its impact on communities, countryside and wildlife, and to determine whether it is compatible with Britain's move towards a low carbon future in 2050 and beyond.

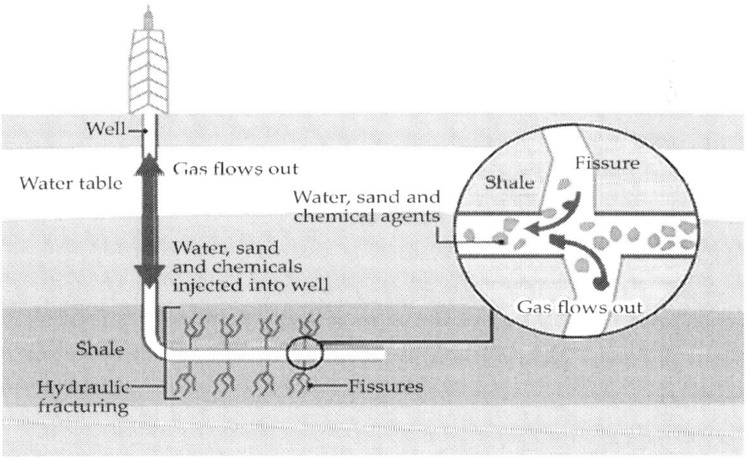

Figure 1. The shale gas 'fracking' process.

Historical Development of Fracking for Shale Gas

The technique of hydraulic fracturing began in the United States of America (USA) [6]in around 1949 when the first two, small-scale commercial vertical wells were initiated in Oklahoma and Texas respectively [7,8]. But it was not until about 1997 that the process known as 'slickwater fracturing' was developed and implemented in the Barnett Shale by the then *Mitchell Energy*. This is a method that involves adding chemicals to water to increase the flowrate at which the fracking fluid can be pumped down a well-bore to fracture extremely dense shale. The fracking fluid is made up of around 98.50% water, 1.00% sand,

and 0.05–0.50% chemical additives [6]. These chemicals are friction reducers, usually a polyacrylamide, together with biocides, surfactants and scale inhibitors. Biocides prevent organisms from blocking the 'downhole' and fissures, whereas surfactants keep the sand grains in fluid suspension. Other chemicals that are sometimes employed include benzene, chromium, and a number of other compounds [6]. North American fracking companies keep the composition of this chemical 'cocktail' secret, claiming commercial confidentiality, although an independent study identified about 650 separate chemicals compounds. However in the UK, companies are obliged under the *Water Resources Act* 1991 to disclose the composition used. Many of these are known to be toxic and widespread concern has been expressed over potential water contamination [6,9–11]. Nevertheless, it was this pressure-induced slickwater fracturing (see again Fig. 1) that made shale gas extraction economical by radically reducing the costs of horizontal fracking [6].

The situation with shale gas development in the UK is quite different from that in the USA, where some 200,000 horizontal gas fracking wells have been in operation over the last two decades or so (Prof. Will Fleckenstein, Colorado School of Mines, USA, private communication 06.11.15) in comparison to just one in Britain – at the Preese Hall site in Lancashire. In addition, the regulatory framework is likely to be tighter in the UK, e.g., 'flowback water' will not be permitted to be reinjected into wells. Environmental limits are also uniformly established across the UK and adherence monitored by national regulators. Water use in the United States (US), by contrast, is regulated on a State-by-State basis.

The Issues Considered

The possible benefits and disbenefits of shale gas fracking include economic, environmental, safety and social consequences [12] for the UK. Here they are discussed as an example of a 'balance sheet' approach: analysis rather than advocacy. In order to draw up an objective and rigorous set of credits and debits for shale gas fracking (or indeed other potentially 'disruptive' technologies) as part of a national dialogue, it is argued that it is desirable to bring together experts from a wide range of disciplines to undertake *energy technology assessments* (ETA) [5] that exhibit balance, objectivity and broad public participation. This should draw on and interact with national and local stakeholders: 'actors' both large and small. Community engagement will consequently be critically important for the adoption of any new energy option that might meet the needs of a low carbon future. This contribution is part of an ongoing research effort aimed at evaluating and optimising the performance of various sustainable energy systems (see, for example, Hammond et al. [13] and Hammond and Hazeldine [14]) in the context of transition pathways [15,16] towards the statutory target of a reduction in UK 'greenhouse gas' (GHG) emissions by at least 80% by 2050 from 1990 levels [17]. It is aimed at illustrating the consequences of shale gas fracking within a UK setting in the light of imperfect, and sometimes contradictory, information. Nevertheless, such assessments provide a valuable evidence base for communities, developers, policy makers, and other stakeholders. They also yield lessons for other European countries attempting to extract significant quantities of shale gas whilst attempting to decarbonise their energy systems, although local circumstances will obviously limit the wider applicability of the present findings.

THE POTENTIAL SHALE GAS RESOURCE IN THE UK

On the positive side of the 'balance equation' is the prospect that fracking could potentially yield significant quantities of shale gas to meet the Britain's energy needs. In contrast (on the negative side), the IEA [2] warn that the significant global development of this gas would put the world on a trajectory towards a long-term temperature rise of over 3.5 °C; well above the widely suggested 'safe' level of 2 °C. The *British Geological Survey* (BGS) [18] has estimated the possible reserves of shale gas in the *Bowland-Hodder* study area or 'play' (encompassing national parks and major cities) and the Weald in the South East for DECC: see Fig. 2. Bassi et al. [19] have also collated several estimates of UK shale gas potential that are presented in Table 1. The great uncertainties inherent in such provisional estimates [total UK technically recoverable shale gas reserves of 150–1130 billion cubic metres (bcm)] can only be refined by extensive investigative drilling, possibly requiring hundreds of wells [20,21]. Notwithstanding the differences between these estimates, they suggest that UK resources are likely to be significant compared to those elsewhere in Europe; given the moratorium on shale gas fracking in France and the limits on extraction in Poland, due to geological constraints [Prof. Danny Reible, Texas Tech University, USA, private communication 05.11.15]. However, making assumptions about recovery rates (based on experience at over 80,000 US horizontal shale gas fracking sites) and the proportion of available resources extractable in the UK, the BGS suggest [18] that recoverable shale gas resources might be equivalent to some 25–50 years of current UK natural gas (NG) demand. That would significantly contribute to Britain's energy security and independence. However, making full use of this resource may not adhere to UK carbon budgets [22] and could risk missing its legally-binding 2050 GHG emissions reduction target.

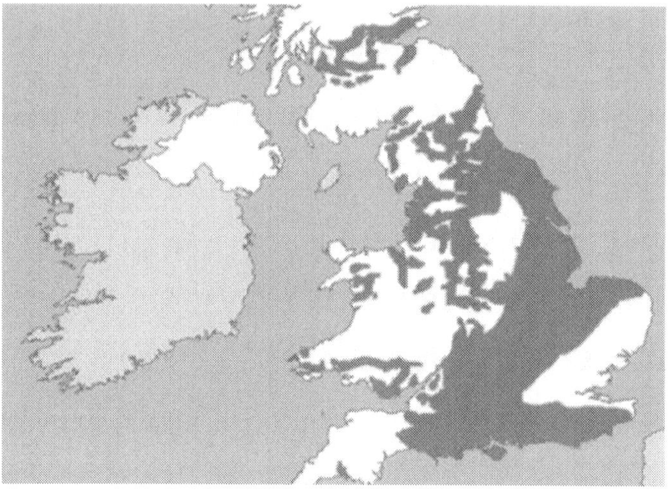

Figure 2. Potential UK shale gas reserves [sites – dark shading].

Table 1. Estimates of shale gas potential in the UK (bcm).

		EIA	BGS/ DECC	Cuadrilla	ECC
Bowland Shale	Gas in place	2690	–	5660	–
	Technically recoverable	540	80–200	900–1200	60–110
Weald Basin (Liassic shale)	Gas in place	60	–	–	–
	Technically recoverable	30	–	–	–

Total UK	Gas in place	2750	–	–	–
	Technically recoverable	570	150	–	1130

Timescale of estimates: 2011–2012.

Units: billion cubic metres (bcm).

Sources: various estimates collated by Bassi et al. [19]: US *Energy Information Administration* (EIA); *Cuadrilla Resources* – a UK unconventional gas exploration company; *British Geological Survey*(BGS); UK *Department of Energy and Climate Change* (DECC); *Energy Contract Company* (ECC) – a UK-based commercial oil and gas consultancy.

SHALE GAS SOCIO-ECONOMIC AND MARKET ISSUES

The UK balance of payments would obviously benefit significantly from the large-scale development of shale gas extraction, although it is unlikely that gas bills for household and industrial consumers would fall dramatically as they have done in North America. This is because the USA is effectively a "natural gas island" with very limited 'Liquefied Natural Gas' (LNG) imports via the gas trading hubs of Europe [23]. In the USA, supplies of conventional NG have been drying up, and unconventional gas (including from shales) has been able to grow rapidly to meet some 60% of marketed production, according to the IEA [2]. Many US energy analysts believe that this fall in gas prices to historically low levels has been caused by advances in extraction techniques, particularly fracking, driving down production costs. Much of this shale gas production occurred as an almost 'free' co-product of unconventional oil extraction. In contrast, the UK is part of the wider European natural gas market [23]where the gas price is determined by the supply and demand for indigenous natural gas, imports from Russia, and LNG from North Africa and Middle East. Shale gas supplies in the UK will only provide a small fraction of those in this wider gas market. So the household economic benefits in Britain are therefore unlikely to live up to the hopes of the UK Prime Minister (David Cameron), who argued in the *Daily Telegraph*newspaper (11/08/13) that it would "see lower energy prices in this country". The British *House of Commons*' Energy and Climate Change Committee [24] argued that domestic shale gas production could reduce the risk that prices would be determined over the longer term by imports (either via natural gas pipeline or by way of LNG). Nevertheless, they concluded that there is substantial uncertainty over the impact of unconventional gas extraction on market prices. David Cameron also cited job creation as another socio-economic benefit. That will undoubtedly follow successful shale gas exploitation, but it is unclear whether this would be any greater than for equivalent programmes aimed at supporting the adoption of energy demand reduction measures (such as thermal insulation or high efficiency lights and appliances) or small-scale low carbon energy options.

A *Task Force on Shale Gas* (TFSG) was established in the UK in September 2014 funded by several companies with commercial interests in the oil, gas and chemicals sectors [*Centrica*, *Cuadrilla Resources*, *Total*, *The Weir Group*, *GDF SUEZ E&P UK Ltd.*, and the *Dow Chemical Company* (until September 2015)]. It was led by the former chair of the *Environment Agency* for England and Wales (Lord Chris Smith) with three other 'independent' panel members. However, environmental *non-governmental organizations* (NGOs), such as *Friends of the Earth*, and various local anti-fracking protest groups have expressed skepticism about its claimed impartiality. Nevertheless, it is useful to compare and contrast the findings of the Task Force with those of the present ETA study. Its last report [25] dealt with the potential economic impacts of shale gas extraction in Britain. The TFSG acknowledged at the outset that it is difficult to judge these effects given the uncertainties around the potential availability of shale gas in the UK. In any event, they recognised that such shale gas extraction would have only a minimal impact on the European market for similar reasons to those suggested above. They therefore recommended the drilling of a number of exploratory wells in order to gain a clearer picture of recoverable shale gas in Britain. Despite the uncertainties, they went on to argue that job creation might amount to thousands of jobs directly and many more in the wider supply chain [25], rather in line with the British Prime Minister's assertion noted above. However, the '*Campaign against Climate Change*' Trade Union Group has estimated (backed by eight national unions and aided by six academic specialists) that far more jobs could be generated via equivalent investment in developments that would mitigate GHG emissions [26]. These 'climate jobs' would be created from the adoption of energy conservation measures in the home and in public buildings, renewable energy technologies, clean public transport, and in the development of 'green skills' that will be required through education and training. Based on case studies on the Fylde (the coastal plain in western Lancashire, northern England) and in Salford (a metropolitan borough of Greater Manchester) they suggest that climate jobs could amount to some 14 times those produced directly from the fracking sector and 80 times nationally, i.e., including indirect employment creation across the supply chain. [A breakdown of the one million 'climate jobs' that might be generated by 2030 are shown in Table 2. In addition, Neale [26] argues that half a million additional, or 'spin-off', jobs could also be created.] Such figures are only indicative, although they provide an important contrast to official rhetoric from David Cameron and others about the prospects of employment creation from shale gas fracking.

Table 2. Estimates of potential 'climate job' creation in the UK

Electricity	400,000
Transport	310,000
Buildings	185,000
Industry	25,000
Education	35,000
Agriculture, waste and forestry	45,000
Total UK 'Climate Jobs'	1,000,000

Perhaps the most important socio-economic issue concerns the distribution of the benefits and costs of shale gas fracking between various communities and demographic groups. Depending on how much shale gas can be exploited, the UK overall could benefit from improved energy security and reduced balance of payments, but it is local communities that will bear most of the risks associated with fracking. The Government intends to offset this potential harm by encouraging (but not requiring) the extraction industry to sign up to a charter that will guarantee payments of some £100,000 to communities located near shale gas exploratory wells. If the gas is ultimately exploited, then they would receive one *per cent* of the resulting revenues: it has been suggested that this might amount to some £10 million. Others have proposed the creation of some form of '*Sovereign Wealth Fund*' (analogous to that generated from Norwegian

North Sea oil and gas revenues) to recompense affected UK regions and communities. It is obviously too early to tell how attractive such financial incentives might be to local communities. The *Task Force on Shale Gas*[25]have recommended that operators explain precisely how they intend to provide the £100,000 of local community payments for exploratory well pads and that the beneficiaries should be clearly defined. Properties directly affected by producing wells would obviously face the greatest disruption. The TFSG [25] therefore believe that there should be community involvement in the development of a "fair and robust" community payments scheme.

INDUCED SEISMICITY

Great public concern over shale gas fracking was triggered in 2011 by two seismic tremors, or minor earthquakes (largest reaching ~2.3 M_L; the local magnitude on the *Richter scale*), caused by exploratory drilling at the *Cuadrilla Resources* site at Preese Hall near Blackpool. Consequently, a moratorium was temporarily placed on shale gas exploration in the UK. Subsequent studies by DECC [27], aided by independent experts, together with a review of the scientific and engineering evidence on shale gas extraction undertaken by the *Royal Society* (RS) and the *Royal Academy of Engineering* (RAEng) [28], found that suitable controls were available to mitigate the risks of undesirable seismic activity. It was argued that the most likely cause of the Preese Hall tremors was 'induced seismicity'; caused by the injection of fracking fluid into and along faults that had already been under stress. The fault then shifts, leading to perceived surface tremors. DECC subsequently announced the introduction of a set of requirements for new controls, permissions and risk assessments on fracking operations in 2012, including oversight by the *Health and Safety Executive* (HSE), at Preese Hall and all future shale gas exploration wells. They included a 'traffic light' seismic monitoring system, as advocated in the RS/RAEng study [28] and subsequently suggested by DECC: *Green* = 0.0 M_L; *Amber* 0.0 ⍰ M_L ⍰ 0.5; and *Red* > 0.5 M_L[29]. Nevertheless, earth scientists (see, for example, Davies et al. [30]) viewed the RS/RAEng fault diagnosis as incomplete, and proposed the additional use of borehole imaging before injection. Recently Westaway and Younger [31] suggested that the existing regulatory limits applicable to quarry blasting could be readily applied to cover such induced seismicity. They argued that future fracking activities in the UK is only likely to cause "minor damage", and that seismic monitoring could be used to 'police' compliance with the regulatory framework. The commercially-sponsored UK *Task Force on Shale Gas* (TFSG) in their second report [32] argued that the DECC 'traffic light' limits as possibly being "unfeasibly low". However, they recommended that independent baseline monitoring should be carried out as early as possible, following the identification of a site to assess seismic risk going forward and also to increase public confidence.

Induced seismic activity has also been linked to the re-injection of large quantities of waste fluid post-fracking, rather than just in the initial hydraulic fracturing process itself [33], which has led to earthquakes of over 5 M_L on the *Richter scale* in the US. A recent study has linked such induced seismicity to disposal wells up to 35 km way [34], much further than previously considered. This practice is unlikely to be carried out in the UK reducing the risk of induced seismicity compared to the USA [29]. It is understood [31] that the reconstituted *Environment Agency* (EA) in England would not grant a permit for this method of wastewater disposal under its current interpretation of the European Union (EU) *Water Framework Directive*. However, the *Task Force on Shale Gas*, again in their second report [32], suggested that there "may be situations and circumstances – where the geology is suitable – where deep injection is sensible, cost effective and popular preferred means of waste disposal". That assertion must be read with an understanding of the commercial interests of the sponsors of the TFSG. In any event, adequate alternative wastewater management systems would need to put in place to safely dispose or reuse of the resultant wastewater [35] away from the fracking site itself.

WATER USE AND CONTAMINATION

Fracking for shale gas typically takes place at depths many hundreds of metres (or several kilometres) below drinking water aquifers, although such wells and acquifers do not co-occur everywhere. Unconventional gas enthusiasts argue that there have been no cases of groundwater contamination due to fracking in the United States, but the US *Environment Protection Agency* is less confident of that and its studies are therefore continuing. Hydraulic fracturing requires large quantities of water dependent on the properties of the shale rock involved. It ranges from 10,000–30,000 m^3 of water per fracking operation or well [36,37]. There has consequently been a lot of publicity in the UK about the large amounts of water used during the fracking process. The primary water demand is for the initial hydraulic fracking process and each subsequent fracturing step, suggesting periods of high water demand, which could put temporal stresses on water resources locally [28]. Excessive water use may lead to a fall in the availability of public water supply, ecosystem degradation and adverse effects on aquatic habitats, erosion, and changes in water temperature [36]. But in the UK abstracting water will require a license from the *Environment Agency* (EA) in England [or the equivalent bodies in the other nations of the UK, i.e., Natural Resources Wales (NRW) and the *Scottish Environment Protection Agency* (SEPA)]. The abstraction of water resources under stress should therefore be avoided via this licensing process. Some of this water may be recyclable, although it could be contaminated, for example, by *Naturally Occurring Radioactive Material* (NORM). The RS/RAEng report [28] suggests that the latter are found in shales at significantly lower levels than safe exposure limits. However, NORMs only give rise to potential hazard if concentrated in scales, for example, which may be precipitated on pipework. In reality, sea water exhibits higher NORM concentrations than deep saline ground waters in the UK. Nevertheless, wastewaters require careful management and monitoring in order to ensure that NORMs do not become concentrated [36].

In the USA some concern has been expressed over the possibility of methane levels in water that might be high enough to be flammable [38]. It has been asserted by DECC [38] that these are normally caused by failures in the well construction or natural background levels of methane rather than fracking *per se*. Indeed the RS/RAEng review [28] considered the possibility of direct groundwater contamination to be very unlikely, although it could result from faulty wells. The RS/RAEng review also warned that environmental contamination, including 'faulty wells, and leaks and spills associated with surface operations', were to be expected as they are common to all oil and gas wells and extractive activities. They argued for integrated operational practices [28], such as recycling and reusing wastewaters, to ensure benign water handling and treatment. DECC [38] have put in place a series of requirements to minimise the risk of groundwater contamination from poorly fabricated wells. These include the need for detailed plans to be submitted to the regulator (the EA in England, NRW in Wales or SEPA in Scotland), together with formal risk assessments [35].

US experience suggests that around 40–80% of the injected fracking fluids will be returned to the surface as 'flowback water' [35,39]. This contaminated 'produced water' also poses a potential risk to groundwater once it reaches the surface [37,38]. In the event of human error or equipment failure, it could potentially leak into streams and seep down to the groundwater. However, all drilling pads are double-lined with impermeable membranes and drainage is intercepted. Operators are required to dispose of flowback fluid from wells in a safe manner. Unlike in the USA, this wastewater is not permitted to be stored in open stores or disposed of by borehole injection [28], reducing the risk of this type of environmental incident in the UK. In addition, all fluids on site must be stored in double-skinned (or integrally 'bunded') tanks in case of spills. Similar hazards arise in other industrial processes, and those associated with hydraulic fracturing should be managed appropriately [10,28,40]. Thus, such risks may be 'designed-out' of the system. Fracking chemicals have to be assessed by the relevant regulator on a case-by-case basis.

ENVIRONMENTAL IMPACTS

Local Environmental Pollution, Health and Related Impacts

There are various local environmental impacts from shale gas fracking: the excessive water use, groundwater contamination and wastewater handling as discussed above, as well as noise, odours, and the disposal of solid wastes. In order to prevent contamination, the integrity of fracking wells must be ensured. Guidelines for achieving this were recommended in the RS/RAEng report [28], which are largely reflected in documents produced by the *American Petroleum Institute* and the HSE that are recommended by DECC [41]. It is believed that properly designed wells should not pose a risk of contamination to underground aquifers [41]. Of course regulation, however good, is ineffective without rigorous enforcement backed by seriously deterrent penalties. Both well design and construction are overseen by an *Independent Well Examiner* and the HSE Wells Inspector. No operation can commence before an inspection by the independent examiner (employed by the operating company or a contractor) who oversees the design, construction and maintenance of a well [42]. However fully independent monitoring of the wells which was recommended by the RS/RAEng report was rejected during the enactment of the UK *Infrastructure Act 2015*[43], primarily leaving the reporting of leakages up to the operating company. Furthermore, the monitoring or management of abandoned wells remains unclear, particularly if an operator becomes insolvent [42].

The (Smith) *Task Force on Shale Gas*, in their second report dealing with local environmental impacts [32], advocated that baseline monitoring of air, land and water should begin as soon as a potential shale gas fracking site had been identified, rather than waiting for planning permission to drill boreholes is granted. They also recommended that the adoption of the recently mandated US process of '*green completions*' or 'reduced emissions completions', whereby the shale gas and associated hydrocarbons is separated from the remaining 'flowback fluid' and the rest of the fluid to be transferred on for further processing, as a compulsory framework for exploratory sites in the UK [32]. This process is claimed to reduce 'fugitive emissions' by around 90%. The TFSG recognised that *green completions* may not be feasible for exploratory wells in Britain, and that some flaring might be required. In that event, operators could convert gas to electricity onsite and link it to the grid as an alternative, more acceptable, option [32].

There are, in addition, aesthetic concerns: visual intrusion of the sort that also results from onshore wind turbine developments. Shale gas fracking requires site operations at the wellhead, as well as the collection and distribution of unconventional gas from the site [44]. Public resistance often focuses on the increased traffic and vehicle exhaust emissions and noise, particularly those emanating from heavy road transport vehicles. Indeed, the first planning application to explore shale gas in the UK was rejected by *Lancashire councillors* [the local authority in that part of the north west of England (see again Fig. 2)] on the grounds of increased noise and visual impact [44]. Drilling often takes place on landscapes of natural beauty that include sensitive wildlife habitats [45]. The *Infrastructure*

Act 2015[43] prohibits hydraulic fracturing from taking place in land at a depth of less than 1000 m, whilst ensuring that communities benefit and that the UK has a robust regulatory regime. The UK government had originally "agreed an outright ban on fracking in national parks, sites of special scientific interest (SSI) and areas of outstanding natural beauty" [45], but have subsequently made changes to these exclusion zones arguing that it could hamper this nascent industry. Surface-level fracking operations are prohibited in environmentally-sensitive areas, such as *National Parks*, *Areas of Outstanding National Beauty* (AONB), the Norfolk Broads, World Heritage Sites, and those where groundwater supplies may be at risk [46]. Sites of *Special Scientific Interest* (SSI) will no longer be excluded from the exploration of shale gas under these new terms. These changes were met by significant concern from environmental groups over the risk posed to some of the UK's most important wildlife sites. Operational environmental permits for shale gas fracking in the UK are issued by the EA, NRW, or the SEPA (as appropriate) on a site-by-site basis in line with the requirements imposed in water abstraction licenses, and actual usage monitored over time.

Climatc Change and Fugitive Emissions

The 2008 UK *Climate Change Act*[17] set a legally binding target of reducing the nation's GHG emissions overall by 80% by 2050 in comparison to a 1990 baseline. In order to meet this reduction target, the Government's independent *Committee on Climate Change* has suggested that the electricity generation sector will effectively need to be decarbonised by 2030. Gas-fired power stations emit far fewer greenhouse gases per unit of electricity output than coal-fired ones and, for this reason, they are favoured by Helm [47] as a transitional energy option. However, if the UK continues to build and operate gas-fired power stations, the power sector will be locked-in to a fossil fuel technology and unable to decarbonise over the lifetime of these gas-fired power plants unless paired with carbon capture facilities. Unfortunately, the UK Government recently cancelled (on 25 November 2015) their £1bn CCS competition, which suggests that this technology may have an uncertain future in Britain. A CCS option would add extra cost to power generation and also constitute a significant risk of climate change policy failure. Many scientists, policy makers and journalists attributed the 7%

reduction in US domestic carbon emissions between 2007 and 2013 to the fuel switch from coal to shale gas. But recent analysis suggested that it in fact only played a small role in this fall, with much of the reduction attributed to the economic recession [48].

Two of the main sources of global warming impact arising from shale gas development are the fugitive methane emissions leaked and vented during extraction processes, and carbon dioxide (CO_2) emissions from the combustion of the shale gas to produce electricity. Like conventional gas, the primary cause of these GHG emissions result from the combustion of the shale gas in boilers. However, there is a greater variation in fugitive methane emissions from the extraction process, depending on the given location, particularly in terms of enforced environmental legislation. Therefore, much of the controversy over the global warming impact of shale gas technology focuses on such fugitive methane. Methane is a much more powerful GHG than CO_2, although it resides in the atmosphere for only 12 years [49]. Some of it may be flared (converting it to CO_2), rather than vented, but this is not or cannot always be done.

The most recent values of *Global Warming Potential* (GWP) for GHGs are provided by the *Intergovernmental Panel on Climate Change* (IPCC) in their *Fifth Assessment Report* (AR5) over three separate time horizons [50]: 20, 100 and 500 year respectively. The application of the three different time horizons are all equally valid for assessing GHGs from a scientific perspective. However, short-lived, more potent GHG emissions have a much higher GWP over the 20 year horizon; consequently methane traps 86 times more heat than carbon over this period, compared to 34 over a 100 year horizon [50]. Some have argued that it might be more pertinent to consider methane emissions over the 20 year horizon to assess the danger it poses to our climate system in the short-term. Nevertheless, the 100 year horizon has been widely used by many, providing a balance between short-term and long-term impact of GHGs on climate change. Furthermore, this is particularly appropriate given that CO_2accumulates over time in the atmosphere, whereas methane dissipates. Accordingly, the results presented in this assessment of the life-cycle GHG emissions from shale gas are over a 100 year time horizon.

Upstream GHG emissions estimated by several studies have been collated [51–57] in order to explore the potential 'carbon footprint' of

UK shale gas (see Fig. 3). The controversial study by Howarth et al. [53] was excluded when averaging this data, because of its relatively high estimates for fugitive emissions. Should rigorous and effective environmental legislation be introduced in the UK, this level of emissions is unlikely to be permitted. Emissions data from the *Ecoinvent* database version 2.2 [58]were used to account for UK NG when generating the latter mix; for both domestic and imported UK fuel routes [from Norway, the Netherlands, the rest of the European Union (EU), and via LNG]. The full life-cycle GHG emissions of shale gas electricity generation are compared to the NG generation using the current UK gas mix, LNG and Russian gas respectively in Fig. 4. The LNG emissions data were taken from a review undertaken previously by Hammond and O'Grady [59]. The operational (or 'stack') emissions are based on current UK technology, but may fall over time as more efficient plants come online. Total life-cycle emissions of shale gas generation were estimated to be in the range of 480–546 $gCO_{2e}/kW\,h_e$, i.e., 4–18% greater than emissions from the current UK gas mix (with a central estimate of 14% greater GHG emissions). Thus, providing effective regulation to curtail fugitive emissions are in place, electricity generation using shale gas could offer significant savings in carbon emissions when displacing coal-fired generation as part of a transitional energy strategy [59]. This result is in keeping with other estimates found in literature [51,54]and a study carried out by DECC [60], which saw a moderate disparity between conventional and unconventional gas. There are large uncertainties associated with these findings, and they should only be considered as 'indicative' until real operational data are available.

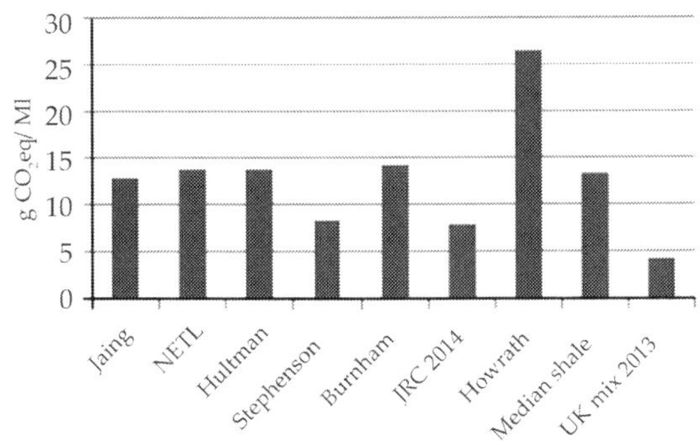

Figure 3. Central estimates for upstream GHG emissions associated with unconventional gas compared to current UK gas mix.

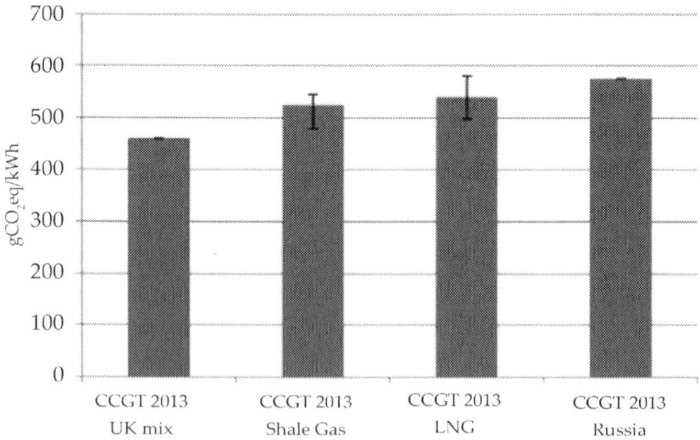

Figure 4. Life-cycle GHG emissions for gas generation with different gas sources.

GHG emissions from shale gas are currently higher than that associated with the UK gas mix, although shale gas may offer less GWP than its counterparts as the UK gas supply mix evolves over time, according to Hammond and O'Grady [59]. Over the coming years, both indigenous and European gas supply will decline, leaving the UK more reliant on imports such as LNG and Russian pipeline gas. LNG was found [59]to be 8–26% greater in terms of GHG emissions than the current UK gas mix, while Russian gas was seen to produce 25% higher emissions. These gas supplies from distant regions require long transportation routes through pipelines, which results in high fugitive emissions [61], or they must undergo liquefaction, shipping and regasification: all energy intensive processes that result in additional GHGs being emitted. The development of a UK shale gas industry could decrease the share of these impactful gases in the future UK gas mix, but are likely to impede the penetration of biomethane, which could still result in increased emissions [59]. Similarly, the growth in UK shale gas may provide a cheap supply of gas that could greatly reduce the investment in renewables, and devalue efforts to adhere to carbon budgets.

Sensitivity analysis performed in connection with shale gas studies have shown large ranges in the impact of shale gas, particularly in terms of fugitive emissions and the estimated ultimate recovery per well [62–65]. Hence, without effective regulation (backed by rigorous enforcement and seriously deterrent penalties) to minimise these fugitive methane emission, many of the notional advantages of shale gas may not be realised. An interesting study in the specific context of the Central Belt of Scotland [65] found that significantly greater methane emissions are likely to arise from shale gas extraction on peatland, in contrast to grassland development typical of the English landscape. However, comparisons between the full chain GHG emissions from conventional and shale gas were taken from the DECC study [60] mentioned above. Legislation to address fugitive methane from shale gas have not yet been specified in the UK, but it is likely that they will be treated in the same manner as fugitive methane from current UK oil and gas production. Consent for venting or flaring in this sector (reserved mainly for maintenance and emergency procedures) must be granted by DECC [66], who are committed to keeping these emissions to a technical and economic minimum. It would be desirable for such 'green completion' techniques, which were also advocated in the second report of the (Smith) *Task Force on Shale Gas*[31], to be mandated during both the explorative and operational phases of the well in order to keep fugitive emissions at an absolute minimum.

The third report of the *Task Force on Shale Gas* (TFSG), published in mid-September 2015, dealt with the climate change impact of shale gas development in the UK [67]. It suggested again that this would be similar to that for conventional gas, provided that the British shale gas sector is "properly regulated" and monitored, and lower than those associated with LNG. They advocated technological innovation and RD&D investment in CCS alongside the extraction of shale gas as a climate change mitigation option. This, they believe, would enable gas to play a transition role in the UK energy mix in the medium-term as advocated by Helm [47] and others [59]. Nevertheless, the TFSG argued that the sector should not prohibit the development of low-carbon energy generation (particularly from renewables), storage and distribution [67]. Just a little after the date that the third report of the TFSG was launched, so too was a review of GHG emissions from conventional and unconventional sources of natural gas over their respective full supply chains [68] by the *Sustainable Gas Institute* (SGI; based at Imperial College London with industrial funding from *BG Group*). These sources included conventional onshore and offshore, shale gas, tight sands, and coal bed methane (CBM). It drew on findings from some 400 papers in order to evaluate the emissions emanating from various extraction, processing and transport routes; albeit mainly based on data from North America. Comparisons were collated between the full chain GHG emissions from conventional and unconventional gas. It found that over the complete range of gas supplies the total GHG emissions associated with electricity generation was 419–636 $gCO_{2e}/kW\ h_e$ (with a central estimate of 496 $gCO_{2e}/kW\ h_e$), which the SGI considered to be well below typical GHG estimates for coal-fired power plants of around 1000 $gCO_{2e}/kW\ h_e$. These embrace the range of full chain emissions found from the present study above, although they are obviously much broader that shale gas data. Unfortunately, the SGI figures were not disaggregated in terms of fuel type (e.g., conventional versus shale gas for the current purposes).

Comparing Environmental Burdens from Different Life-Cycle Impact Categories

Climate change is the primary focus of most of these studies of shale gas, with little attention given to its wider environmental implications. The first life-cycle assessment was only recently conducted by Stamford and Azapagic [64] but has generated some controversy in the way their results have been represented in the media: as, for example, "fracking trumps renewables" [according to a Media Release by the UK *Institution of Chemical Engineers* in their members' magazine ('The Chemical Engineer') http://aboutdatajournalism.org/tcetoday-news-lca-shows-fracking-trumps-renewables/]. This is because the authors examined a variety of life-cycle impact categories in addition to climate change (for which their central estimate was 462 $gCO_{2e}/kW\ h_e$) with varying results between other environmental burdens. Shale gas was comparable or superior to conventional gas, nuclear power and renewables in terms of the depletion of abiotic resources and eutrophication, as well as freshwater, marine and human toxicities. In contrast, they found shale gas to be more environmentally damaging when photochemical smog and terrestrial toxicity were examined; both, of course, are associated with increased human mortality. Nevertheless, carbon footprints have become the 'currency' of debate in a climate-constrained world [59], where the UK is seeking to dramatically reduce its carbon emissions by 2050. It is therefore of greater significance than these other (important, although perhaps not critical) impact categories. In that regard, shale gas fracking certainly does not "trump renewables".

Stamford and Azapagic's study [64] has attracted other criticisms, including that by Westaway et al. [69]. The latter expressed doubt over some assumptions taken in a UK context in regards to drilling waste disposal, well completion, and the *Estimated Ultimate Recovery* (EUR) of the wells. Westaway et al. argued that Stamford and Azapagic assumed that practices that carried out in the USA, that have long been illegal in the UK (and EU generally), would potentially be used by the shale gas industry in Britain. However, Stamford and Azapagic stressed that the large uncertainties involved in their work was due to the nascent nature of the British shale gas industry. Despite the criticisms, their study [64] demonstrates the high sensitivity of the LCA results to particular parameters of shale gas, reporting large ranges in life-cycle impacts for UK shale gas. It is critical that robust environmental data is collected as an industry grows in Britain, with strong collaboration between operators and the EA.

In view of the early stage of UK shale gas development and of the consequent scarcity of country-specific data, baseline monitoring studies are being planned and undertaken in order to provide GHG emissions and related data over the British supply chain employing airborne, remote sensing, sampling, and sensor network methods. Thus, Allen et al. [70] have obtained and validated trace-gas-concentration and thermodynamic profiles throughout the troposphere and planetary boundary layer using data from aircraft campaigns over and around London. Similarly, Sommariva et al. [71] have taken shale samples from the Bowland-Hodder formation (in northern England) in order to determine, using mass spectrometry, methane and non-methane hydrocarbons (NMHCs). Their results indicate that high temperatures significantly increase the amount of NMHCs released from shale, whilst humidity tends to suppress them. A large fraction of the gas is also released within the first hour after the shale has been fractured. Clearly, much more needs to be done in terms of basic data gathering for real-world hydraulic fracturing operations in the UK. Such data would help reduce uncertainty and allow the potential environmental impact of UK shale gas to be more accurately determined.

PUBLIC AND STAKEHOLDER ENGAGEMENT

UK Government Ministers have indicated their concern over the fierce resistance to their shale gas fracking policies; particularly from rural communities in both the south and north of England. Potential sites stretch all the way from Dorset to the Kent borders (across the south), via the Bowland-Hodder (in northern England), to at least the Midland Valley of Scotland (see again Fig. 2). There are also potential shale gas reserves in Wales, although there is a relative paucity of data for the Principality. Initial concerns were raised in northern England as a result of induced seismicity caused by exploratory drilling at Preese Hall. Such communities need to be engaged in a two-way dialogue aimed at clarifying the impacts of the shale gas fracking process, along with its potential costs and benefits. Challenges to energy infrastructural developments in the form of local activism have typically been led by groups, such as the Green Parties (of the nations of the UK) and environmental campaigning organisations, like *Friends of the Earth* and *Greenpeace*, as well as various nature conservation bodies. Public opposition could prove to be a 'showstopper' for this energy option unless the various stakeholders are engaged in an appropriate consultation. Pigdeon et al. [72] recently examined some of the critical issues concerning the design and conduct of public deliberation processes on energy policy matters of national importance. In order to develop their argument, they employed as an illustrative case study, some of their earlier work on public values and attitudes toward future UK energy system change. They note that national-level policy issues are often inherently complex; involving multiple interconnected elements and frames, analysis over extended scales, and different (often high) levels of uncertainty. It is their view that facilitators should engage the public in terms of 'whole systems' thinking at the problem scale, provide balanced information and policy framings, and use different approaches that encourage participants to reflect and deliberate on the issues. This is similar to what is often referred to as interactive, participatory methods by the technology assessment community [3–5].

DECC has engaged in a process of evaluation of public attitudes to various energy technologies [73], including shale gas fracking. They have commissioned periodic surveys of just over 2000 face-to-face in-home interviews with adults over 16 years of age by specialists from the *Office of National Statistics* (ONS) in a series, or wave, of studies. The most recent (August 2015) survey [73] indicated that around ¾ of the British public were aware of fracking, although only 14% indicated that they knew a significant amount about the process. Of the sample, some 46% were neither supporters nor opponents of extracting shale gas. Those that offered an opinion indicated that around 28% were opposed, whereas 21% supported of the exploitation of this technology. DECC have suggested, on the basis of the ONS survey, that support for fracking is related to awareness with 54% of those who claim a lot of knowledge being opposed to the technology (in contrast to 32% in favour of shale gas fracking) [73].

An academic study of UK public perceptions of shale gas fracking by O'Hara et al. [74] has attracted a lot of attention amongst social scientists and policy makers in Britain [37]. It included analysis of policy documents and media sources, semi-structured interviews, and a series of seven nation-wide surveys over the period 2012–2013. The initial survey indicated that 37% of respondents were familiar with shale gas extraction, but this rose to over 60% and then flattened out. This latter phenomenon occurred in spite of a significant increase in media coverage of fracking over this period, particularly via the BBC. O'Hara et al. [74] found that respondents under 25 years of age were least aware of shale gas extraction and its implications. Over 58% of people thought that it would aid energy security, although respondents answered 'don't know' to questions about how shale gas development would impact in terms of climate change. The majority of people who were familiar with the process felt that it should be permitted. Another recent, detailed experimental (online) survey of public perceptions of shale gas fracking in the UK (N = 1457) by Whitmarsh et al. [75], included analysis of the effects of different messages on support for, or risk perceptions of, shale gas fracking. They found that the public were generally ambivalent about shale gas, but perceived more risks than benefits. This was strongly influenced by demographic, political and environmental considerations and values. The study [75] discovered that prior knowledge of shale gas extraction had the greatest impact on those respondents who were initially ambivalent or 'undecided', which suggested an important role for information and awareness raising.

The (Smith) *Task Force on Shale Gas* also considered matters around local engagement in their first report [76]. They proposed the establishment of a community engagement plan by operators (albeit with local community involvement) and the full disclosure of information to the local community before an application is submitted to the appropriate regulator. The latter should include logistic and broad site access issues [76], such as the likely number and size of transport movements over the life of a potential drill pad. The *British Geological Survey* has been involved in practical activities associated with public engagement about shale gas fracking [Dr. Robert Ward, BGS, UK, private communication 06.11.15]. They have advocated interaction much in line with that recommended by TFSG [76] although, in addition, the BGS stress the need to utilise tailor-made approaches for different areas/groups, as well as the importance of keeping "information up-to-date, fresh and understandable". Further elaboration of such surveys, employing the deliberative framework proposed by Pigdeon et al. [72], might go some way towards securing better awareness and understanding of hydraulic fracturing by the public in general.

PLANNING, REGULATION AND MONITORING

One set of issues for which politicians of different persuasions and community groups agree on is the important need for adequate measures in the area of unconventional gas planning and regulation. Nevertheless, it has yet to be determined whether what community groups consider 'effective regulation' will be accepted and upheld by the government. For instance, many local groups are opposed to recent moves by the UK Government to facilitate planning permission for fracking by preventing landowners objecting to the process taking place under their land [77]. DECC issues licenses to onshore oil and gas operators for exclusive drilling rights, and have listed a long set of pre-drilling approvals that are needed from the various regulators [1]. Operators are required to obtain planning permission from the appropriate UK minerals planning authority (county council or unitary authorities in England and the planning authorities in Scotland and Wales) and seek access to the site from landowners [1].

The *Environment Agency* (EA) in England is currently undergoing a consultation to define new standard permits for onshore oil and gas [78]. The process of applying for permits to drill and carry out preliminary testing of wells would be streamlined by removing public consultation. Although a separate permit for hydraulic fracturing will still be required, this has been seen by many as a move to simplify the process and reduce costs for operators. It has also been argued that prior independent [28,32] evaluations of well integrity should be undertaken before drilling can commence, followed by mandated disclosure of hazardous incidents, ongoing process monitoring and contamination assessments. *Public Health England* (PHE) [79], for example, have recently proposed that baseline environmental monitoring be instigated in order to facilitate the impact assessment of shale gas extraction on the environment and public health, that the fracking chemicals (including NORMs) should be publicly disclosed and their risks assessed before use, and that the type and composition of the extracted gas should be determined on a site-by-site basis. This was criticised by a group of medical specialists from the US [80], who argued that it was based on the idea that many of the public health problems experienced in the USA would be replicated in a more densely populated country like Britain. Law et al. [80] suggest that these impacts as yet remain undetermined and require further scientific study using rigorous, quantitative epidemiological methods. However, the UK *Task Force on Shale Gas*[31] generally agreed with the PHE recommendations, but also proposed that the Government should establish a 'National Advisory Committee' of independent academic experts to monitor data from shale gas operations in order to evaluate health impacts.

The TFSG examined planning and regulatory issues in their first report [76]. Their main recommendation was that the UK Government should explore the possibility of creating a new, bespoke regulator for onshore underground energy (unconventional oil and gas, including CBM) that would take over the current regulatory responsibilities of the *Environment Agency* (and its national counterparts), the *Health & Safety Executive*, and DECC. The (Smith) Task Force also suggested [76] that full-scale 'environmental impact assessments', which they view as being not easily accessible, should be replaced by 'environmental risk assessments' [35]. The TFSG argued [76]that the latter methodology would be “more succinct and approachable” for use by the new regulator in a way that could be made readily available to local communities.

CONCLUDING REMARKS

An energy technology assessment (ETA) has been undertaken [3–5,13,14] in order to evaluate the credit and debit ‘columns’ of the shale gas fracking ‘balance sheet’. The adoption of this extraction technology is at a very early stage in the UK with great uncertainty over the scale of the potential shale gas resource. An extensive programme of investigative drilling across the country will therefore be needed in order to provide reliable estimates; possibly requiring hundreds of exploratory wells [20,21]. Nevertheless, the successful exploitation of large-scale development of shale gas extraction in the UK might contribute positively in terms of fuel security and independence, as well as jobs and growth, providing the potentially harmful ‘side-effects’ outlined here can be satisfactorily resolved. But it is unclear whether job creation would be any greater than that arising from equivalent programmes aimed at supporting the adoption of energy demand reduction measures or small-scale low carbon energy options. Similarly, the UK balance of payments would benefit, although it is unlikely that gas bills for household and industrial consumers would fall dramatically as they have done in North America. This is because the UK is part of the wider European natural gas market [23,37] where the gas price is determined by the supply and demand for indigenous natural gas, imports from Russia, and LNG from North Africa. Unconventional gas supplies in the UK will be only be a small fraction of those in this wider market. Lessons from the present ETA study will therefore also apply in general across the European Union, albeit tailored by local circumstances.

Hydraulic fracturing requires significant quantities of water which may potentially lead to a fall in the availability of public water supply, ecosystem degradation and adverse effects on aquatic habitats, erosion, and changes in water temperature [28]. However, abstracting water in the UK will require a license from the *Environment Agency* (EA) in England [or the equivalent bodies in the other nations of the UK, i.e., NRW and SEPA]. The abstraction of water resources under stress should therefore be avoided via this licensing process. Recycled fracking fluid could be used for ongoing fracking operations [35,36], except that a proportion of this is not recovered. In the USA some concern has been expressed over the possibility of methane levels in water that might be high enough to be flammable. It has been asserted by DECC [38] that the high methane levels found in some US drinking water supplies were caused by failures in the well construction or natural background levels of methane, rather than fracking *per se*. They have put in place minimum requirements to avoid groundwater contamination from poorly fabricated wells. Fracking chemicals have to be assessed by the appropriate regulator (EA, NRW or SEPA) on a case-by-case basis [79]. Operators are required to dispose of 'flowback fluid' from well in a safe manner.

The life-cycle carbon footprint of shale gas has been shown to be lower than that of coal-fired power generators providing stringent regulation is implemented to minimise fugitive methane emissions. On the other hand, the life-cycle carbon footprint was found to be slightly higher than conventional gas, and considerably higher than nuclear power and renewables. It could therefore form part of a transitional UK energy strategy [47,59], although this might jeopardise the attainment of a low (near zero) carbon transition pathway by 2050. The penetration of shale gas into the UK energy mix would likely lead to the lock-in of gas-fired power generation for some decades. Furthermore, without the large-scale use of *carbon capture and storage* (CCS) [81], such a transition would be incompatible with meeting legislated carbon budgets and limiting GHG concentrations to a 'safe' level [82].

The socio-economic benefits and costs of shale gas fracking are not evenly distributed between various communities and demographic groups. Thus, the UK overall might benefit from improved energy security and reduced balance of payments, whilst it will be local communities that bear any adverse environmental and health risks of fracking. Induced seismicity caused by the injection of fracking fluid into and along faults that are already under stress can lead to minor earthquakes or surface tremors. DECC have introduced a set of requirements for new controls, permissions and risk assessments on fracking operations in 2012, based on a 'traffic light' system [27–29] to monitor unusual seismic activity. However, several prominent UK earth scientists have argued [30,31] that future fracking activities in the UK are only likely to cause 'minor damage'; yet again, provided a robust regulatory framework is put in place. Local environmental impacts are critical to neighbouring communities near the wellhead. They focus on shale fracking site operations, as well as the collection and distribution of unconventional gas from the wellhead [44,47]. Public resistance has been largely concerned about increased traffic, which causes vehicle exhaust emissions and noise [79], particularly those emanating from heavy road transport vehicles. In addition, drilling places environmental burdens on landscapes that are often in areas of natural beauty with sensitive wildlife habitats [47].

In order to draw up an objective and rigorous 'balance sheet' for the fracking of shale gas (or indeed other critical technologies) as part of a national dialogue, it would be desirable to bring together experts from a range of disciplines in order to carry out the necessary ETAs. They would need to interact with national and local stakeholders: 'actors' both large and small. That work should be seen by the wider community as *analysis* and not *advocacy*. The UK Coalition Government, when it came into office in 2010, unfortunately closed down or withdrew funding from a number of independent, 'arms-length' bodies established by government departments (sometimes known by the term 'quasi-autonomous non-governmental organisations', or 'Quangos') who might have been capable of conducting studies of this type. Two such bodies were the *Royal Commission on Environmental Pollution* and the *Sustainable Development Commission*. It is unlikely that the present government would re-establish these organisations, but perhaps a future government might consider establishing an alternative. One model might be the *Office of Technology Assessment at the German Bundestag* (TAB), or the equivalent bodies in the Scandinavian countries [4]. In the present context, a new UK agency might look wider than just the scientific and engineering issues, bringing together the technical with the social science perspectives. The latter would be critically important in obtaining insights from various stakeholder groups. That would guard against unwanted side-effects by identifying them in advance of deployment, and could go some way towards engaging and reassuring the community. Constraints on the exploration for unconventional gas are likely to be as much about public acceptance as they are about the various technical issues [72–76]. Community engagement in a genuinely participative process will consequently be critically important for the adoption of any new energy option [3,4]that might meet the needs of a low carbon future. That is certainly one lesson from the shale gas fracking controversy.

ACKNOWLEDGEMENTS

This is an extended and updated version of a paper originally presented at the *7th International Conference on Applied Energy* (ICAE2015) held in Abu Dhabi, UAE over the period 28–31 March 2015 (denoted then as paper ICAE2015-600). The work reported forms part of a programme of research at the University of Bath on the technology assessment of low carbon energy systems and transition pathways that is supported by a series of UK research grants and contracts awarded by various bodies. In the present context, the first author (GPH) is jointly leading a large consortium of university partners funded by the UK *Engineering and Physical Sciences Research Council* (EPSRC) entitled 'Realising Transition Pathways: Whole Systems Analysis for a UK More Electric Low Carbon Energy Future' [under Grant EP/K005316/1]. The second author (AO'G) is wholly funded as a Research Associate via this grant. Both authors are grateful for the interaction with other members of the Consortium (and its predecessor) made up of participants from nine UK universities.

Since the ICAE2015 conference version of this paper was delivered, the first author (GPH) participated in a specialist *Joint US-UK Workshop on Unconventional Hydrocarbons* held at Virginia Tech Research Center in Washington DC over 5–6 November 2015. Its technical programme was arranged by Prof. Richard Davies of Newcastle University (UK) and Prof. Danny Reible of Texas Tech University (USA). The British participants were supported by the UK *Natural Environment Research Council* (NERC) and those from the US by the *National Science Foundation* (NSF). Insights from this workshop, particularly in regard to the exploration for shale gas extraction in the UK, have been incorporated into the revised paper as appropriate. Both authors have also greatly benefitted from critical comments on an early version of this paper by their University of Bath colleague Dr. David E. Packham (Emeritus Senior Lecturer in Materials Science). Finally, they are grateful to an anonymous reviewer ("an experienced well engineer") for elaborating on some of the mining engineering and geological aspects of the paper. However, the views expressed here are those of the authors alone, and do not necessarily reflect the opinions of the collaborators or the policies of the funding body.

REFERENCES

1. Department of Energy and Climate Change (DECC). Onshore oil and gas exploration in the UK: regulation and best practice – England. London, UK: DECC; 2013.
2. International Energy Agency (IEA). World Energy Outlook: Are we entering a golden age of gas? Special Report. Paris, France: IEA; 2011.
3. Tran TA, Daim T. A taxonomic review of methods and tools applied in technology assessment. Technol Forecast Soc Chang 2008;75(9):1396–405.
4. Vig NJ, Paschen H, editors. Parliaments and technology: the development of technology assessment in Europe. New York, USA: Suny Press; 2000.
5. Manne AS. ETA: a model of energy technology assessment. Bell J Econ 1976;7 (2):379–406.
6. Chivers D. The big story – fracking. New Internationalist 2013;468:12–24.
7. Rogers H. Shale gas – the unfolding story. Oxford Rev Econ Policy 2011;27 (1):117–43.
8. Zuckerman G. The frackers: the outrageous inside story of the new energy revolution. London, UK: Portfolio Penguin; 2013.
9. Theodori GL, Luloff AE, Willits FK, Burnett DB. Hydraulic fracturing and the management, disposal, and reuse of frac flowback waters: views from the public in the Marcellus Shale. Energy Res Soc Sci 2014;2:66–74.
10. Broderick J, Anderson K, Wood R, Gilbert P, Sharmina M, Footitt A, et al. Shale gas: an updated assessment of environmental and climate change impacts. [A report commissioned by the Co-operative and undertaken by researchers at the Tyndall Centre, University of Manchester. Manchester], UK: Tyndall Centre; 2011.
11. Lechtenböhmer S, Altmann M, Capito S, Matra Z, Weindrorf W, Zittel W. Impacts of shale gas and shale oil extraction on the environment and on human health. [A study by the Wuppertal Institute for Climate, Environment and Energy and Ludwig-Bölkow-Systemtechnik GmbH requested by the European Parliament's

Committee on Environment, Public Health and Food Safety. Report IP/A/ENVI/ST/2011-07]. Brussels, Belgium: European Parliament; 2011.

12. Hammond GP. Energy, environment and sustainable development: a UK perspective. Trans. IChemE Part B. Process Saf Environ Prot 2000;78 (4):304–23.
13. Hammond GP, Jones CI, Spevack R. The 'Shoots Barrage': an indicative energy technology assessment of a tidal power scheme. J. Sustain Dev Energy Water Environ Syst 2014;2(4):388–407.
14. Hammond GP, Hazeldine T. Indicative energy technology assessment of advanced rechargeable battery technologies. Appl Energy 2015;138:559–71.
15. Foxon TJ, Hammond GP, Pearson PJ. Developing transition pathways for a low carbon electricity system in the UK. Technol Forecast Soc Chang 2010;77 (8):1203–13.
16. Hammond GP, Pearson PJG. Challenges of the transition to a low carbon, more electric future: from here to 2050 (Editorial). Energy Policy 2013;52:1–9.
17. Climate Change Act 2008, Chapter 27, London, UK: The Stationary Office Limited; 2008.
18. Harvey T, Gray J. The unconventional hydrocarbon resources of Britain's onshore basins – Shale gas. [Report by the British Geological Survey (BGS)]. London, UK: DECC; 2013.
19. Bassi S, Rydge J, Khor CS, Fankhauser S, Hirst N, Ward B. A UK 'dash' for smart gas, Policy Brief. London, UK: Centre for Climate Change Economics & Policy and Grantham Research Institute for Climate Change & the Environment, London School of Economics and Political Science; 2013.
20. Parliamentary Office of Science and Technology [POST]. POSTBOX: UK shale gas potential. London, UK: POST; 2013.
21. White E, Fell M, Smith L, Keep M. Shale gas and fracking. Standard Note SN/SC/ 6073. London, UK; House of Commons Library; 2014.
22. Committee on Climate Change. Meeting Carbon Budgets – Progress in reducing the UK's emissions, Report to Parliament. London, UK: CCC; 2015.

23. Stern JP. Continental European long-term gas contracts: is a transition away from oil product-linked pricing inevitable and imminent? Oxford, UK: Oxford Institute for Energy Studies; 2009.
24. House of Commons' Energy and Climate Change Committee. The Impact of Shale Gas on Energy Markets. HC 785. London, UK: The Stationary Office; 2013.
25. Task Force on Shale Gas [TFSG]. Economic Impacts. Final Report. London, UK: TFSG; 2015.
26. Neale J, editor. One million climate jobs: tackling the environment and economic crises. London, UK: Campaign against Climate Change; 2014.
27. Green CA, Styles P, Baptie BJ. Preese hall shale gas fracturing: Review & recommendations for induced seismic mitigation. London, UK: DECC; 2012.
28. Royal Society/Royal Academy of Engineering (RS/RAEng). Shale gas extraction in the UK: a review of hydraulic fracturing. DES2597. London, UK: RS/RAEng; 2012.
29. Department of Energy and Climate Change [DECC]. Fracking UK shale gas: understanding earthquake risk. London, UK: DECC; 2014.
30. Davies R, Foulger G, Bindley A, Styles P. Induced seismicity and hydraulic fracturing for the recovery of hydrocarbons. Mar Pet Geol 2013;45:171–85.
31. Westaway R, Younger PL. Quantification of potential macroseismic effects of the induced seismicity that might result from hydraulic fracturing for shale gas exploitation in the UK. Q J Eng Geol Hydrogeol 2014;47:333–50.
32. Task Force on Shale Gas [TFSG]. Assessing the Impact of Shale Gas on the Local Environment and Health. Second Interim Report. London, UK; TFSG; 2015.
33. McGarr A. Maximum magnitude earthquakes induced by fluid injection. J Geophys Res: Solid Earth 2014;119(2):1008–19.
34. Keranen KM, Weingarten M, Abers GA, Bekins BA, Ge S. Sharp increase in central Oklahoma seismicity since 2008 induced by massive wastewater injection. Science 2014;345(6195):448–51.

35. Prpich G, Coulon F, Anthony EJ. Review of the scientific evidence to support environmental risk assessment of shale gas development in the UK. Science of The Total Environment 2015: available online 26 November http://dx.doi.org/ 10.1016/j.scitotenv.2015.11.026.
36. The Charted Institution of Water and Environmental Management (CIWEM). Hydraulc fracturing (fracking) of shale in the UK. Policy Position Statement. London, UK: CIWEM; 2012.
37. Bradshaw M. Global energy dilemmas: energy security, globalization, and climate change. Cambridge, UK: Polity Press; 2014.
38. Department of Energy and Climate Change [DECC]. Fracking UK shale gas: water. London, UK: DECC; 2014.
39. Engelder T, Cathles LM, Bryndzia LT. The fate of residual treatment water in gas shale. J Unconventional Oil Gas Resour 2014;7:33–48.
40. Lawrence KW, Hung Y-T, Lo HH, Yapijakis C, editors. Handbook of industrial and hazardous wastes treatment. Boca Raton, USA: CRC Press; 2004.
41. Department of Energy and Climate Change [DECC]. Fracking UK shale gas: safety from design to decommissioning. London, UK: DECC; 2014.
42. Environment Agency [EA]. Review of assessment procedures for shale gas well casing installation, Bristol, UK: EA; 2012.
43. Infrastructure Act 2015, Chapter 7: Part 6 – Energy. London, UK: The Stationery Office Limited; 2015.
44. Lancashire County Council (LCC), Shale gas developments in Lancashire. Preston, UK: LCC; 2015. .
45. Rudd A. House of Commons debate: Infrastrucure Bill .HC Deb 26 January 2015 Vol 591 c586.
46. The Onshore Hydraulic Fracturing (Protected Areas) Regulations 2015.Draft Regulations laid before Parliament under section 4B(5) of the Petroleum Act 1998, for approval by resolution of each House of Parliament. London, UK: The Stationery Office Limited. 2015.
47. Helm D. The carbon crunch: how we're getting climate change wrong – and how to fix it. New York and London: Yale University Press; 2012.
48. Feng K, Davis SJ, Sun L, Hubacek K. Drivers of the US CO2

emissions 1997–2013. Nat Commun 2015;6:7714. http://dx.doi.org/10.1038/ncomms8714.

49. Intergovernmental Panel on Climate Change [IPCC]. Climate Change 2007 – The Physical Science Basis. Cambridge, UK: Cambridge University Press; 2007.
50. Intergovernmental Panel on Climate Change [IPCC]. Climate change 2013 – the physical science basis summary for policymakers.. Cambridge, UK: Cambridge University Press; 2013.
51. Jiang M, Griffin WM, Hendrickson C, Jaramillo CP, VanBriesen J, Venkatesh A. Life cycle greenhouse gas emissions of Marcellus shale gas. Environ Res Lett 2011;6(3):034014.
52. Hultman N, Rebois D, Scholten M, Ramig C. The greenhouse impact of unconventional gas for electricity generation. Environ Res Lett 2011;6 (4):044008.
53. Howarth R, Santoro R, Ingraffea A. Methane and the greenhouse-gas footprint of natural gas from shale formations. Climatic Change 2011;106(4):679–90.
54. Stephenson T, Valle JE, Riera-Palou X. Modeling the relative GHG emissions of conventional and shale gas production. Environ Sci Technol 2011;45 (24):10757–64.
55. Burnham A, Han J, Clark CE, Wang M, Dunn JB, Palou-Rivera I. Life-cycle greenhouse gas emissions of shale gas, natural gas, coal, and petroleum. Environ Sci Technol 2011;46(2):619–27.
56. Skone TJ, Littlefield J, Marriott J. Life cycle greenhouse gas inventory of natural gas extraction, delivery and electricity production. DOE/NETL-2011/1522. Pittsburgh, USA: US Department of Energy [DOE]., National Energy Technology Laboratory [NETL]; 2011.
57. Edwards R, Hass H, Larivé J-F, Lonza L, Maas H, Rickeard D. Well-to-Wheels analysis of future automotive fuels and powertrains in the European context. Joint Research Centre (JRC) WELL-TO-TANK Report Version 4a. Luxembourg: Publications Office of the European Union; 2014.
58. Hischier R, Weidema B, Althaus H-J, Bauer C, Doka G, Dones R, et al. Implementation of life cycle impact assessment methods. Ecoinvent Report No. 3 (version 2.2). Dübendorf, Switzerland: Swiss Centre for Life Cycle Inventories; 2010.

59. Hammond GP, O'Grady A. The threat of an evolving fuel supply to the carbon targets for the UK electricity sector in a more electric dependent future. In: Proc. 9th Conference on sustainable development of energy, water and environment systems [SDEWES], Venice-Istanbul, 20–27 September 2014, Paper SDEWES2014-0365.
60. MacKay DJ, Stone TJ. Potential greenhouse gas emissions associated with shale gas extraction and use. London, UK: DECC; 2013.
61. Hammond GP, O'Grady A. The implications of upstream emissions from the power sector. Proc Inst Civ Eng: Energy 2014;167(1):9–19.
62. Laurenzi IJ, Jersey GR. Life cycle greenhouse gas emissions and freshwater consumption of Marcellus shale gas. Environ Sci Technol 2013;47 (9):4896–903.
63. Heath GA, O'Donoughue P, Arent DJ, Bazilian M. Harmonization of initial estimates of shale gas life cycle greenhouse gas emissions for electric power generation. Proc Natl Acad Sci 2014;111(31):E3167–76.
64. Stamford L, Azapagic A. Life cycle environmental impacts of UK shale gas. Appl Energy 2014;134:506–18.
65. Bond CE, Roberts J, Hastings A, Shipton ZK, João EM, Tabyldy Kyzy J, et al. Lifecycle assessment of greenhouse gas emissions from unconventional gas in Scotland. Aberdeen/Glasgow, Scotland, UK: ClimateXChange; 2014. .
66. Harvey T. Guidance notes for onshore oil and gas field development plans. London, UK: DECC; 2014.
67. Task Force on Shale Gas [TFSG]. Assessing the Impact of Shale Gas on Climate Change. Third Report. London, UK: TFSG; 2015.
68. Balcombe P, Anderson K, Speirs J, Brandon N, Hawkes A. Methane and CO2 emissions from the natural gas supply chain: an evidence assessment. London, UK: Sustainable Gas Institute, Imperial College London; 2015.
69. Westaway R, Younger PL, Cornelius C. Comment on 'Life cycle environmental impacts of UK shale gas' by L. Stamford and A. Azapagic. Applied Energy, 134, 506–518, 2014. Applied Energy 2015; 148: 489–95.
70. Allen G, Illingworth SM, O'Shea SJ, Newman S, Vance A,

Bauguitte SB, et al. Atmospheric composition and thermodynamic retrievals from the ARIES airborne TIR-FTS system – Part 2: Validation and results from aircraft campaigns. Atmos Measur Tech 2014;7(12):4401–16.

71. Sommariva R, Blake RS, Cuss RJ, Cordell RL, Harrington JF, White IR, et al. Observations of the release of non-methane hydrocarbons from fractured shale. Environ Sci Technol 2014;48(15):8891–6.
72. Pidgeon N, Demski C, Butler C, Parkhill K, Spence A. Creating a national citizen engagement process for energy policy. Proc Natl Acad Sci 2014;111(Suppl. 4):13606–13.
73. Department of Energy and Climate Change [DECC]. DECC Public Attitudes Tracker – Wave 14. London, UK: DECC; 2015.
74. O'Hara S, Humphrey M, Jaspal R, Nerlich B, Poberezhskaya M. Public perception of shale gas extraction in the UK: how people's views are changing. Nottingham, UK: University of Nottingham; 2013.
75. Whitmarsh L, Nash N, Upham P, Lloyd A, Verdon JP, Kendall JM. UK public perceptions of shale gas hydraulic fracturing: the role of audience, message and contextual factors on risk perceptions and policy support. Appl Energy 2015;160:419–30.
76. Task Force on Shale Gas [TFSG]. Planning, Regulation and Local Engagement. First Interim Report. London, UK: TFSG; 2015.
77. Department of Energy and Climate Change [DECC]. Underground Drilling Access: Government Response to the Consultation on Proposal for Underground Access for the Extraction of Gas, Oil or Geothermal Energy. London, UK: DECC; 2014.
78. Environment Agency [EA]. Standard rules consultation No. 11 – new standard rules for onshore oil and gas activities. Bristol, UK: EA; 2015. .
79. Public Health England [PHE]. Review of the Potential Public Health Impacts of Exposures to Chemical and Radioactive Pollutants as a Result of the Shale Gas Extraction Process. [PHE-CRCE-009]. London, UK: PHE; 2014.
80. Law A, Hays J, Shonkoff SB, Finkel ML. Public Health England's draft report on shale gas extraction (Editorial). BMJ 2014;348:g2728.

81. Hammond GP, Ondo Akwe SS, Williams S. Techno-economic appraisal of fossil-fuelled power generation systems with carbon dioxide capture and storage. Energy 2011;36(2):975–84.
82. Committee on Climate Change [CCC]. The need for a carbon intensity target in the power sector [Letter from CCC to the UK Secretary of State for Energy and Climate Change, 13 September]. London, UK: CCC; 2012.

9 BUILDING ENERGY METERING AND ENVIRONMENTAL MONITORING – A STATE-OF-THE-ART REVIEW AND DIRECTIONS FOR FUTURE RESEARCH

Muhammad Waseem Ahmad[a], Monjur Mourshed[a], David Mundow[b], Mario Sisinni[c], Yacine Rezgui[a]

[a] BRE Centre for Sustainable Engineering, School of Engineering, Cardiff University, Cardiff CF24 3AA, United Kingdom

[b] SMS Plc – Energy & Utilities Division, Prennau House, Copse Walk, Cardiff Gate Business Park, Cardiff CF23 8XH, United Kingdom

[c] SMS Plc, Via Gaudenzio Ferrari 21, 21047 Saronno, Italy

ABSTRACT

Buildings are responsible for 40% of global energy use and contribute towards 30% of the total CO_2 emissions. The drive to reduce energy consumption and associated greenhouse gas emissions from buildings has acted as a catalyst in the increasing installation of meters and sensors for monitoring energy use and indoor environmental conditions in buildings. This paper reviews the state-of-the-art in building energy metering and environmental monitoring, including their social, economic, environmental and legislative drivers. The integration of meters and sensors with existing building energy management systems (BEMS) is critically appraised, especially with regard to communication technologies and protocols such as ModBus, M-Bus, Ethernet, Cellular, ZigBee, WiFi and BACnet. Findings suggest that energy metering is covered in existing policies and regulations in only a handful of countries. Most of the legislations and policies on energy metering in Europe are in response to the Energy Performance of Buildings Directive (EPBD), 2002/91/EC. However, recent developments in policy are pointing towards more stringent metering requirements in future, moving away from voluntary to mandatory compliance. With regards to metering equipment, significant developments have been made in the recent past on miniaturisation, accuracy, robustness, data storage, ability to connect using multiple communication protocols, and the integration with BEMS and the Cloud – resulting in a range of available solutions, selection of which can be challenging. Developments in communication technologies, in particular in low-power wireless such as ZigBee and Bluetooth LE (BLE), are enabling cost-effective machine to machine (M2M) and internet of things (IoT) implementation of sensor networks. Privacy and data protection, however, remain a concern for data aggregators and end-users. The standardization of network protocols and device functionalities remains an active area of research and development, especially due to the prevalence of many protocols in the BEMS industry. Available solutions often lack interoperability between hardware and software systems, resulting in vendor lock-in. The paper provides a comprehensive understanding of available technologies for energy metering and environmental monitoring; their drivers, advantages and limitations; factors affecting their selection and future directions of research and development – for use a reference, as well as for generating further interest in this expanding research area.

INTRODUCTION

Globally, Heating, buildings account for 40% of total energy used and contribute towards 30% of the total CO_2 emissions [1,2]. Buildings are the largest consumer of energy in the European Union, accounting for up to 40% of the total energy consumption and approximately 36% of the greenhouse gas emissions [3]. Buildings' share of CO_2 emissions is higher in some countries; e.g., the sector represents 50% of the total of 570 $MtCO_2e$ emissions in the UK in 2013 [4]. Energy is used in buildings for heating and cooling, hot water, lighting and appliances, and the majority of this energy come from the burning of fossil fuel, which amounted to 81.23% of global energy consumption in 2011 [5]. Associated GHG emissions from the burning of fossil fuels have been attributed as the extremely likely cause of anthropogenic climate change [6]. According to the International Energy Agency (IEA) [7], global greenhouse gas emissions are rapidly increasing and it will be difficult to limit the long-term rise in global average temperature to 2 °C below pre-industrial levels. Addressing the issue of global climate change and reversing the trend of rising energy consumption is essential to reduce the impact of climate change to a 2 °C rise in global average temperature [7]. The building sector, therefore, plays a significant role in mitigating the impacts of climate change – first, through reducing the demand; i.e. energy conservation, and second, by maximising the use of renewable energy – both aimed at reducing GHG emissions [8].

Developed and developing countries alike are faced with the challenges of decarbonisation of their respective building stocks over the next 30–40 years [9]. Proposed measures for decarbonisation include but not limited to increasing the share of low- and zero-carbon buildings; smarter and more efficient energy systems and behavioural change among consumers. Gathering data on and evaluation of energy and environmental performance are at the heart of decarbonisation strategies to enhance energy efficiency and to reduce energy use in buildings. Advanced metering and control, and environmental monitoring technologies have a significant role to play in data-driven energy efficiency measures. There is also an increase in energy-related building legislations and regulations around the world, including stringent energy efficiency requirements that act as a motivator for increased level of metering, sub-metering and automated control, especially in non-domestic buildings. As an example, the European Energy Performance of Buildings Directive (EPBD) also requires EU members to encourage intelligent metering and active control systems in new or renovated buildings through building regulations [10]. In the USA, the Energy Policy Act of 2005 has metering requirements for federal buildings. Most US States, cities and districts have adopted the International Energy Conservation Code (IECC) [11], with enhanced requirements for metering and control.

Building energy metering and environmental monitoring give stakeholders valuable information regarding how buildings are performing. Knowledge gleaned from analytics can also be used to improve the performance further. A comprehensive energy metering and environmental monitoring system has the potential to get all stakeholders (tenants, building owners, energy managers) on board to take energy-efficiency measures [12]. Energy metering can also help in identifying cost-cutting opportunities by detecting inefficiencies; benchmarking buildings internally and externally; improving load planning and energy usage; and managing demand to minimise exposure to system reliability and price volatility risks [12]. There is currently a myriad of metering and sensor technologies that are available on the market with varying degrees of sophistication and functionality. Choosing a correct metering and sensor solution for a building is a challenging and non-trivial task as it depends on different factors. These factors need to be considered to ensure the metering and monitoring solution can meter and monitor the relevant quality (power consumption, gas consumption, air-temperature, etc.). Therefore, it is important to be aware of different factors that influence the selection of these technologies. In future, smart buildings infrastructure will support bi-directional communication standards, which will allow continuous interaction between the utility, the consumer and the controllable electrical load [13]. According to Ras [14], a highly reliable communication will be required to transfer a high volume of data. Therefore, communication and network technologies will play a significant role in the integration of smart buildings and grids.

The increasing installation of energy meters and environmental monitoring sensors has acted as a catalyst towards the drive to reduce energy consumption and associated green gas emissions. This papers reviews drivers for installing these technologies; including social, economic, environmental and legislative. The recent developments in policy are pointing towards more stringent metering installation requirements and in future these requirements will become mandatory. The selection of metering and environmental solution can be a challenging task due to recent developments on accuracy, robustness, data storage, miniaturisation, ability to connect using multiple communication protocols, integration with BEMS and the Cloud. Different factors that influence this decision making process are also covered in this article.

Different widely used communication protocols including low-power wireless (e.g. ZigBee) are also reviewed in the paper. Data protection and privacy is a main concern for both end users and data aggregators. Other challenges including cost, interoperability, lack of ICT infrastructure are also reviewed. In this paper, it is aimed to provide necessary information to energy managers, academic researchers, energy suppliers, owners, tenants, manufacturers, and technicians. The rest of the article is structured as follows. Section 2 details social, economic environmental and legislative drivers for energy metering and environmental monitoring. This section also highlights some of the legislations that are currently in place in different countries. In Section 3, classification of metering and monitoring devices is presented along with their working principle, types and cost comparison between different widely available meters and sensors. Section 4 details various factors that needs to be considered while selecting a metering and monitoring solution. The following section outlines widely used communication and network technologies along with their advantages and limitations. Section 6 focuses on the directions for future development and challenges including interoperability, security, cost, MEMS sensor technology and lack of ICT infrastructure. Concluding remarks are presented in the end.

DRIVERS FOR ENERGY METERING AND ENVIRONMENTAL MONITORING

There are many existing drivers for the installation of meters and sensors in buildings, but the overriding driver is usually to enable better management of energy within the building and ultimately identifying and achieving potential energy and cost savings and to verify savings that have been achieved [15]. More specific drivers are summarised below. However, there is a considerable overlap between these drivers and often it is difficult to assess which driver takes precedence over others.

Mitigating Climate Change and Enhancing Energy Efficiency

In order to limit the projected global temperature rise to 2 °C, the IEA has proposed four emissions reduction policy solutions, of which the policies related to specific energy efficiency measures have been found to yield in greater savings – about 49% of the total emissions savings identified in the World Energy Outlook 2013 report [7]. In Europe, energy efficiency within the overall energy systems and infrastructure has been recognised as one of the five closely related and mutually reinforcing dimensions of the EU Energy Union agenda, aimed at reducing energy consumption to reduce pollution and preserve domestic energy sources [16]. In recent years, the overarching drive towards greater energy efficiency has resulted in renewed interests in heating and cooling as distinct areas of focus [17], in recognition of their share (50%) in the European Union's final energy consumption [18].

Electric and thermal smart energy grids with distributed and dynamic generation, transmission and distribution of energy are increasingly being seen as essential for the transition towards low carbon energy systems [19,20]. Data-driven control and management are the basis for smart energy grids that relies on the collection, aggregation and analysis of high resolution energy usage data from consumer premises, as well as the grid itself [20]. Advanced metering infrastructure (AMI); i.e. smart meters are the backbone of smart grids enabling wider integration of energy and information and communication technologies (ICT). AMI and associated technologies enable realtime, two-way communication between suppliers and consumers – thus creating more dynamic control and interaction of energy flows. It is expected that all 26 million households in England, Wales and Scotland will be fitted with smart meters by 2020 [21]. On the other hand, energy efficiency measures is a broad term and can include different strategies or initiatives within buildings. Chalmers [22] has discussed mitigation technology options including using daylighting, high-performance building envelope, efficient lighting, building automation and control system, smart meters and grids. Studies in the USA have also suggested that building efficiency measures that rely on sensors and meters, such as building energy management systems, lighting control, automatic fault detection and diagnosis (AFDD) and demand control ventilation, can reduce energy use by 5% up to 60% [23]. Therefore, it is evident that metering and sensing devices will play an important role to mitigate climate change.

Compliance with Regulations and Legislations

Recognising the need for tacking climate change, many governments have introduced legislation aimed at reducing greenhouse gas emissions. Different countries have implemented a range of legislations that are summarised in Table 1; with either having specific energy metering and controls requirements or indirectly encouraging building owners/users for their wider application in buildings. In Europe, most of the legislations have been introduced in response to the EU Directives, specifically the Energy Performance of Buildings Directive (EPBD 2002 [24] and 2010 [25]) and the Energy Efficiency Directive (2012). EPBD 2002 introduced building energy labelling, and EPBD 2010 included specific requirements for metering and controls while the Energy Efficiency Directive also required the provision of smart meters to all customers and the creation of energy audit regimes for businesses.

Table 1. Energy legislation and regulations in different countries

Country	Instrument	Building type	Metering requirements and drivers	Control requirements and drivers
JP	CCREUB (1999)	Commercial	None	Requires *proper controls* for HVAC, lighting, hot water and lifts
EU	EPBD 2002/91/EC (2002)	All new and refurbished buildings	Energy certification encourages the installation of smart meters and sub-meters	None
USA	Energy Policy Act (2005)	All Federal	Meter electricity where practicable, hourly interval data collected daily by 1 October 2012	None
UK	The Companies Act (2006)	GHG emissionsa	Data collection requirements encourage installation of smart meters	None
USA	Energy Independence and Security Act (2007)	All Federal	Provide equivalent metering as EPA 2005 for natural gas and steam	None

UK	The Energy Performance of Buildings (Certificates and Inspections) Regulations (2007)	All (with exceptions)	Data collection requirements for DECs encourage installation of smart meters. Sub-metering may be required in some buildings (e.g., on campuses, multi-use or with excluded energy uses) to calculate accurately DEC rating	Controls will improve energy performance (EPC) asset rating and *in-use* DEC rating
UK	Carbon Reduction Commitment (2007)	Electricity and gas supplies	Reporting requirements plus 10% uplift on use of estimated data encourage roll-out of smart meters to ensure timely and accurate data collection	Control measures that reduce energy use will reduce carbon allowance purchase costs.
USA	International Energy Conservation Code (2009)	Allb	None	Thermostatic control of HVAC systems including zone control. Minimum 7-day time control. Automatic shut-off of supply and exhaust dampers when systems/spaces not in use. Demand control ventilation of spaces greater than 50 m^2
USA	Seattle Energy Code (2009)	New and refurbished (excluding single family homes)	End-use metering of HVAC systems, lighting, plug and miscellaneous loads plus new systems for buildings >1858 m^2. Automatic daily data acquisition and 36 months data storage. Energy displays for building managers	As IECC 2009 plus: Humidistat control for each humidity control system. Optimum start control for systems with design air capacities exceeding 2000 CFM. Speed control of supply fans for AHUs with cooling capacity
JP	Energy Conservation Law (2009)	All	No specific metering/control required but buildings >2000 m^2must submit an energy conservation report to the local authority before any new construction or major alteration, to demonstrate compliance with law and to implement recommended improvements. Local governments offer incentives such as relaxed height and size restrictions for energy efficient buildings to encourage inclusion of greater energy efficiency measures. [31]	

EU	EPBD 2010/31/ EU (2010)	All	*Encourage* the use of intelligent metering systems for new or renovated buildings	*Encourage* the installation of active control systems such as automation, control and monitoring systems that aim to save energy
UK	Building Regulations Part L2A and L2B (2010)	All	Assign end-use categories (heating, lighting, small power, etc.) to 90% of building energy use. Separately monitor the output of any renewable energy systems. Include automatic meter reading &data collection capability in buildings over 1000 m^2	Separate zone control for areas with different solar exposure, occupancy or use type. Prevent simultaneous heating &cooling. Demand led control of central plant (default to off)
EU	Energy Efficiency Directive 2012/27/EU (2012)	General EEc	Free and simple customer access to real-time and historical energy consumption data through more accurate individual metering. Introduce energy audit obligations for large energy organisations including annual energy reporting	Set and achieve energy efficiency targets. Renovate 3% of central government buildings and procure energy efficient buildings, products and services
UK	Energy Saving Opportunities Scheme (2013)	Private organisations	No regulatory requirement but the need for reporting encourages the roll-out of smart meters to ensure timely and accurate data collection	Audit reports likely to include controls recommendations
UK	UK Smart Metering Implementation Plan (2014)	Small non-domestic &domestic	Roll-out of *smart meters*on smaller non-domestic and domestic electricity supplies by 2020 including wireless communications ability and real-time data collection and visualisation	None

AUS	Building Energy Efficiency Disclosure Act (2014)	Commercial	No specific requirements but sub-metering enable exclusion of tenant energy use from *Base Building* energy rating on Building Energy Efficiency Certificate that is required when office space is leased.	No specific requirements but reductions in energy use from good controls will improve the *Base Building* energy rating
AUS	National Construction Code vol. 1 (2014)	Commercial and multi-occupancy dwellings	Individual gas and electricity metering for all buildings with a floor area exceeding 500 m^2(includes campus site). Sub-metering of main energy end-uses in buildings >2500 m^2	Timer control of main HVAC plant. Independent thermostatic control of zones with differing thermal characteristics and occupancy requirements. Occupancy linked control of *room conditioners*. Provision of an *outdoor air economy cycle* in air conditioning systems. Use of variable speed fans when variable air supply rates are required

a GHG emissions.

b Adopted in 46 US States, District of Columbia, NYC, Puerto Rico and US Virgin Islands.

c General Energy Efficiency.

In the UK, sub-metering and control requirements are included in the Building Regulations Part L while building energy labelling is covered by the Energy Performance of Building Regulations 2007. The UK Smart Metering Implementation Plan covers the roll-out of smart gas and electricity meters to all homes and businesses by the end of 2020. The Energy Saving Opportunities Scheme (ESOS) will require large private sector enterprises to conduct energy audits on a sample of their buildings by 5th December 2015 and every four years after that [26]. In 2010, the UK government also introduced the Carbon Reduction Commitment (CRC) energy efficiency scheme independently of any EU directives. This was originally intended as a carbon trading scheme for large and non-intensive energy users and was not included in the EU Emissions Trading Scheme. However, now it is a simpler carbon allowance purchase scheme where participants must purchase and surrender carbon allowances annually to cover carbon emissions from their electricity and gas supplies. As detailed in Table 1, the reporting regime does encourage the wider use of automatic metering to aid data collection and implementation of energy saving strategies to reduce carbon emissions.

In the USA, the Energy Policy Act 2005 and the Energy Independence and Security Act 2007 introduced metering requirements for federal buildings. Most of US states, cities and districts have adopted the International Code Council's International Energy Conservation Code (IECC) as their basis for building energy regulations and have also introduced enhanced requirements. These various legislations are summarised in Table 1 including one example of enhanced regulations in the USA. The Australian Building Energy Efficiency Disclosures Act allows commercial landlords to exclude tenant energy use if separately metered, to produce a "Base Building" energy rating. This act encourages sub-metering of tenant energy consumption, as the improved energy rating can be a useful marketing tool. Like UK and USA, Australian building regulations also have very stringent and specific requirements for sub-metering and controls. In Japan, legislations and regulations on building energy use is less prescriptive with regards to sub-metering and controls and only relies on voluntary compliance in many cases. However, compliance rates are quite high, and the requirement to provide an energy conservation report demonstrates that compliance with the Energy Conservation Law encourages the use of controls and also to have some degree of sub-metering. With regards to control, in the Japanese Commercial Building Code it is only stated that *a proper control system* should be adopted for controlling a range of energy supplying plant. Table 1 lists more details on different regulations.

Feedback on Energy Consumption and Cost Reduction

Researchers have done an extensive amount of research on providing feedback to users on their energy consumption and energy savings that can be achieved through feedback and it was concluded that feedback can reduce energy consumption. Darby [27] has reviewed three types of feedback for residential sector i.e. direct feedback (providing real-time meter readings), indirect feedback (billing) and inadvertent feedback (a by-product of technical, household or social change). It was found that energy savings from direct and indirect feedback were in the range of

5–15% and 0–10% respectively. Therefore, metering solutions are better options for providing feedback to occupants. We acknowledge that most of the studies conducted on providing feedback and its effect on energy savings are performed on residential buildings, and fewer studies were carried out on non-domestic buildings. Also, the impact of occupants behaviour on energy consumption in non-domestic buildings in less when compared to claimed 20% reduction in energy use [28]. The reasons for this variation is actual and simulated energy savings in non-domestic buildings as reviewed in Cox et al. [29]. The evaluation of behavioural initiative in non-domestic buildings is poorly understood because of the variety of organisational processes and lack of sub-metering. Employers can also provide feedback to their employees by displaying energy metered data on a plasma screen in a communal area inside their building. A UK trial of advanced energy meters in small and medium-sized businesses (SMEs) between 2004 and 2006 demonstrated that SMEs can identify an average of 12% carbon savings by using advanced metering [30]. The report concluded that low-cost metering services provided by automated systems could be effective in reducing energy use and if implemented on UK level can potentially save £ 650 million [30]. Shaikh et al. [1] have also mentioned that sensor based intelligent building automation systems could deliver average 30% reduction in building energy use worldwide.

Corporate Social Responsibility

Many companies promote their actions on energy efficiency through corporate social responsibility (CSR). The UK retailer Marks and Spencer's *Plan A* is a good example and includes concrete measures such as the use of automatic meters to validate energy efficiency savings for its UK operations, requirements for suppliers to report energy data and achieve savings and the introduction of an Eco-Factory standard for vendors. This standard includes requirements for sub-metering major energy end-uses and report them daily and to introduce controls for heating and cooling systems and electrical motors. 53 of Marks and Spencer's top 100 suppliers had achieved this rating by the end of 2013, and 85 had adopted temperature control measures [32,33]. Similarly, the international bank Santander gives prominence to energy metering and control measures in their CSR reports. This includes the roll-out of smart electricity meters and gas loggers to track energy performance and installation of controls for HVAC and lighting [34]. Therefore, it can be concluded that policies on a government level as well as corporation level are currently in place and are currently being implemented.

CLASSIFICATION OF METERING AND MONITORING DEVICES

Fugate et al. [35] proposed three categories for building sensors and meters for measuring and sensing different building performance parameters, occupants comfort perception and machinery characteristics. Temperature, occupancy, humidity, $CO_2$and air quality sensors are used to sense occupant comfort and activity. On the other hand; building energy meter, sub-metering, plug-load measurements, natural gas meters, and other sources of energy meters are used to measure energy consumption. Measurements of machinery characteristics can be done by using refrigerant temperature, machinery electrical current, machinery vibration, and return and supply airflows sensors. The focus of this article will only be on devices used for the measurement of primary energy (electricity and gas), thermal comfort, indoor air quality (IAQ) and lighting and occupancy control.

According to Fanger [36], the evaluation of the thermal environment requires the knowledge of six quantities: four physical parameters (air humidity, air temperature, mean radiant temperature and air velocity) and two subjective (clothing thermal insulation and metabolic rate). These parameters have to be measured or estimated to evaluate and monitor thermal comfort [37]. While subjective quantities are generally estimated according to benchmarks for specific activities and types of buildings, physical ones can be directly or indirectly measured [38]. The sensors used to evaluate thermal comfort, sensors used for daylighting control or occupancy detection serves different purposes. These sensors are installed in building to make sure that occupants have comfortable indoor air conditions, minimise energy consumption during unoccupied periods, provide detailed building energy consumption information, provide input for security, fire alarm system, etc.

In the following sections, each parameter has been characterised on the basis of the main sensing technologies used for metering and monitoring purposes, with a description of their operative principles, advantages, and disadvantages. Then for each parameter the main commercial products exploiting the above sensing technologies have been characterised on the basis of their main meteorological parameters (such as accuracy and operational range), usage and cost. The sensing and metering considered in this article are commercially available electronic devices that measure and record data for further analysis and not only individual the sensing components.

Electricity Metering

Electricity meters operate by continuously sensing the instantaneous values of current and voltage to provide a measurement of energy used (and possibly power demand). The oldest and most common type of electricity meter is the electromechanical watt-hour meter (a 2-phase inductive motor). It is composed of an electrically conductive metal disc whose rotational speed is proportional to the power passing through the meter. The disc is influenced by two sets of coils properly connected to generate magnetic fluxes proportional to the load current and the line voltage respectively. This produces eddy currents in the disc that exert a force in proportion to the active power. Contemporary a permanent magnet produces an opposing force proportional to the rotational speed of the disc. The interaction between the two forces results in the disc rotating at a speed that is proportional to either the power or the rate of energy usage [39]. In the first mechanical meters the energy data had to be acquired using direct dial readings; modern ones, instead, can rely on encoders and other registering mechanisms to produce electronic or pulse output [40]. Although electromechanical meters represent the majority of the currently installed meters in both residential and commercial buildings, the future of the utility industry is represented by electronic (solid-state/digital) advanced meters, which do not require moving parts and are capable of storing and managing data. The most conventional current and voltage sensor circuits are typically current and voltage transformers. Due to their higher accuracy, data storage and communication capability and the consequent possibility to manage energy and power data, electronic meters are gradually replacing the old electromechanical models. Different types of current sensors consist of resistive shunt and Hall-effect circuits [41]. Furthermore, advanced meters have led in recent years to the development of automatic meter reading (AMR) solution and services [42]. Next generation meters (called smart meters) will make full use of AMR's features, enabling the creation of automatic meter infrastructure (AMI) with bidirectional communication between the meter (consumer's side) and the utility provider [14,43].

The accuracy of these devices depends on their application and the metering class they belong to (as will be discussed in coming sections). Typical accuracy values for meters in Europe are in the range of between ±0.5% to ±2%. The price of these devices is mainly dependent on their communication protocol, ease of deployment and AMR compatibilities and range between $300–700.

Gas Metering

Gas consumption must be adequately measured to comply with European and local codes. Several types of gas meters are currently present on the market; though diaphragm, rotary and turbine ones are the most commonly installed. The diaphragm gas meter is divided into two or more chambers formed by movable diaphragms. The chambers alternately receive and expel gas, expanding and contracting the diaphragms; the linear motion of the diaphragms is converted into rotary motion using cranks and the gas-flow information is thus transferred to a counter mechanism. Rotary meters are used when high gas-flow levels and higher accuracy are required. This type of meter consists of two rotors that spin in precise alignment and let a specific quantity of gas pass at each complete turn. The gas flow information correspondent to each rotation is then recorded using electrical pulses or mechanical counters. In turbine gas meters a rotor with helical vanes freely rotates with low friction around an axis parallel to the flow and thus to measure the gas flow speed. Volumetric flow and speed are linked by the continuity equation and are characterized by linear proportion once that the internal meter area is known. The information is then transmitted electronically or mechanically to a counter. Turbine gas meters require the non-turbulent flow condition to work properly [44].

All of the three types of meters described above use dynamic mechanisms to measure the gas flow. Although dynamic meters (and, in particular, diaphragm meters) represent the near totality of flow meters used by distribution companies. However, their construction simplicity and low cost is often accompanied by wear problems, high-pressure losses and inability to indicate instantaneous flow rate value [45]. In recent years, many static methods for the measurement of gas flow have gained popularity both in the commercial and research fields. Among the static devices, ultrasonic flow meters have shown significant results regarding accuracy and non-intrusiveness [45]. They typically use narrowband piezoelectric transducer arrangements to interrogate the flow of gas in a pipe [46] and use variation to velocities of the sound pulse in the acoustic path between transducers to measure the gas flow [47]. Other static gas flow meters are the fluidic meters that are based on fluidic dynamic oscillation principle [48] and those that use the Coriolis effect [49].

Table 2. Summary of gas metering technologies

Type	Use	Measuring range	Output	AMR compatible	Approx. price (US$)
Diaphragm	Residential	up to 7 m^3/h	Odometer counter	✓	125–185
			Pulse		150–265
	Commercial	11.33–28.32 m^3/h	Odometer counter	✓	400–550
			Pulse		500–650
Fluidic	Commercial	up to 28.32 m^3/h	Pulse	✓	2000
		up to 84.95 m^3/h	Pulse	✓	2600

Like electricity meters, gas metering has also experienced a broad development of smart models and functionalities in recent years. Although the measurement principles of the meters do not change, smart meters contain electronic devices that convert the pulse information into an energy consumption reading and then storing the reading or send it to a server, enabling forecast and energy consultancy services [50]. Due to a large number of traditional meters that are already installed, many manufacturers have also developed AMR solution to be directly connected to the pulse output of traditional meters. In this way, the conventional meter can be upgraded to a smart meter in a simple, quick and non-invasive way. An example of such upgraded smart meters was developed by Smart Metering Systems Plc. [51], that can be connected to a gas meter in a non-invasive way to provide half-hourly consumption reading using GPRS at communication medium. Table 2 summarises different types of gas meters currently widely available in the market for both domestic and non-domestic sectors. The cost of gas meters depends mainly on the measuring technology, pulse output option, AMR compatibility and measuring range. Generally, static methods-based meters are more expensive than dynamic ones.

Air Temperature

Air temperature can be defined as "the temperature of the air surrounding the human body, which is representative of the aspect of the surroundings that determines the heat flow between the body and air" [37]. The range of techniques used for the measurement of temperature are wide and can rely on many phenomena, such as thermoelectricity, fluorescence and spectral characteristics, a temperature-dependent variation of the resistance of electrical conductors, thermal expansion, etc. [52].

For the measurement of air temperature, however, most of the sensors are based on mechanical and electrical operative principles. Mechanical devices are in general cheap and reliable but are usually used for direct control purposes and/or visualisation rather than for metering and monitoring. The most common example is the domestic room thermostat, where the output is the temperature-dependent physical displacement of a bimetal to operate directly a switch [53]. Another example is the mercury-in-glass or liquid-filled thermometer, where the temperature information can be only visualised on a graduated scale [37]. Electrical sensors, instead, have the capability to convert the measured value into a digitally encoded signal for direct communication for control and measurement purposes. Electrical devices are mainly based on thermoelectric and electrical resistance principles. Thermocouples are the most widespread example of thermoelectric devices and are commonly used because of their low cost, simplicity, robustness, temperature range and size and because there is no requirement of the power supply. As a matter of fact, their functioning is based on the See-beck effect, that is the production of an electromotive force in a circuit of two dissimilar conductors experiencing a thermal gradient [52]. Resistance temperature devices (RTD), instead, use the dependency between temperature and electrical resistivity of a conductor. Although any conductor could, in theory, be used as an RTD, practical considerations of cost, reliability and manufacturing constraints limit the choice to copper, gold, nickel, platinum and few other semiconductors [52]. Examples of RTDs used in the building sector are platinum resistance thermometer (PRT) and thermistor. The former can rely on great accuracy, reliability and stability, but has the disadvantage of the high cost and the very high time response. The latter, which consists of a semiconductor (usually a mixture of oxides of nickel, manganese, iron, copper and others [52], whose resistance is sensitive to temperature, is cheaper and has good accuracy, but has problems of drift and requires proper calibration [53].

Approximate range of values of sensing parameters, cost and relevant information of air temperature sensors are reported in Table 3. Prices of different sensors are taken from datasheets and user manuals of some of the main manufacturers. It is evident that thermocouples are cheap and can measure a wide range of temperatures, but have low accuracy (system errors of less than 1 °C can be difficult to achieve). PRTs have higher accuracy (though it would require a 4-wire connection) and stability than thermocouples, but longer response time and cost are the drawbacks of these types of sensors. Thermistors are the most accurate and sensitive type of sensors for temperature reading. They have shorter response time as compared to RTDs but nearly have same response time when compared with thermocouples.

Table 3. Summary of air temperature, air humidity, mean radiant temperature, air velocity, CO_2, CO and VOC concentrations sensors

Measured parameter	Type of sensor	Accuracy	Response time	Comm tech.a	Use	Approximate price (US$)	Measuring range
Room temperature	Thermistor	±0.1–0.5 °C	10–30 s	Wired Wireless	Direct visualisation BMS HVAC	20–70	−50 to 180 °C
	Thermocouple	±0.8–4 °C	10–80 s	Wired Portable	BMS HVAC	10–50	−100 to 300 °C
	PRT	±0.25–0.6 °C	3–8 min	Wired Wireless	Direct visualisation BMS HVAC	35–100	−50 to 100 °C
	Capillary thermostat	–	5–12 min	Closed control	HVAC	20–60	−35 to 65 °C

Relative humidity	Capacitive polymer	±2–4.5% RH	10–50 s	Wired Wireless	Direct visualisation BMS HVAC	50–200	0–100% RH
	Ceramic resistance	±2–5% RH	10–50 s	Wired Wireless	Direct visualisation BMS HVAC	40–150	10–90% RH
Mean radiant temperature	Globe-thermometer	±1–3 °C	8–30 minb	Wired Portable	Thermal comfort tests	40–150	20–120 °C
Indoor air velocity	Hotwire	±2–5% of reading	0.2–5 s	Wired Wireless	HVAC Airflow control Thermal comfort test	50–180	0.05–20 m/s
CO_2 concentration	NDIRc	±30–80 ppm	30–50 s	Wired Wireless	Air-flow control	200–500	0–2000 ppm
CO and VOC concentration	MOSFETd	±50–100 ppm	<60 s	Wired Wireless Portable	Air-flow control	200–500	400–2000 ppm CO_2 eq.

a Communication Technology.

b Adjustment time.

c Non-dispersive infrared.

d Metal–oxide–semiconductor field–effect transistor.

Mean Radiant Temperature

The mean radiant temperature is defined by ISO 7726 as "the uniform temperature of an imaginary enclosure in which radiant heat transfer from the human body is equal to the radiant heat transfer in the actual non-uniform enclosure" [54]. It also suggested by ISO 7726 [54] that it can be calculated from the temperature of surrounding surfaces and their orientation with respect to the body (subject) [37]. However, besides calculation methods, ISO 7726 proposes a comprehensive series of methodologies and instruments for the indirect measurement of the mean radiant temperature and globe thermometer is widely used among them. This type of instruments has low cost and high traceability but also have high response times (that limit real-time measures) and overestimation of radiative contribution related to horizontal surfaces (floors and ceilings). Other usable instruments are contact and radiation thermometers and unlike globe thermometers both these devices can evaluate the asymmetry of the radiant temperature [54] within a room. However, to provide a reliable measurement, contact thermometers need significant attention regarding contact resistance evaluation, while radiometers require the knowledge of the emissivity of the surface to be measured. Finally, the most accurate sensors are the net radiometers, but their usage in the building is usually limited due to their high cost and the need for multiple sensors [55]. Most of the available devices in the market used to measure radiant temperature are globe thermometers. These types of probes for the measurement of mean radiant temperature are part of expensive portable instruments that are capable of the measurement of all the parameters that define thermal comfort.

Indoor Air Velocity

Air movement around the subject (occupants) can increase/decrease heat flow from and to the body and hence body temperature [37]. Indoor air velocity plays a significant role in the definition of thermal comfort conditions [36,56]. Indoor air velocity can be considered as "the mean air velocity intensity over an exposure time of interest and integrated over all directions" [37]. This means that any measuring device must be able to measure the effective velocity whatever the direction of the air is.

A commonly used principle is to calculate the air velocity from the cooling (or heating) produced by air moving across the instrument. The kata thermometer [57], for example, is an alcohol thermometer with a large bulb that is heated and exposed to the environment. Although these thermometers provide average practical values of air velocity, they are inadequate for significant fluctuations and cannot be used for recording and computer analysis purposes [37]. Another type of sensors, which also belongs to the same category and are widely used in the building sector, are hot wire anemometers [58]. These devices measure the cooling capacity of air moving across a 'hot' wire and relate this to the air velocity using the value of air temperature within the environment [59]. Compared to kata thermometer, hot wire anemometers measure fluctuations in the air temperature and measurements can be easily recorded for later analysis. They are, however, directional in response and can be inaccurate in low air velocities due to the natural convection of the hot wire [37]. An improvement of the hot wire anemometer is the low-speed hot sphere anemometer. It consists of two sphere/sensors: an electrically heated sensor that registers the air velocity and a cold sensor that measures the air temperature to allow for the correction of the heated sensor temperature. With respect to the hot wire anemometer, the hot sphere sensor is omnidirectional and, therefore, it is the most practical device that can be used for the measurement of indoor air velocity [60].

Air flow can be also measured by using more advanced (and thus expensive) sensors with directional sensitivity, like ultrasonic and laser Doppler anemometers. The former consists of 2, 4 or 6 prongs (about 100 mm apart) that emit a sonic signal in 1, 2 or 3 directions respectively. The velocity of the air in one direction influences the time of flight of pulsed sound waves between the corresponded prongs. As a consequence, pulsing two ultrasonic signals in opposite direction between each set of prongs and comparing the time of flights of the out signal and the back signal, it is possible to calculate easily the air velocity [61]. In the latter, the velocity measurement is measured by focusing a laser beam onto the measuring point and recording the frequency shift of the light scattered by moving particles [61]. These instruments are mostly used to room air velocity measurement both in laboratories and in field applications [62], their high cost, size and sensitiveness have limited their commercial development. There are other sensors that are based on principles involving pressure measurements (like Pitot tubes) or rotating element immersed in the air-flow, but sensors based on these principles are suitable for air speed measurement only when flow direction is determined, such as in air ducts. Their use for indoor air velocity measurement is therefore very limited.

Relative Humidity

Humidity is defined as the water vapour content in the air or other gases. It is generally measured in terms of absolute humidity, dew point or relative humidity (RH) [63], but all three measures can be calculated from each other if the air temperature is known. One of the oldest used instrument for determining relative humidity is the whirling hygrometer (or sling psychrometer). Whirling hygrometer consists of two thermometers that are turned by vigorously swinging the handle and exposing the thermometers to rapid air movement. The bulb of one thermometer is placed in direct contact with the room air to measure dry-bulb temperature. The bulb of the other thermometer is covered with a silk or muslin sleeve that is kept moist to record the wet bulb temperature. The atmospheric humidity is determined by calculations, steam tables, or using a psychrometric chart (considering the difference in reading between the two thermometers) and, therefore, the use of these devices for monitoring purposes is very limited [37].

Modern electronic devices use change in electrical capacitance or resistance to measure humidity differences. Capacitive type sensors consist of a substrate (typically glass, ceramic, or silicon) on which a thin film of polymer or metal oxide is deposited between two conductive electrodes. The incremental change in the dielectric constant of the sensor is nearly directly proportional to the relative humidity of the surrounding environment. Resistive sensors usually consist of noble metal electrodes either deposited on a substrate by photoresist techniques or wire-wound electrodes on a plastic or glass cylinder. Resistive humidity sensors measure the change in electrical impedance of a hygroscopic medium such as a conductive polymer, salt or treated substrate [59]. The materials used in humidity sensors that exploit variations of electrical parameters can be classified into electrolytes, organic polymers and ceramics [64]. Commercially developed humidity sensors are mainly made of polymer films and porous ceramic [65]. Thick film technology, in particular, offers advantages of cost efficiency, robustness and flexibility in device design [66]. Thermal conductivity sensors (also known as absolute humidity sensors) are mostly used to measure absolute humidity even at high temperatures or in polluted environments by means of a system that employs two thermistors in a bridge connection [59].

Indoor Air Quality (IAQ)

According to Fanger [67], there are three basic requirements for human occupancy: temperature and relative humidity acceptability, maintenance of normal concentrations of pollutants gasses (mainly carbon dioxide (CO_2) and volatile organic compounds (VOCs) and secondly carbon monoxide (CO) and others) and distribution of adequately ventilated air. Internal air quality (IAQ) control aims to fulfil all these three requirements. As discussed above in this paper, temperature and humidity sensor technologies and measurement techniques have been widely developed and deployed in the industrial and civil sectors in recent decades. In contrast, carbon dioxide (CO_2), carbon monoxide (CO) and air pollutants control, though thoroughly studied in terms of concentration acceptability and related air-change rates for the building sector [68], has shown a massive technological development only in recent years.

According to the measurement principle, CO_2 sensors can be mainly classified into two categories: the first type uses chemical methods, which can ensure high accuracy, but lack of stability and durability due to heterogeneous gases poisoning effects; the second uses a non-dispersive infrared (NDIR) method based on the physical principle of gas-absorption at a particular wavelength. NDIR-based sensors have higher energy consumption but still are the most widely applied for the real-time monitoring, due to their higher durability, reliability and accuracy [69]. CO_2 sensor information can be directly used in building automation to increase the natural or mechanical airflow, but many studies have also suggested to use visual *traffic-light*-type sensors to inform the occupants of the increasing CO_2 concentration in a room [70]. CO_2 concentration can also be used to estimate occupancy within a building and this information can be used to implement a more efficient control of the HVAC system [71].

CO and VOCs detection in building, instead, is mainly achieved by using the physical principle of gas-adsorption on semiconductors. Simple metal oxides such as SnO_2, WO_3 and TiO_2 are well known for their high sensitivity to changes in the air composition [72]. Their operation relies on the establishment of adsorption equilibrium between the molecules of the measured gas and the grains of semiconductor, resulting in a change of electrical conductivity of the sensor›s material [73]. Metal oxide sensors show high and stable sensitivity, but they also require an electric heater to keep the sensing element to approx. 300 °C, resulting in high electric consumptions. This aspect and the very short life of these devices have led to the recent development of new gas sensors based on electrochemical methods. Their operative principle does not different from a fuel cell, which in this case is designed to produce a current output proportional to the concentration of the particular pollutant in the air. Due to their very low power consumptions, quick response time, high sensitivity and longer life, electrochemical gas sensors have been increasingly used for CO detection [74]. The same method can also be used for VOCs monitoring, but provides inaccurate results when sensing mixtures of chemicals; its exploitation is therefore generally limited to the industrial sector for the monitoring of individual pollutants.

Table 3 reports examples CO_2, CO and VOCs concentration measurement devices currently available in the market. Technical data is taken from the datasheets of different manufacturers while costs have been deduced from information found on the Internet.

Occupancy and Daylight Sensors

Lighting energy consumption can be significantly reduced by switching off or dimming lamps based on factors like occupancy and availability of daylight [75]. Technologies that perform this type of lighting control differ in their input parameters, their control method, control algorithm, cost of installation, complexity of commissioning, etc. [76]. In below sections, the focus of the discussion will be on different occupancy and lighting sensors that can be used in building for controlling lighting systems.

Occupancy Sensors

Occupancy sensors employ motion sensing techniques to detect the presence of occupants in a given range of space; the lights are turned on when the sensor detects any occupant while are turned off when there is no occupant within a pre-fixed delay period [76]. Although many studies have demonstrated the importance of occupancy information for whole building climate control [77], occupancy sensors in buildings are currently mainly used for lighting control purposes.

Among the available sensing systems passive infrared (PIR) detection systems and ultrasonic sensors are the most common. PIR sensors rely on detecting a change in the temperature pattern in the observed space using a pyroelectric detector [78]. The sensitivity of these type of sensors depends on the distance of the subject from the sensor, and this can lead to *False-off* errors when movements of occupants very far from the sensor are not accurately detected [79]. Ultrasonic sensors work based on the principle of Doppler effect; emitting ultrasonic waves and comparing them to reflected signals. Ultrasonic waves are reflected by room surfaces, and, therefore, these sensors do not require a fixed field of vision like the PIR sensors. As a consequence, ultrasonic sensors have been proven to be more reliable in detecting presence over long distances [80], though movements coming from activities in the room other that occupants can lead to *False-on* errors [76]. Some other types of sensors that are also gaining popularity include sensors based on radio frequency identification (RFID) and imaging techniques such as camera and body recognition to detect occupancy [76]. RFID sensors are mostly used in offices for personnel identification [76].

Daylight Sensors

Daylight-linked lighting control can be used either to switch lights ON or OFF when daylight level inside the occupied space is above or below a threshold value or in combination with dimmable electronic ballasts to adjust the artificial lighting level according to the daylight contribution [81]. Daylight-linked lighting control can be divided into *Switching systems* and *Dimming systems*, according to the way the lighting system is controlled [76]. A daylight responsive system is composed of a sensitive element, such as a photosensor, a lighting controller and an electronic dimming ballast [82].

Table 4 lists different types of occupancy and lighting sensors. Data comes directly from datasheets of the various manufacturers while cost is collected from information found on the web. It can be clearly seen that price mainly depends on the type of the system – Standalone system with direct load control are cheaper. Price also depends on the accuracy of different sensors and type of communication medium used. Sensors which offer wireless functionality are more expensive as compared to wired connected sensors. In the case of occupancy sensors, Ultrasonic sensors are more expensive as compared to PIR sensors. For lighting sensors, photosensors are more expensive than photodiodes.

Table 4. Summary of daylight and occupancy sensors

Parameter	Type of sensor	Accuracy/ range	Response time	Communication	Approx. price (US$)	Dimmer compat-ibility
Occupancy	PIR	3–5 m radius or 5–12 m front and 3–8 m lateral	10 s–15 min	Wireless Wired Standalone[a]	25–80	N/A
	Ultrasonic	$185\ m^2$	30 s–30 min	Wireless Wired Standalone[a]	150–300	N/A

Light	Photo-diode	±8–10% of illum.	N/A	Wireless Wired Standalone[a]	50–150	Yes
	Photo-sensor	±8–10% of illum.	N/A	Wireless Wired Standalone[a]	100–250	Yes
Daylight	Photo-diode	±8–10% of illum.	N/A	Wireless Wired Standalone[a]	50–120	Yes
	Photo-sensor	±8–10% of illum.	N/A	Wireless Wired Standalone[a]	100–250	Yes

a Standalone sensors are directly connected to an actuator to switch on/off or dim artificial lighting.

SELECTION OF SENSING SOLUTIONS

In this section, different factors that influence the selection of sensing and metering solutions will be discussed. The key factors are listed below:

- Accuracy
- Ease of deployment
- Communication protocol
- Granularity
- Cost
- Availability

Accuracy

In Europe, there are different bodies to govern meter accuracy and one of these is the Measuring Instrument Directive (MID) (2004/22/EC) [83]. MID is a European Directive to harmonize the requirements of different measurement instrument types (including electricity and gas meters). In the UK, only MID approved electricity and gas meters can be used for both supplier–consumer billing and tenant billing [84]. By reviewing available sub-meters, it is found that although none of them are MID approved however they do comply with other standards. MID approved meters also comply with those standards as well. Some example of the meters' accuracy standards for kWh are EN 62052-11 [85], EN 62053-

21 [86], EN 62053-23 [87], BS 8431 [88], IEC 1036 [89], IEC 687 [90], ANSI C12.20 [91]. One of the common feature of most of these standards is to categorise meters into Classes representing their percentage accuracy e.g., Class 0.5 = ±0.5% accuracy, Class 1.0 = ±1% accuracy and Class 2.0 = ±2% accuracy. In Table 5, most widely available fiscal and non-fiscal meters are categorised according to their classes and whether they are MID approved or not. It is clear shown that some of Class 0.5 and Class 1.0 meters are MID approved. Class 1.0 meters are most suitable for sub-metering purposes, and Class 2.0 meters' accuracy is considered as the minimum standards for sub-metering.

Table 5. Accuracy classes of some common fiscal and non-fiscal electricity meters.[a]

Manufacturer	MID approved	Class 0.5	Class 1.0	Class 2.0	Connectivity					
		±0.5%	±1.0%	±2.0%	Ethernet	WiFi	ModBus	M-Bus	GPRS	PLC
									GSM	
Fiscal meters										
Elster A1120/40	•				•		•	•	•	
Elster A1170	•		•		•		•	•		
Elster AS230	•		•		•	•	•	•	•	
EDMI MK10A	•		•			•	•	•	•	•
Landis &Gyr ZMD405	•	•			•	•	•	•	•	
Landis &Gyr E650	•	•			•	•	•	•	•	
Non-fiscal meters										
ND Meters Cube/Rail 350 IP			•		•		•	•		
Schneider Electric PM750		•								
Schneider Electric PM850		•			•		•			
Selec MFM 384			•		•		•			
Carlo Gavazzi EM21-72R				•	•		•			
Carlo Gavazzi EM24	•		•		•		•	•		

Carlo Gavazzi WM4-96			•		•		•		•	
Hager EC310				•						
Hager EC320				•						

a This table is based on most widely available fiscal and non-fiscal meters in the UK.

For electricity meters, it is also necessary to consider the accuracy of current transformers (CTs), as there are most commonly used current sensors in high current solid state electricity meters [92]. CTs are used in fiscal metering when a high operating current (typically above 100 A) is applied directly to a meter [93]. CTs are also widely used in retrofit electricity sub-metering installations as they allow sub-meters to be fitted safely with minimal interruption of power provision. Their use in fiscal meters is evident of their acceptability as an accurate measurement instrument.

Ease of Deployment

Deployment of metering and sensing solution can offer different installation and networking challenges that must need to be tackled. Wireless sensors reduce the installation cost and also provide greater flexibility. A study conducted by [94] showed that wireless sensors have more advantages rather than just replacing wires. Milenkovic and Amft [95] have demonstrated that it is feasible to use existing BEMS sensors and controls to improve energy efficiency and reduce installation requirements. Their study also demonstrated the effectiveness of self-powered sensors (incorporating a battery and small solar panel) with wireless communications ability. On the other hand, the deployment of energy meters offers less flexibility as these needs to be connected to the energy distribution system. In general practice, sub-metering of electrical services is carried out by using either a panel or rail meters installed in electricity distribution board using CTs to minimise cost, have safety and ease of deployment. Installation of mechanical energy meter for gas, oil and heat metering also offer less flexibility as these need to be installed on the relevant pipe that serves devices, area or building. General practice is to install these mechanical meters based on finding a suitable location, minimising disruption to supply and compliance with safety regulations. These restrictions are to some degree offset by the fact that typically far fewer devices are served by the utilities monitored by mechanical meters so limiting the number of sub-meters is required. Different factors that should be considered while deploying sensors and meters in building are listed below:

- New sensors and meters need to be interoperable with the existing installation in the building including BEMS, IT network, communication protocol and transport layers.
- The wireless capability of any new sensors should be examined to make sure that they will satisfy the coverage limitation.
- Meters location should be easily accessed for the installers.
- Making sure that there is minimal disruption in the supply.
- Safety legislations and regulations regarding metering installation should be complied.

Communication Protocol

A wide range of communication and network technologies are discussed in Section 5to transfer sensor and metering data. Table 5 listed fiscal and non-fiscal meters along with the available communication protocol those meters offer. Fiscal meters offer the widest range of network and communication protocol options but are not ideal for multiple sub-metering installations. ModBus and Ethernet are commonly available for sub-meters and offer the most attractive option due to ease of integration with existing communication networks and the potential to collect other energy data. In summary, commercially available meters do offer a wide choice of communications and network compatibility that gives flexibility for integration with most existing systems.

Granularity

Granularity or resolution of measurement determines the possible level of analysis that can be performed on the metered or sensed data. It can be considered as a function of equipment's resolution, data collection interval and the number of equipment connected to a data collection device or system. Resolution of equipment has an apparent impact on granularity as it determines the minimum level of data that can be collected. Pulsed outputs on many of the electricity meters can be scalable and configured by the user. However, a pulsed output is only generated when a threshold value is exceeded, for example if a device is scaled at 1 pulse per kWh there will be no information available on consumption between 0 and 999 Wh. Typical data collection intervals in non-domestic buildings for fiscal metering and data logging in BEMS are 30 min and 15 min; and are considered as sufficient for building level collection and lighting, HVAC motors, and pumps due to infrequent changes in power modes. For devices with more frequent changes in power states or with very low levels of energy use (e.g., computers, screens, and printers) more frequent sample rates may be required. A study by Lanzisera et al. [96] concluded that date sampling at 1 min provided sufficient accuracy to determine power states for electrical loads in an office building. However, this study was performed for a limited period (6–16 months) and concluded that two months was sufficient to generate representative data for the particular study. Data resolution as frequent as 1 min is useful for some specific purposes but it may not be required for monitoring over extended periods. Also, data resolution at this level is very rarely available in existing commercial meter and sensor systems.

Granularity is also affected by the network structure adopted for the data collection system and specifically the number of sensor devices connected to a data collection device or system due to the time required to communicate with each sensor and retrieval of data from the sensing point. This time can be approx. up to 30 s per sensor. Therefore, in a data collection network with a large numbers of sensors the time delay between retrieving data from the first and last sensor can be few minutes. This issue can be addressed by a combination of increasing the number of installed data collection devices and also by installing sensors with built-in storage so that data collectors can recover historical data as opposed to live data. This solution has a drawback of increased cost and, therefore, there should be a trade-off between granularity requirements and installation costs.

Cost

Equipment cost is identified as a primary driver in deciding on investing in metering and sensing equipment; operating cost is more important to those stakeholders who are choosing to integrate the system, on the other hand, initial cost reduction is more beneficial to those who do not want to integrate the system [23]. Cost perceived as a financial risk factor by building owners or design teams. The data to enable a clearer understanding of the balance between the cost of control systems and the savings those systems might achieve is often lacking. As a result, the number of sensors and meters installed in existing buildings is usually limited to the minimum to provide adequate control of HVAC systems and ensure compliance with any relevant legislations.

Availability

The availability of a particular manufacturer's meters and/or sensor will vary geographically and from a practical point of view it is necessary to have reasonable lead in times for delivery and provision of technical support and spare parts. Product availability makes one of the most critical factors in deciding which devices to deploy, and it is a factor that can only be assessed at the time of deployment. As a result, similar monitoring and sensing requirements are met by a range of devices in different countries depending on the geographical availability of products.

COMMUNICATION AND NETWORK TECHNOLOGIES

Communication and network technologies are mainly divided into two categories, i.e., wired and wireless technologies. These technologies can be used for data transmission between utility meters and utility providers, building energy management system and indoor environmental monitoring sensors, building users and smart appliances, etc. Both of these technologies (wired and wireless) have their advantages and disadvantages e.g., wireless technologies are low-cost solutions as compared to the wired technologies, and have low installation and operating costs. On the other hand, wired technologies are a more secure way of data transfer as compared to the wireless technology. The choice of the technology depends on the requirements of the problem.

Communication network will play a significant role in the success of smart grids, and smart buildings and cities. According to Depuru et al. [97], the selection of the communication network and design of the communication devices are critical and must satisfy complex requirements. As we are moving towards smart buildings and cities, the amount of data transfer will be enormous and different challenges needs to be tackled to handle sensitive and confidential data. There are different communication networks' guidelines and standards to make sure that the data transmissions between various components are secured. Sood et al. [98] mentioned that the communication links will be using different kinds of communication technologies i.e. hard-wired links to fibre optics, wireless, satellites and microwave links; due to the fact that smart grids will be covering a vast geographical territory. In the below section, we will discuss some of the widely used communication and networks technologies for metering and sensory purposes in the buildings.

ZigBee

ZigBee was built on the IEE 802.15.4 standard and operates in the 2.4 GHz and 900 MHz bands [99], this standard defines the physical layer and medium access control sub-layer specifications for a low rate and cost personal area networks. ZigBee supports the star, tree and mesh networking topologies. ZigBee provides a low cost, reliable data transfer, short range, low power consumption home area wireless network solution. ZigBee is mostly used for security system, building automation and remote meter readings. ZigBee has application limitations due to its low power and rate, and also short-distance nature. Therefore, it is not recommended to be used for the applications with strong real-time and large amount of data [100].

This low cost and power consumption are also selling points for ZigBee. It is also an automatic dynamic network and with high security [101]. ZigBee meets the requirements of a seven-layer protocol layer and also has an additional module for security purposes. A ZigBee network can be extended to a maximum of 255 nodes; this makes it a highly expandable network [102]. ZigBee network is based on Master/Slave and facilitates two-way communication. The media storage control (MAC) layer operates on talk-when-ready collision prevention mechanism [103], which means the data is transferred when it is required, and the receiver also confirms when the data packet is being received. This feature increases the ability of the ZigBee network to make sure the transmission data packet. Although Zigbee HAN devices have a very low energy consumption (<1 mW), ZigBee network may not be suitable for rural areas due to having a 100 m line of sight range [104].

Different studies developed and experimentally validated ZigBee-based smart meters. Luan et al. [105] used ZigBee communication technology to transmit not only the detailed power consumption data but also outage event data to rear-end processing system. Experimental results showed a high potential for using ZigBee-based solution to build the area-based advanced metering infrastructure (AMI). Han and Lim [106] proposed and implemented a smart home control system using ZigBee and IEEE 802.15.4. The authors also developed a new routing protocol, DMPR (Disjoint Multi Path based Routing), to enhance the performance of ZigBee sensor network [107]. Luan et al. [108] designed a ZigBee-based smart power meter with outage recording function to develop an automatic reliability system for AMI. It was concluded that the proposed system has great potential to be integrated into the AMI to improve the accuracy of reliability calculation. Cao et al. [109] presented a remote wireless AMR reading system based on wireless sensor networks and embedded technology. ZigBee was used to transmit data from the sensor nodes to the data collector, and Ethernet was used to transmit data from the data collector to the server. It was found that the average receive rate of the sensor node using ZigBee was 92.81%, therefore, it was concluded that it is possible to implement the proposed system for remote AMR applications. Kang et al. [110] implemented a smart loading monitoring and control system with ZigBee wireless network. The integration of ZigBee smart loading control module and digital meter was able to monitor and control the electricity consumption of household appliances and also manage standby power. ZigBee technologies have also been used for monitoring photovoltaic and wind energy systems [111]. The authors also used ZigBee for building and home energy management. The experimental results demonstrated the proficiency of ZigBee devices applied in smart metering systems and distributed renewable generation.

Power Line Carriers

Power line carriers (PLCs) uses existing electricity grid, cellular network, mesh network, licensed and unlicensed radio, wireless modem, power line communication, RS-232/485, WiMAX, Ethernet and Wi-Fi to upload data using IEC DNP [97]. PLCs are more suitable for remote locations, where cellular or GPRS coverage is not available or where the number of consumers are relatively low [14]. The data transmission near electricity transmission can be problematic in PLC as there is data distortion and in most cases it is ideal to avoid transformer points. PLCs also have low bandwidth (up to 20 Kbps) but the advantages of PLCs are that they have low running cost and can be installed using current infrastructure, there also have low initial cost. For utility providers, PLCs are better options due to having full control over the network and their low cost [112]. According to Depuru et al. [97], power line carrier is an efficient technology for automation of data in smart meter applications. PLCs can be used to transmit data to and from the building along low and high voltage power line [113]. As PLCs have lower bandwidth, therefore, they are mostly suitable for lower bandwidth applications e.g., changing traffic light signals, switching on and off street lights, etc. Kim et al. [114] used PLC to transmit control data, however, its integration with other systems (e.g. window mounted security insurance system) was difficult to perform. Choi et al. [115] also used PLC as a communication protocol for their smart metering system. It was acknowledged by the authors that wireless mode of communication without repeaters could be challenging in buildings that are constructed from ferroconcrete as most of the signals will be blocked.

ModBus

ModBus was introduced by Modicon Corporation and used it as a point-to-point connections with EIA-232C interfaces of Programmable logic controllers). It is one of the widely used communication protocol due to its simplicity and reliability. It includes RTU (Remote Terminal Unit), TCP (Transmission Control Protocol) and ASCII mode of transmission and supports RS-232, RS-422, RS-485 and Ethernet based equipment [116]. ModBus is based on Master/Slave method for data communication; an information request is sent by the Master and Slave produces the response to the requested information. ModBus, because of its simplicity and open source availability, is more popular for local communication in buildings and also has become standard for industrial SCADA systems [117]. ModBus is a client-server protocol that was initially designed to communicate low-speed serial in process control network, and security issues were not addressed [118]. One of the primary concerns of ModBus is that it does not support authentication nor encryption [119], which means it is less secure and more vulnerable to cyber attacks. ModBus has been used in different research works; Bhatt and Verma [120] developed ModBus based intelligent Building Automation System (BAS). The authors wired sensors and ModBus compliant I/O modules to form an RS-438 network at field level. A ModBus compliant submeter was also used by [121] to monitor real current consumed by up to 21 devices.

GPRS and GSM

GPRS and GSM stand for general packet radio service and global system for mobile. GSM is well known as one of the world's most deployed technology and is present even in the remote parts of the world. Reliability of short message service (SMS) during network congestion times can be a challenging issue for GSM/GPRS technologies [122]. GSM based smart meters are already being used and available in the market. Tan et al. [123] and Ashna and George [124] proposed a GSM based meter for energy billing; SIM (subscriber identity module) are embedded in the utility meters, and the data is sent over to the utility provider through SMS or GPRS. However, as discussed previously missing SMS can be a serious issue for GSM/GPRS based technologies. Two other issues from supplier perspective could be the hacking by the consumer to alter their energy consumption data, and someone can also cut the power remotely (this can be cutting the power of an entire cities or cities).

There are other applications of GSM/GPRS apart from automatic energy metering. Technologies based on these communication protocols are also used for indoor environment and load control purposes. Al-Ali et al. [125] developed a GSM based home appliances monitoring and control system. Alheraish [126] also proposed a similar system, which shows the popularity of GSM/GPRS based system due to their extensive coverage. However, before deploying a GSM/GPRS based solution, the strength (quality) and availability (quantity) needs to be checked/ determined [127].

M-Bus

M-Bus stands for Meter-Bus and is a European standard (EN 13757-2 physical and link layers [128], EN 13757-3 application layer [129]) and is used for transmitting gas, heat, oil, etc. meter readings to utility providers. M-Bus was specifically designed for meter reading and is less popular than ModBus protocol. By using M-Bus, a large number of devices can be connected with each other, and there can also be the possibility of network expansion. Metering bus also have wireless applications, and recent research has been focussed on Wireless M-Bus e.g., [130,131]. Open Metering System group has proposed the use of Wireless M-Bus for metering scenarios and recommended to be used in smart meters [130].

M-Bus is been used by researchers in different R&D projects; REMPLI project [132]included the support for the M-Bus protocol in the automatic metering domain. Hosek et al. [133] also used M-Bus in their study to investigate the feasibility of cost-efficient measurement architecture based on already deployed home and network access technologies. M-Bus was also used in Intelligent Metering project [134] to transmit energy and water consumption data to monitor occupants' behaviour. Spinsante et al. [135] analysed the adoption of the Wireless Metering-Bus (WM-Bus) in future smart water grids. It was found that the WM-Bus protocol at 169 MHz ensured adequate transmission capabilities.

Ethernet

Ethernet is a low-cost communication method and is widely used for communication between PLCs and supervisory control and data acquisitions (SCADA). Ethernet is available on three media: i.e. optical fibre, shielded twisted pairs and coaxial cables. Among these three mediums, optical fibre is more secure and popular because of the absence of electromagnetic interference and electrical current [136]. Ethernet uses carrier sense multiple access collision detection (CSMA-CD) methods for sending data [137]. According to Bertoluzzo et al. [138], Ethernet is not suitable for real-time application because the a priori estimation of the data packet's maximum transmission time is impossible. Different solutions can be used for real-time applications of Ethernet and are known as real-time Ethernet (RTE) e.g., Powerlink [139], EtherCAT [140], PROFINET IO [141], etc. Ethernet requires cabling and is often less favourable option. Ethernet is a robust and common solution with no running cost, but it can be interfered easily and also require the involvement of IT team.

WiFi

In recent years, wireless sensing technology has gained more popularity because wireless sensors are easy to install and cheaper in price. However, its use for the metering technology is still at very early stage because of security issues. Mushtaq [142] has discussed in detail the security vulnerabilities concern of wireless networks. WiFi can use existing wireless local area network (WLAN) and there is no addition need of a dedicated network. This communication technology is more expensive as compared to other technologies because of its high power consumption and model price. Wireless technology is mostly used for building automation, remote control, meter reading, security systems, etc.

A WiFi based home automation system was designed and implemented by ElShafee and Hamed [143]. The proposed system can be scalable and one server was able to manage many hardware interface modules (as long as they existed on WiFi network coverage). The proposed system has the benefits of low cost, secure, remotely controllable, auto-configurable and ubiquitously accessible system. Martani et al. [144]proposed a method to measure occupants' activity by using WiFi connections as a proxy for human occupancy. Although this study was not particularly focussed on WiFi communication protocol, however, WiFi was used to study the correlation between energy consumption and occupancy. Balaji et al. [145] performed an occupancy based HVAC actuation by using the existing WiFi infrastructure in commercial buildings. It was found that the system accurately determined the occupancy in office spaces and resulted in energy savings of 17.8%.

BACnet

BACnet stands for building automation and control network and was initiated by ASHRAE (American Society of Heating, Refrigeration and Air-conditioning Engineers) and is now an American National Standard (ANSI/ASHRAE 135-1995) and ISO Standard (ISO 16484-5) standard protocol [146]. The advantage of using BACnet is its ability to be integrated with different types of equipment without having any need for additional hardware. BACnet has the capabilities from field layer (bottom layer) and BAS layer (top layer) for building control application and is considered as the full-fledged international standard [146]. A wireless version of BACnet using short range wireless networks (e.g., ZigBee) has also been developed [14]. The use of using ZigBee as a communication protocol for BACnet was announced in addendum 135-2008q to take advantage of reduced installation cost, flexible extension and mobile device connectivity of ZigBee [147]. BACnet's popularity can be judged by the fact that ASHRAE SSPC 135 has established the Smart Grid Working Group to enable buildings as full participants in the Grid [148]. In that regard, the Smart Grid Working Group has released some BACnet objects that can be used for demand reduction in buildings [149]. The BACnet Web Services issued by ASHRAE can allow external application to have interaction with building energy management or automation system for advanced control and/or optimisation of HVACoperation.

An effort has been done to couple building energy management and control system (EMCS) with different simulation programmes [150]. Haves and Xu [150] has developed a software called building controls virtual bed (BCVTB) to couple EnergyPlus[1] with control systems. A BACnet interface [151] is also been developed allowing the simulation programme to communicate with BACnet-compatible building control system. This interface was also used by Pang et al. [152] to compare actual and expected building energy performance in real time. Hong et al. [149]experimentally evaluated BACnet-ZigBee smart grid gateway for demand response (DR) applications in buildings. The gateway was used to map application of ZigBee to that of BACnet and vice versa. Another study by Li and Hong [153] also introduced a gateway architecture for mapping the DR applications of BACnet to those of EnOcean,[2] and vice versa. A study by Krioukov et al. [154] demonstrated a personalised lighting control system by using BACnet protocol to control different zones and several switches.

FUTURE DIRECTIONS AND CHALLENGES

Interoperability

Interoperability, in its broadest sense, can be defined as the ability of products and services to work together and to exchange information during all lifecycle stages without the loss of semantics [155]. Standardised representation of information is at the heart of interoperable solutions. In the case of metering and sensing devices, it refers to the standardized exchange of data so that the components of building energy monitoring and metering system can communicate with each other irrespective of different manufacturers and physical medium. Advanced sensing and monitoring solutions are available in the market from different manufacturers; with a different range, communication protocol and capabilities; therefore, interoperability needs to be maintained between these technologies for their consistent communication.

This challenge can be tackled by developing frameworks that should include standards and protocols to achieve interoperability between metering, sensing devices and also with the control system. An example of such framework for Smart grids is known as NIST framework [156]. NIST framework includes protocols and standards for information management for interoperability of smart grid systems and devices [157]. Such kind of frameworks needs to be developed on building level as well. Also, there is a need to introduce, develop and deploy vendor-independent metering and sensing solutions to increase flexibility.

Security

The security of the advanced metering infrastructure (AMI) will be one of the biggest challenge for and high priority of consumers, suppliers and policy makers. AMI will be able to transfer all kind of data e.g., socio-economic status, work habits, which and when appliance are used, food consumption patterns, etc., which can be assessed both legitimately and illegitimately) [43]. Attackers can alter the transmission and can also get access to the information. This issue needs to be discussed not only for deliberate attacks but also for unintentional user errors and equipment failures [157]. The privacy of customers' energy and environmental data should be respected as the customers do not want to share their energy related information to unauthorised people or marketing firms [158].

Wireless communication protocols are more vulnerable because of air (open medium) being used for data transmission. Different standards bodies are working towards making securing data transfer links. Wireless protocols will be secured by using technologies such as 802.11i [159] and 802.16e [160]. On the other hand, wired links will be made more secured by using virtual private networks (VPN), firewall and also by using secure shell (SSH), security socket layer (SSL) [157].

Cost

Cost is one of the deciding factors for selecting sensing and/or metering solution in buildings. Wired communication solutions are costly due to their initially cost, on the other hand, mobile phone technology based solution such as GSM and GPRS are expensive as well due to their high power consumption. Wireless solutions are cheaper because of the absence of the wires, but there are other drawbacks. Despite implementing different policies and legislations, there are not many incentives from government's side for building owners to pay towards their smart solutions. There are other incentive schemes e.g., Warm front, The Green Deal, etc., but unfortunately, there are not many incentive schemes for metering and sensing technologies. In the UK, householders can get an optional payment of £230 per year for purchasing a metering and monitoring solution on the installation of heat pumps through renewable heat incentive (RHI) [161]. The smart grid is a key step towards sustainable energy future [137], Stanislav et al. [162] defined smart grids as the power system will be *smart* by using meter, controls, sensors and analytical tools to control and monitor the energy flow from plant to plug. Therefore, there should be more incentives rather than just policies to reduce energy consumption and to move towards sustainable future. Also, cost related to a dedicated communication network for the transmission of data will be a very expensive solution and according to MacDonald [163] 'piggybacking' on existing communication infrastructure may be less expensive.

MEMS Sensor Technologies

Sensors and processing devices are getting smaller and cheaper, microelectromechanical (MEMS) sensor technology is one the advancements in this regard. In future, more advancements are expected in MEMS technology to develop sensors that are more capable and versatile. One of the many applications of MEMS in buildings is using MEMS based 'Smart Dust motes' wireless platforms by [164]. The authors used the platform for controlling daylighting systems in commercial buildings and concluded that the proposed solution enabled less expensive and disruptive retrofitting of commercial buildings. MEMS sensors show immense potential towards energy saving and smart sensing solutions.

Lack of ICT Infrastructure

The European Commission has plans to replace 80% of electricity meters with smart meters by 20%. It is projected that the installation of smart meters will lead to a reduction of 9% in emissions in the EU. The cost of installing a smart energy meter will provide a savings of £160 for gas and £309 for electricity per metering point. However, the installation of smart meters and moving towards smart grids and smart cities will also require the deployment of advanced measurement and control devices along with different sensors [165]. Strbac [165] mentioned that the key ingredients for smart grids and cities do exist, however, target trials are required to gain more experience and most importantly a commitment to implementation would be a key factor towards smart grids and smart cities. We can also learn from the trials/implementation that were carried out in other countries. One of the examples of such implementations, the example of implementation of smart meters in the Australian state of Victoria needs to be studied before moving forward. A report by Auditor-General [166] criticised the deployment of smart meters due to the risk of accelerated installation schedule in advance of national and international standards.

CONCLUSIONS

In this article, we presented a comprehensive review discussing metering and sensing technologies for buildings. The study suggests that there has been active research and technological advancements in building energy metering and environmental monitoring. There are a broad range of technologies available and used by the researchers and industry people to overcome different challenges. In home automation and HVAC system monitoring, wireless area networks show promising trends towards deploying a low-cost metering and monitoring solution. Wireless networks are easy to retrofit the exciting system as compared to the wired system, and they are also more preferred because of their lower installation costs. However, choosing a proper metering and monitoring solution is a challenging task as it depends on different factors such as granularity, accuracy, cost, availability, ease of deployment and communication protocol. This paper also overviews different drivers for energy metering and environmental monitoring. The paper also reviewed some of the policies and regulations that are in place in the various countries and it is evident that only a few countries address the need to install metering and sensing technologies in buildings.

Communication and network technologies should be cost benefit and be able to provide good transmittable range along with secured data transmission and better power quantity and bandwidth. One of the biggest challenges of AMI is the protection of metering data against unauthorized access. The privacy and quality of the transferred data are important to both clients and service providers. To detect forged or hacked smart meters an end to end secured communication protocols can be used and also encryption of data needs to be ensured for protecting against leakage of information. Also, future research should focus on developing multi-protocol devices, or further standardisation is required to address the issue of interoperability, which is one of the biggest challenges for AMI. This article also highlighted some other challenges and future directions that need to be addressed. It can be concluded that the area of building energy metering and indoor environmental technologies has witnessed many technological advancements in recent years, and it will be continued because of the developments in information and communication technologies.

ACKNOWLEDGEMENT

The authors acknowledge financial support from the European Commission (FP7). Grant reference – 609154.

REFERENCES

1. P.H. Shaikh, N.B.M. Nor, P. Nallagownden, I. Elamvazuthi, T. Ibrahim, A review on optimized control systems for building energy and comfort management of smart sustainable buildings, Renew. Sustain. Energy Rev. 34 (2014) 409–429.
2. A. Costa, M.M. Keane, J.I. Torrens, E. Corry, Building operation and energy performance: monitoring, analysis and optimisation toolkit, Appl. Energy 101 (2013) 310–316.

3. J. Grözinger, T. Boermans, A. John, F. Wehringer, J. Seehusen, Overview of Member States information on NZEBs: Background paper – final report, ECOFYS GmbH, Cologne, Germany, 2014.
4. DECC, 2013 UK Greenhouse Gas Emissions, Provisional Figures and 2012 UK Greenhouse Gas Emissions, Final Figures by Fuel Type and End-User, Department of Energy and Climate Change, London, UK, 2014.
5. The World Bank, World Development Indicators 1960–2013, The World Bank, Washington, DC, 2014.
6. IPCC, Climate Change 2013 – The Physical Science Basis: Working Group I Contribution to the Fifth Assessment Report of the Intergovernmental Panel on Climate Change, Cambridge University Press, New York, NY, 2013.
7. International Energy Agency, Redrawing the Energy Climate Map – World Energy Outlook Special Report, 2013 http://www.iea.org/publications/ freepublications/publication/WEO Special Report 2013 Redrawing the Energy Climate Map.pdf (accessed 01.10.15).
8. R.M. Dowd, M. Mourshed, Low carbon buildings: sensitivity of thermal properties of opaque envelope construction and glazing, Energy Procedia 75 (2015) 1284–1289, http://dx.doi.org/10.1016/j.egypro.2015.07.189.
9. T. Oreszczyn, R. Lowe, Challenges for energy and buildings research: objectives, methods and funding mechanisms, Build. Res. Inf. 38 (1) (2010) 107–122, http://dx.doi.org/10.1080/09613210903265432.
10. European Commission, Energy Efficiency Directive, http://ec.europa.eu/ energy/efficiency/eed/eed en.htm (accessed 01.10.15), 2014.
11. Internal Code Council, International Energy Conservation Code.
12. J. Genet, C. Schubert, Designing a metering system for small and medium-sized buildings, Technical Report SEMED310007EN, SE, 2011.
13. N.M.G. Strategy, Advanced Metering Infrastructure, US Department of Energy Office of Electricity and Energy Reliability.
14. , in: A survey on advanced metering infrastructure, Int. J. Electr. Power Energy Syst. 63 (2014) 473–484.

15. S. Smith, How sub-metering changed the way Nissan, Smyrna plant does business, in: Future of Instrumentation International Workshop (FIIW), vol. 20, 2011.
16. EC, State of the Energy Union 2015, COM(2015) 572, European Commission, Brussels, BE, 2015.
17. M.W. Ahmad, M. Mourshed, B. Yuce, Y. Rezgui, in: Computational intelligence techniques for HVAC systems: a review, Build. Simul. (2016), http://dx.doi.org/10.1007/s12273-016-0285-4.
18. EU, Common Vision for the Renewable Heating & Cooling Sector in Europe, European Union, Brussels, BE, 2011.
19. B. Juul-Kristensen, Efficiency in Energy Supply, High Efficiency CHP and Heating/Cooling, CA-EED, Brussels, BE, 2014.
20. M. Mourshed, S. Robert, A. Ranalli, T. Messervey, D. Reforgiato, R. Contreau, A. Becue, K. Quinn, Y. Rezgui, Z. Lennard, in: Smart grid futures: perspectives on the integration of energy and ICT services, Energy Procedia 75 (2015) 1132–1137, http://dx.doi.org/10.1016/j.egypro.2015.07.531.
21. DECC, Smart Meters – A Guide, Department of Energy and Climate Change, London, UK, 2016.
22. P. Chalmers, Climate Change: Implications for Building, 2014, pp. 1–20 http://www.cisl.cam.ac.uk/business-action/low-carbon-transformation/ ipcc-briefings/pdfs/briefings/IPCC AR5 Implications for Buildings Briefing WEB EN.pdf.
23. M.R. Brambley, P. Haves, S.C. McDonald, P.A. Torcellini, D. Hansen, D. Holmberg, K. Roth, Advanced Sensors and Controls for Building Applications: Market Assessment and Potential R&D Pathways, Citeseer, 2005.
24. European Council, Directive 2002/91/EC of the European parliament and of the council of 16 December 2002 on the energy performance of buildings, Official Journal of the European Communities. DIRECTIVE 91.
25. EU Parliament, Directive 2010/31/EU of the European Parliament and of the council, 2010.
26. Department of Energy & Climate Change, Energy Savings Opportunity Scheme, 2013 https://www.gov.uk/government/

consultations/energysavings-opportunity-scheme (accessed 01 September).

27. S. Darby, in: The effectiveness of feedback on energy consumption, a review for DEFRA of the literature on metering, Billing Direct Disp. 486 (2006) 1–24.
28. S. Junnila, in: The potential effect of end users on energy conservation in office buildings, Facilities 25 (7/8) (2007) 329–339, http://dx.doi.org/10. 1108/02632770710753352.
29. A. Cox, T. Higgins, R. Gloster, B. Foley, A. Darnton, The Impact of Workplace Initiatives on Low Carbon Behaviours, Scottish Government Social Research, Edinburgh.
30. The Carbon Trust, Advanced metering for SMEs – Carbon and cost savings.
31. M. Evans, B. Shui, T. Takagi, Country report on building energy codes in Japan, 2009 http://www.pnl.gov/main/publications/external/technical reports/PNNL-17849.pdf.
32. Mark and Spencer, M&S Plan A Report, 2014 http://planareport.marksandspencer.com/downloads/M&S-PlanA-2014.pdf (accessed 01 September).
33. P. Marian, 14 Steps to becoming an M&S eco-factory, 2011 http://www.juststyle.com/analysis/14-steps-to-becoming-an-ms-eco-factory id112528. aspx (accessed 01.09.14).
34. Santander, Corporate Social Responsibility, 2014 http://www.santander.co. uk/uk/about-santander-uk/csr.
35. D. Fugate, P. Fuhr, T. Kuruganti, Instrumentation systems for commercial building energy efficiency, in: Future of Instrumentation International Workshop (FIIW), 2011, pp. 21–24.
36. P.O. Fanger, et al., Thermal comfort. Analysis and applications in environmental engineering.
37. K. Parsons, Human Thermal Environments: The Effects of Hot, Moderate, and Cold Environments on Human Health, Comfort, and Performance, Making Sense of, Taylor and Francis, 2014, ISBN 9781466595996.
38. F.R.d. Alfano, M. Del′ lIsola, B.I. Palella, G. Riccio, A. Russi, in: On the measurement of the mean radiant temperature and its influence

on the indoor thermal environment assessment, Build. Environ. 63 (2013) 79–88.

39. Edison Electric Institute, Handbook for Electricity Metering, EEI Publication, Edison Electric Institute, 2002, ISBN 9780931032523, http://books.google. co.uk/books?id=ZmPyPQAACAAJ.
40. G. Sullivan, W. Hunt, R. Pugh, W. Sandusky, T. Koehler, B. Boyd, Metering Best Practices: A Guide to Achieving Utility Resource Efficiency, http:// energy.gov/sites/prod/files/2013/10/f3/mbpg.pdf.
41. E. O'Driscoll, G.E. O'Donnell, in: Industrial power and energy metering – a state-of-the-art review, J. Clean. Prod. 41 (2013) 53–64.
42. T. Khalifa, K. Naik, A. Nayak, in: A survey of communication protocols for automatic meter reading applications, IEEE Commun. Surv. Tutor. 13 (2) (2011) 168–182.
43. M.P. McHenry, in: Technical and governance considerations for advanced metering infrastructure/smart meters: technology, security, uncertainty, costs, benefits, and risks, Energy Policy 59 (2013) 834–842.
44. F. Cascetta, P. Vigo, Flowmeters: A Comprehensive Survey and Guide to Selection, Instrument Society of America, 1988, ISBN 9781556170997.
45. G. Buonanno, in: On field characterisation of static domestic gas flowmeters, Measurement 27 (4) (2000) 277–285.
46. I. ÓSullivan, W. Wright, in: Ultrasonic measurement of gas flow using electrostatic transducers, Ultrasonics 40 (1-8) (2002) 407–411.
47. J.G. Drenthen, G. de Boer, in: The manufacturing of ultrasonic gas flow meters, Flow Meas. Instrum. 12 (2) (2001) 89–99.
48. K. Sakai, M. Okabayasi, K. Yasuda, in: The fluidic flowmeter – a gas flowmeter based on fluidic dynamic oscillation, Flow Meas. Instrum. 1 (1) (1989) 44–50.
49. M. Anklin, W. Drahm, A. Rieder, in: Coriolis mass flowmeters: overview of the current state of the art and latest research, Flow Meas. Instrum. 17 (6) (2006) 317–323.
50. S. Kearney, The age of advanced metering arrives, in: Rural Electric

Power Conference, 2005, pp. 1–4, http://dx.doi.org/10.1109/REPCON.2005. 1436328.

51. A. Foy, Remote metering device, http://google.com/patents/EP2318809A1?cl=zh-cn, eP Patent App. EP20,090,785,727, 2011.
52. P.R.N. Childs, J.R. Greenwood, C.A. Long, in: Review of temperature measurement, Rev. Sci. Instrum. 71 (8) (2000) 2959–2978.
53. CIBSE, Guide H: Building control systems, ISBN: 9781906846008.
54. ISO, ISO Standard 7726: Ergonomics of the Thermal Environment – Instruments for Measuring Physical Quantities, 1998, pp. 1–52.
55. X. Chen, K. Kamimura, A. Katoh, M. Miura, in: Measurement of the indoor thermal environment using a radiant temperature sensor, ASHRAE Trans. 108 (1) (2002) 341–348.
56. D. McIntyre, in: Preferred air speed for comfort in warm conditions, ASHRAE Trans. 84 (2) (1978) 263–277.
57. L. Hill, H. Vernon, D. Hargood-Ash, in: The kata-thermometer as a measure of ventilation, Proc. R. Soc. Lond. Ser. B, Contain. Pap. Biol. Character 93 (651) (1922) 198–206.
58. S. Corrsin, in: Extended applications of the hot-wire anemometer, Rev. Sci. Instrum. 18 (7) (1947) 469–471.
59. J. Wilson, Sensor Technology Handbook Electronics & Electrical, vol. 1, Elsevier, 2005, ISBN 9780750677295.
60. M. Loomans, A.v. Schijndel, in: Simulation and measurement of the stationary and transient characteristics of the hot sphere anemometer, Build. Environ. 37 (2) (2002) 153–163.
61. H.B. Awbi, Ventilation of Buildings, Taylor & Francis, 2003.
62. H. Koskela, J. Heikkinen, R. Niemel, T. Hautalampi, in: Turbulence correction for thermal comfort calculation, Build. Environ. 36 (2) (2001) 247–255.
63. J. Fraden, Handbook of Modern Sensors: Physics, Designs, and Applications, Handbook of Modern Sensors, Springer, 2004, ISBN 9780387007502.
64. E. Traversa, in: Ceramic sensors for humidity detection: the state-of-the-art and future developments, Sens. Actuators B: Chem. 23 (2-3) (1995) 135–156.
65. K. Arshak, K. Twomey, in: Investigation into a novel humidity

sensor operating at room temperature, Microelectron. J. 33 (3) (2002) 213–220.
66. L.J. Golonka, B.W. Licznerski, K. Nitsch, H. Teterycz, in: Thick-film humidity sensors, Meas. Sci. Technol. 8 (1) (1997) 92.
67. B. Brooks, Understanding Indoor Air Quality, Telford Press S, Taylor & Francis, 1991, ISBN 9780849388460.
68. CEN, Ventilation for Buildings: Design Criteria for the Indoor Environment, Brussels, European Committee for Standardization.
69. J. Kwon, G. Ahn, G. Kim, J.C. Kim, H. Kim, A study on NDIR-based CO2 sensor to apply remote air quality monitoring system, in: ICCAS-SICE, 2009, pp. 1683–1687.
70. M. Griffiths, M. Eftekhari, in: Control of CO2 in a naturally ventilated classroom, Energy Build. 40 (4) (2008) 556–560.
71. D. Preethichandra, Design of a smart indoor air quality monitoring wireless sensor network for assisted living, in: IEEE International Instrumentation and Measurement Technology Conference (I2MTC), 2013, pp. 1306–1310.
72. K. Zakrzewska, in: Mixed oxides as gas sensors, Thin Solid Films 391 (2) (2001) 229–238.
73. J. Gebicki, ˛ A. Kloskowski, in: Electrochemical sensor for measurement of volatile organic compounds employing square wave perturbation voltage, Metrol. Meas. Syst. 17 (4) (2010) 637–649.
74. A. Yasuda, T. Shimidzu, in: Electrochemical carbon monoxide sensor with a Nafion® film, React. Funct. Polym. 41 (1–3) (1999) 235–243.
75. M.W. Ahmad, M. Mourshed, J.-L. Hippolyte, Y. Rezgui, H. Li, Optimising the scheduled operation of window blinds to enhance occupant comfort, in: Proceedings of BS2015: 14th Conference of International Building Performance Simulation Association, Hyderabad, India, 2015, pp. 2393–2400.
76. M.A. ul Haq, M.Y. Hassan, H. Abdullah, H.A. Rahman, M.P. Abdullah, F. Hussin, D.M. Said, in: A review on lighting control technologies in commercial buildings, their performance and affecting factors, Renew. Sustain. Energy Rev. 33 (2014) 268–279.

77. F. Oldewurtel, D. Sturzenegger, M. Morari, in: Importance of occupancy information for building climate control, Appl. Energy 101 (2013) 521–532.
78. X. Guo, D. Tiller, G. Henze, C. Waters, in: The performance of occupancy-based lighting control systems: a review, Light. Res. Technol. 42 (4) (2010) 415–431.
79. New Buildings Institute, Inc, J. Benya, D.E. Weigand, Advanced Lighting Guidelines, New Buildings Institute, 2003.
80. V. Magori, H. Walker, in: Ultrasonic presence sensors with wide range and high local resolution, IEEE Trans. Ultrason. Ferroelectr. Freq. Control 34 (2) (1987) 202–211.
81. D.H. Li, K. Cheung, S. Wong, T.N. Lam, in: An analysis of energy-efficient light fittings and lighting controls, Appl. Energy 87 (2) (2010) 558–567.
82. A.-S. Choi, K.-D. Song, Y.-S. Kim, in: The characteristics of photosensors and electronic dimming ballasts in daylight responsive dimming systems, Build. Environ. 40 (1) (2005) 39–50.
83. M.I. Directive, Directive 2004/22/EC, European Council.
84. National Measurement Office, MID Approved Gas and Electricity Meters, 2014 https://www.gov.uk/mid-approved-gas-and-electricity-meters (accessed 01.09.15).
85. International Electrotechnical Commission, Electricity Metering Equipment (AC) – General Requirements, Tests and Test Conditions – Part 11: Metering Equipment, Geneva, Switzerland.
86. International Electrotechnical Commission, Electricity Metering Equipment (AC) – Particular Requirements – Part 21: Static Meters for Active Energy Class 1 and 2, Geneva, Switzerland.
87. International Electrotechnical Commission, Electricity Metering Equipment (AC) – Particular Requirements, Tests and Test Conditions – Part 11: Metering Equipment, Geneva, Switzerland.
88. British Standards Institution, Electrical Static Meters for Secondary Metering and Sub-metering. Specification, ISBN: 978 0 580 68095 3.
89. International Electrotechnical Commission, Alternating Current Static Watt-hour Meters for Active Energy (Classes 1 and 2).

90. International Electrotechnical Commission, International Standard for Alternating Current Static Watt-hour Meter for Active Energy (Classes 0.2 S and 0.5 S).
91. ANSI, American National Standard for Electricity Meters – 0.2 and 0.5 Accuracy Classes.
92. W. Koon, Current sensing for energy metering, in: IIC-China/ESC-China, 2002, pp. 321–324.
93. EDF Energy, Understanding changes to the meter class, 2014 http://www. edfenergy.com/products-services/large-business/PDF/B2B ePublications/ B2B-MBC-040.pdf (accessed 01.10.14).
94. M. Kintner-Meyer, in: Opportunities of wireless sensors and controls for building operation, Energy Eng. 102 (5) (2005) 27–48.
95. M. Milenkovic, O. Amft, An opportunistic activity-sensing approach to save energy in office buildings, in: Proceedings of the Fourth International Conference on Future Energy Systems, 2013.
96. S. Lanzisera, S. Dawson-Haggerty, H.I. Cheung, J. Taneja, D. Culler, R. Brown, in: Methods for detailed energy data collection of miscellaneous and electronic loads in a commercial office building, Build. Environ. 65 (2013) 170–177.
97. S.S.S.R. Depuru, L. Wang, V. Devabhaktuni, in: Smart meters for power grid: challenges, issues, advantages and status, Renew. Sustain. Energy Rev. 15 (6) (2011) 2736–2742.
98. V. Sood, D. Fischer, J. Eklund, T. Brown, Developing a communication infrastructure for the smart grid, in: Electrical Power Energy Conference (EPEC), Montreal, 2009, pp. 1–7, http://dx.doi.org/10.1109/EPEC.2009. 5420809.
99. IEEE, Approved Draft Amendment to IEEE Standard for Information technology-Telecommunications and information exchange between systems-PART 15.4:Wireless Medium Access Control (MAC) and Physical Layer (PHY) Specifications for Low-Rate Wireless Personal Area Networks (LR-WPANs): Amendment to add alternate PHY (Amendment of IEEE Std 802.15.4), IEEE Approved Std P802.15.4a/D7, 2007, pp. 1–26.
100. Q. Zhang, Y. Sun, Z. Cui, Application and analysis of ZigBee technology for smart grid, in: International Conference on Computer and Information Application (ICCIA), 2010, pp. 171–174.

101. D. Xu, J. Liu, in: IPv6-based smart metering network for monitoring building electricity, Adv. Mech. Eng. (2013) 1–8.

102. S. Lin, J. Liu, Y. Fang, ZigBee based wireless sensor networks and its applications in industrial, in: IEEE International Conference on Automation and Logistics, 2007, pp. 1979–1983.

103. W.-T. Sung, Y.-C. Hsu, in: Designing an industrial real-time measurement and monitoring system based on embedded system and ZigBee, Expert Syst. Appl. 38 (4) (2011) 4522–4529.

104. J. Cheng, T. Kunz, A survey on smart home networking, Carleton University, Systems and Computer Engineering, Technical Report, SCE-09-10.

105. S.-W. Luan, J.-H. Teng, S.-Y. Chan, L.-C. Hwang, Development of a smart power meter for AMI based on ZigBee communication, in: International Conference on Power Electronics and Drive Systems (PEDS), 2009, pp. 661–665, http://dx.doi.org/10.1109/PEDS.2009.5385726.

106. D.M. Han, J.H. Lim, in: Smart home energy management system using IEEE 802.15.4 and ZigBee, IEEE Trans. Consum. Electron. 56 (3) (2010) 1403–1410, http://dx.doi.org/10.1109/TCE.2010.5606276, ISSN 0098-3063.

107. D.M. Han, J.H. Lim, in: Design and implementation of smart home energy management systems based on ZigBee, IEEE Trans. Consum. Electron. 56 (3) (2010) 1417–1425, ISSN 0098-3063.

108. S.-W. Luan, J.H. Teng, S.-Y. Chan, L.-C. Hwang, Development of an automatic reliability calculation system for advanced metering infrastructure, in: 8th IEEE International Conference on Industrial Informatics (INDIN), 2010, pp. 342–347, http://dx.doi.org/10.1109/INDIN.2010.5549395.

109. L. Cao, J. Tian, Y. Liu, Remote wireless automatic meter reading system based on wireless mesh networks and embedded technology, in: Fifth IEEE International Symposium on Embedded Computing, 2008. SEC '08, 2008, pp. 192–197, http://dx.doi.org/10.1109/SEC.2008.57.

110. M.S. Kang, Y.L. Ke, J.S. Li, Implementation of smart loading monitoring and control system with ZigBee wireless network, in: 6th IEEE Conference on Industrial Electronics and Applications

(ICIEA), 2011, pp. 907–912.

111. N. Batista, R. Melício, J. Matias, J. Catalao, in: Photovoltaic and wind energy systems monitoring and building/home energy management using ZigBee devices within a smart grid, Energy 49 (2013) 306–315, http://dx.doi.org/10. 1016/j.energy.2012.11.002, ISSN 0360-5442.

112. G. Deconinck, An evaluation of two-way communication means for advanced metering in Flanders (Belgium), in: Instrumentation and Measurement Technology Conference Proceedings, 2008. IMTC 2008. IEEE, ISSN 1091-5281, 2008, pp. 900–905, http://dx.doi.org/10.1109/IMTC.2008. 4547164.

113. S. Marvin, H. Chappells, S. Guy, in: Pathways of smart metering development: shaping environmental innovation, Comput. Environ. Urban Syst. 23 (2) (1999) 109–126, http://dx.doi.org/10.1016/S0198-9715(99)00011-3, ISSN 0198-9715, http://www.sciencedirect.com/science/ article/pii/S0198971599000113.

114. D.S. Kim, S.Y. Lee, K.Y. Wang, J.C. Choi, D.J. Chung, in: A power line communication modem based on adaptively received signal detection for networked home appliances, IEEE Trans. Consum. Electron. 53 (3) (2007) 864–870, http://dx.doi.org/10.1109/TCE.2007.4341558, ISSN 0098-3063.

115. T.S. Choi, K.R. Ko, S.C. Park, Y.S. Jang, Y.T. Yoon, S.K. Im, Analysis of energy savings using smart metering system and IHD (in-home display), in: Transmission Distribution Conference Exposition: Asia and Pacific, 2009, 2009, pp. 1–4, http://dx.doi.org/10.1109/TD-ASIA.2009.5356956.

116. D. gang Peng, H. Zhang, L. Yang, H. Li, Design and realization of ModBus protocol based on embedded Linux system, in: International Conference on Embedded Software and Systems Symposia, 2008. ICESS Symposia '08, 2008, pp. 275–280.

117. S. Le Blond, T. Lewis, M. Sooriyabandara, Towards an integrated approach to building energy efficiency: drivers and enablers, in: 2nd IEEE PES International Conference and Exhibition on Innovative Smart Grid Technologies (ISGT Europe), 2011, pp. 1–8.

118. Y. Mo, T.-H. Kim, K. Brancik, D. Dickinson, H. Lee, A. Perrig, B. Sinopoli, in: Cyber – Physical security of a smart grid infrastructure,

Proc. IEEE 100 (1) (2012) 195–209, http://dx.doi.org/10.1109/JPROC.2011.2161428, ISSN 0018-9219.

119. I. Modicon, Modicon ModBus Protocol Reference Guide, North Andover, Massachusetts, 1996, pp. 28–29.

120. J.G. Bhatt, H. Verma, RS-485/MODBUS based intelligent building automation system using LabVIEW, in: 4th International Conference on Computer Applications in Electrical Engineering-Recent Advances (CERA-09), IIT Roorkee, Roorkee, India, 2010, pp. 19–21, Section-A1: instrumentation, Paper no. 04 (abstract).

121. Collaborative recommender systems for building automation, in: 42nd Hawaii International Conference on System Sciences, 2009, HICSS '09, 2009, pp. 1–10, http://dx.doi.org/10.1109/HICSS.2009.114, ISSN 1530-1605.

122. A. Usman, S.H. Shami, in: Evolution of communication technologies for smart grid applications, Renew. Sustain. Energy Rev. 19 (2013) 191–199.

123. A. Tan, C. Lee, V.H. Mok, Automatic power meter reading system using GSM network, in: International Power Engineering Conference (IPEC), 2007, pp. 465–469.

124. K. Ashna, S. George, GSM based automatic energy meter reading system with instant billing, in: International Multi-Conference on Automation, Computing, Communication, Control and Compressed Sensing (iMac4s), 2013, pp. 65–72.

125. A.R. Al-Ali, M. Rousan, M. Mohandes, GSM-based wireless home appliances monitoring control system, in: International Conference on Information and Communication Technologies: From Theory to Applications, 2004, pp. 237–238.

126. A. Alheraish, in: Design and implementation of home automation system, IEEE Trans. Consum. Electron. 50 (4) (2004) 1087–1092.

127. P.-K. Cuvelier, P. Sommereyns, Proof of concept smart metering, in: 2009 20th International Conference and Exhibition on Electricity Distribution, 2009.

128. CEN, Communication systems for meters and remote reading of meters, Part 2: Physical and link layer, in: EN 13757-2, 2004, pp. 1–32, ISBN: 0 580 44959 9.

129. CEN, Communication systems for meters and remote reading of meters, Part 3: Dedicated application layer, in: EN 13757-3, 2004, pp. 1–58, ISBN: 0 580 44960 2.
130. S. Squartini, L. Gabrielli, M. Mencarelli, M. Pizzichini, S. Spinsante, F. Piazza, Wireless M-Bus sensor nodes in smart water grids: the energy issue, in: Fourth International Conference on Intelligent Control and Information Processing (ICICIP), 2013, 2013, pp. 614–619, http://dx.doi.org/10.1109/ ICICIP.2013.6568148.
131. A. Flammini, S. Rinaldi, A. Vezzoli, The sense of time in open metering system, in: 2011 IEEE International Conference on Smart Measurements for Future Grids (SMFG), 2011, pp. 22–27.
132. T. Sauter, M. Lobashov, in: End-to-end communication architecture for smart grids, IEEE Trans. Ind. Electron. 58 (4) (2011) 1218–1228, http://dx. doi.org/10.1109/TIE.2010.2070771, ISSN 0278-0046.
133. J. Hosek, P. Masek, D. Kovac, M. Ries, F. Kropfl, Universal smart energy communication platform, in: 2014 International Conference on Intelligent Green Building and Smart Grid (IGBSG), 2014, pp. 1–4, http://dx.doi.org/10. 1109/IGBSG.2014.6835232.
134. H. Lunzer, Intelligent metering, in: 5th IEEE International Conference on Industrial Informatics, vol. 2, 2007, pp. 1215–1219, http://dx.doi.org/10. 1109/INDIN.2007.4384949, ISSN 1935-4576.
135. S. Spinsante, S. Squartini, L. Gabrielli, M. Pizzichini, E. Gambi, F. Piazza, in: Wireless M-Bus sensor networks for smart water grids: analysis and results, Int. J. Distrib. Sens. Netw. 2014 (2014).
136. O. Strobel, J. Lubkoll, Fiber-optic communication — an overview, in: 20th International Crimean Conference on Microwave and Telecommunication Technology (CriMiCo), 2010, 2010, pp. 16–20.
137. E. O'Driscoll, G.E. O'Donnell, in: Industrial power and energy metering – a state-of-the-art review, J. Clean. Prod. 41 (2013) 53–64.
138. M. Bertoluzzo, G. Buja, S. Vitturi, Ethernet networks for factory automation, in: Proceedings of the 2002 IEEE International Symposium on Industrial Electronics, 2002, vol. 1, 2002, pp. 175–180.
139. EPSG, POWERLINK, 2014 http://www.ethernet-powerlink.org/en/ powerlink/technology/ (accessed 01 September).

140. ETG, EtherCAT – The Ethernet Fieldbus, 2016 http://www.ethercat.org (accessed 01 March).
141. PI, PROFIBUS, http://www.profibus.com/technology/profibus (accessed 01 September), 2014.
142. A. Mushtaq, Wireless network security vulnerabilities and concerns, in: Security Technology, Disaster Recovery and Business Continuity, Communications in Computer and Information Science, vol. 122, 2010, pp. 207–219.
143. A. ElShafee, K.A. Hamed, in: Design and implementation of a WIFI based home automation system, World Acad. Sci. Eng. Technol. 68 (2012) 2177–2180.
144. C. Martani, D. Lee, P. Robinson, R. Britter, C. Ratti, in: ENERNET: Studying the dynamic relationship between building occupancy and energy consumption, Energy Build. 47 (2012) 584–591, http://dx.doi.org/10.1016/j. enbuild.2011.12.037, ISSN 0378-7788, http://www.sciencedirect.com/ science/article/pii/S0378778811006566.
145. B. Balaji, J. Xu, A. Nwokafor, R. Gupta, Y. Agarwal, Sentinel: occupancy based HVAC actuation using existing WiFi infrastructure within commercial buildings, in: Proceedings of the 11th ACM Conference on Embedded Networked Sensor Systems, SenSys '13, ACM, New York, NY, USA, 2013, pp. 17:1–17:14, http://dx.doi.org/10.1145/2517351.2517370, ISBN 978-1-4503-2027-6.
146. S. Wang, Z. Xu, J. Cao, J. Zhang, in: A middleware for web service-enabled integration and interoperation of intelligent building systems, Autom. Constr. 16 (1) (2007) 112–121.
147. C. Reinisch, W. Kastner, G. Neugschwandtner, W. Granzer, Wireless technologies in home and building automation, in: 5th IEEE International Conference on Industrial Informatics, vol. 1, 2007, pp. 93–98, http://dx.doi. org/10.1109/INDIN.2007.4384737, ISSN 1935-4576.
148. D.G. Holmberg, S.T. Bushby, BACnet® and the Smart grid, ASHRAE.
149. S.H. Hong, S.H. Kim, G.M. Kim, H.L. Kim, in: Experimental evaluation of BZ-GW (BACnet-ZigBee smart grid gateway) for demand response in buildings, Energy 65 (2014) 62–70, http://

dx.doi.org/10.1016/j.energy.2013. 12.008, ISSN 0360-5442, http://www.sciencedirect.com/science/article/pii/ S0360544213010657.

150. P. Haves, P. Xu, The building controls virtual test bed – a simulation environment for developing and testing control algorithms, strategies and systems, in: Building Simulation '07, Bejing, China, 2007 http://www.ibpsa. org/proceedings/BS2007/p748 final.pdf.

151. T. Nouidui, M. Wetter, Z. Li, X. Pang, P. Bhattacharya, P. Haves, BacNet and analog/digital interfaces of the building controls virtual testbed, in: Building Simulation Conference, 2011.

152. X. Pang, M. Wetter, P. Bhattacharya, P. Haves, in: A framework for simulation-based real-time whole building performance assessment, Build. Environ. 54 (2012) 100–108, http://dx.doi.org/10.1016/j.buildenv.2012.02. 003, ISSN 0360-1323.

153. Y.-C. Li, S.H. Hong, in: BACnet-EnOcean Smart Grid Gateway and its application to demand response in buildings, Energy Build. 78 (2014) 183–191, http://dx.doi.org/10.1016/j.enbuild.2014.04.022, ISSN 0378-7788, http://www.sciencedirect.com/science/article/pii/S0378778814003326.

154. A. Krioukov, S. Dawson-Haggerty, L. Lee, O. Rehmane, D. Culler, A living laboratory study in personalized automated lighting controls, in: Proceedings of the Third ACM Workshop on Embedded Sensing Systems for Energy-Efficiency in Buildings, BuildSys '11, ACM, New York, NY, USA, 2011, pp. 1–6, http://dx.doi.org/10.1145/2434020.2434022, ISBN 978-1-4503-0749-9.

155. M. Mourshed, Interoperability-based optimisation of architectural design, Ph.D. thesis, National University of Ireland, Cork, Ireland, 2006.

156. U.S. Departmemt of Commerce, NIST Framework and Roadmap for Smart Grid Interoperability Standards, 2012, Release 2.0, http://www.nist.gov/ smartgrid/upload/NIST Framework Release 2-0 corr.pdf.

157. Y. Yan, Y. Qian, H. Sharif, D. Tipper, in: A survey on smart grid communication infrastructures: motivations, Requir. Chall. IEEE Commun. Surv. Tutor. 15 (1) (2013) 5–20.

158. F. Cleveland, Cyber security issues for advanced metering infrasttructure (AMI), in: 2008 IEEE Power and Energy Society

General Meeting – Conversion and Delivery of Electrical Energy in the 21st Century, 2008.

159. IEEE, IEEE Standard for Information technology-Telecommunications and information exchange between systems Local and metropolitan area networks-Specific requirements. Part 11: Wireless LAN Medium Access Control (MAC) and Physical Layer (PHY) Specifications, IEEE Std 802. 11-2012 (Revision of IEEE Std 802.11-2007), 2012, pp. 1–2793.

160. IEEE, IEEE Standard for Local and Metropolitan Area Networks. Part 16: Air Interface for Fixed and Mobile Broadband Wireless Access Systems Amendment 2: Physical and Medium Access Control Layers for Combined Fixed and Mobile Operation in Licensed Bands and Corrigendum 1, IEEE Std 802.16e-2005 and IEEE Std 802.16-2004/Cor 1-2005 (Amendment and Corrigendum to IEEE Std 802.16-2004), 2006, pp. 1–822.

161. DECC, Domestic Renewable Heat Icentive, 2013 https://www.gov.uk/ government/uploads/system/uploads/attachment data/file/212089/ Domestic RHI policy statement.pdf (accessed 01.10.15).

162. P. Stanislav, K. Bryan, M. Tihomir, Smart grids better with integrated energy system, in: 2009 IEEE Electrical Power Energy Conference (EPEC), 2009, pp. 1–8.

163. M. MacDonald, Appraisal of costs & benefits of smart meter roll out options, Final Report. Report for Department of Business Enterprise and Regulatory Reform, London, accessed 17, 2007, pp. 08–11.

164. Y.-J. Wen, J. Granderson, A. Agogino, Towards embedded wireless-networked intelligent daylighting systems for commercial buildings, in: IEEE International Conference on Sensor Networks, Ubiquitous, and Trustworthy Computing 2006, vol. 1, 2006, pp. 1–6.

165. G. Strbac, in: Demand side management: benefits and challenges, Energy Policy 36(12)(2008)4419–4426, http://dx.doi.org/10.1016/j.enpol.2008.09. 030, ISSN 0301-4215.

166. V. Auditor-General, Towards a 'Smart Grid' – The Roll-out of Advanced Metering Infrastructure, Victorian Auditor-General's Office, Melbourne, Australia, 2009, pp. 1–49.

10 INDUSTRIALIZED WATERSHEDS HAVE ELEVATED RISK AND LIMITED OPPORTUNITIES TO MITIGATE RISK THROUGH WATER TRADING

Sheila M.W. Reddy[a], Robert I. McDonald[b], Alexander S. Maas[ac], Anthony Rogers[adh], Evan H. Girvetz[ei], Jennifer Molnar[b], Tim Finley[f], Gená Leathers[f], Johnathan L. DiMuro[g]

[a] Office of the Chief Scientist, The Nature Conservancy, 334 Blackwell St., Suite 300, Durham, NC 27701, USA

[b] Office of the Chief Scientist, The Nature Conservancy, 4245 N. Fairfax Dr, Suite 100, Arlington, VA 22203, USA

[c] Department of Agricultural and Resource Economics, Colorado State University, Clark B-320, Ft. Collins, CO 80523, USA

[d] Nicholas School of the Environment, Duke University, Durham, NC 27708, USA

[e] Office of the Chief Scientist, The Nature Conservancy, 1917 1st Avenue, Seattle, WA 98101, USA

[f] Texas Operations, The Dow Chemical Company, 2301 Brazosport Rd., APB Building, Freeport, TX 77541, USA

[g] Global EH&S and Sustainability, The Dow Chemical Company, 2020 Dow Center, Midland, MI 48674, USA

[h] Pew Charitable Trust, 901 E. St., NW, Washington, DC 20004-2008, USA

[I] International Center for Tropical Agriculture, CGIAR, Kasarani Rd., ICIPE Complex, Nairobi, Kenya

ABSTRACT

Businesses are increasingly concerned about water scarcity and its financial impacts, as well as competing needs of other stakeholders and ecosystems. Industrialized watersheds may be at more serious risk from water scarcity than previously understood because industrial and municipal users have inelastic demand and a high value for water. Previous water risk assessments have failed to sufficiently capture these economic aspects of water risk. We illustrate how hydro-economic modeling can be used to improve water risk assessments at a basin scale and we apply the methodology to the industrialized Brazos River Basin (85% municipal and industrial withdrawals) and consider implications for The Dow Chemical Company's Freeport Operations in Texas, US. Brazos water right holders pay only operating and maintenance costs for water during normal periods; however, when shortages occur, leasing stored water or reducing production may be the only mitigation option in the short-run. Modeling of water shortages and the theoretical cost of leasing water under nine combined scenarios of demand growth and climate change suggests that water lease prices to industry could increase by 9–13X. At best, a more developed water rights and storage lease market could result in lower lease prices (2–3X); however, given that transactions would be limited it is more likely that prices would still increase by 4–13X. These results suggest that markets are unlikely to be a robust solution for the Brazos because, in contrast to other watersheds in the Western US, there is little reliable water to trade from low value users (agricultural) to high value users (industry and municipalities). Looking at demand trends across the contiguous US as an indicator of water risk, 2% of watersheds have municipal and industrial demands that outstrip total surface and ground water supplies and in these watersheds industry has historically paid higher lease prices for water. This study provides new ways for businesses to characterize water risk and forecast water prices that uncovers hidden water risk and highlights the positive but diminished mitigating effects of water markets in a highly industrialized basin.

Keywords: Corporate sustainability, Hydro-economic model, Water scarcity, Climate change, Environmental market, Ecosystem service

INTRODUCTION

Businesses are increasingly using a variety of tools to report and manage their impacts and dependencies on water resources. As of 2012, 76% of the 250 largest businesses are reporting on water and 44% report plans to reduce water use [1]. Developing strategic plans for water presents new challenges to businesses. Currently, there are a variety of tools to help businesses assess water risk across sites, through their value chain, and within their supply chain [2–5]. Yet, these tools do not provide information on the economic value of water to the business or to other stakeholders. This means that businesses may underestimate risk or mischaracterize risk, especially in highly developed or industrialized river basins. A number of businesses have begun to make efforts to better understand the value of water and the approaches have been varied [6]. For example, Puma estimated the cost to society of water used in their supply chain and reported it in their Environmental Profit and Loss Account and Hitachi assessed the financial and social costs and benefits of a new water supply and treatment plant [7]. Here we demonstrate the use of basin-scale hydro-economic modeling to characterize water risk, forecast water prices, and assess market solutions. We apply this approach to inform business strategies for future scenarios of climate change and water demand at The Dow Chemical Company's largest facility in Freeport, TX at the mouth of the Brazos River.

Water is a unique resource that provides both private (e.g., bottled drinking water) and public (e.g., recreation) goods and services. As a result, markets do not provide good information on the value of water nor do they result in optimal allocation of water [8,9]. Hydro-economic models, or linked hydrological and economic models, therefore have been important tools to evaluate water policy and management decisions over the last 25 years [8]. Hydro-economic models vary in their particular design and complexity. However, the models are typically designed to represent spatially distributed water resource systems, infrastructure, and water demand by different types of users. Harou et al. [10] and Booker et al. [8] provide comprehensive and recent reviews of hydro-economic models and their applications. Although these models have been primarily used by governments, they may be of great use to businesses that want to better understand future trends in water shortages, economic losses, and prices at a basin scale in order to develop new water strategies and solutions.

Historic trends in economic development led to an increase in total water use and a reallocation of water from agriculture to municipalities and industry (M&I) [9,11]. Now the majority of the world's river basins are classified as water scarce [12] and, on average, 70% of water use in high-income countries is from M&I as compared to 18% for low to middle-income countries [13]. Previous analyses of water use show that M&I water users have relatively inelastic demand and are willing to pay (WTP) high prices for water as compared to agricultural users [14–18]. In river basins dominated by M&I, this means that M&I water demand may drive-up short-term water prices and thereby spur large economic losses during droughts. For example, the median price of water rights purchased by M&I users from agricultural users in the Reno/Truckee Basin, US, between 2002–2009 was $\$15.30/m^3$ ($18800/acre-ft), while the median price paid by agricultural users was $\$1.30/m^3$ ($1600/acre-ft) (2012 USD, value in perpetuity) [19]. During droughts, high-value, water-intensive businesses, such as clothing, food and beverage, biotech, petrochemical, electronics, mining, and electric utilities may have no recourse except to pay high prices or reduce production and thus risk substantial economic losses if no water is available for trading [20].

The risk associated with trends in climate change and M&I demand has not been easy for businesses to account for in regular financial planning, in part, because water prices rarely reflect the forces of supply and demand [9]. Instead, water prices are often set by utilities or governments, or they do not exist at all [21]. When water trading does occur and market prices exist, these prices may be highly distorted, in part, because current institutions restrict water trading in order to ensure that any water trade is conducted with an "absence of damage" [22]. More critically, even when institutions enable efficient trading, market prices during normal years do not provide a good indication of the value of water during drought years when relatively inelastic demand from M&I users could cause water prices to rise quickly. Water risk assessments that measure water scarcity (total water use/total water available) [e.g., [12]] or water stress (total water withdrawals/total water available) [e.g., [23]] have been helpful in shedding light on where future shortages are most likely to occur. However, these risk assessments do not provide sufficient information on the potential economic impacts of shortages or the ability of water markets to mitigate these shortages because they do not adequately account for the effects of water demand from different water users.

To address this challenge, we show how simple hydro-economic modeling can be used to assess water risk to business, using The Dow Chemical Company's Freeport Facility on the Brazos River as a case study. The Brazos River spans 840 miles across the state of TX from northwest TX to the Gulf of Mexico. Along its lower reaches, it is an important source of water to the petrochemical industry and various municipalities. The most recent estimates of water use in the Brazos show that 85% of water use is by M&I users and total demand from M&I is expected to increase in the future (Fig. 1). In addition, the Brazos is a river characterized by highly variable flows where storage is needed to make water rights reliable through the year. After the drought of record in the 1950s, reservoirs were built along the river but analyses of recent drought conditions suggest the current period may be a new drought of record and that there is insufficient stored water [24]. In 2011–2012, the Brazos experienced an extreme drought where some stretches of the river ran dry and water right holders relied heavily on stored water, leased water when available, or decreased water use [25,26].

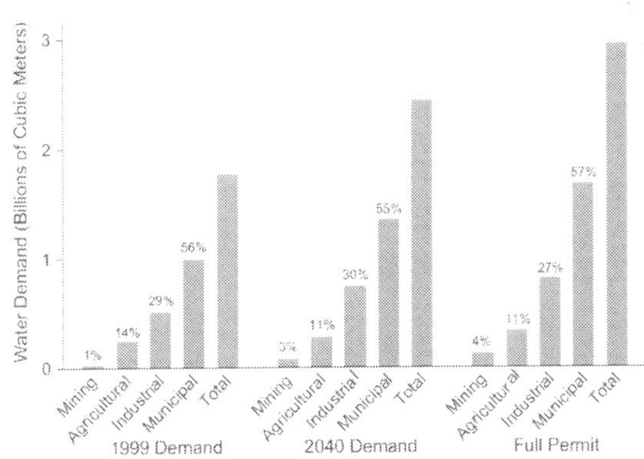

Figure 1. Water demand by scenario and sector. The 1999 demand level was based on measured historic use. Demand for 2040 was modeled based on trends in population growth, economic development, technology change, and historic demand. Fully permitted demand is a scenario in which water users divert their entire legal water right.

We used hydro-economic modeling to examine the effects of future climate change and M&I demand-driven water shortages on economic losses, willingness to pay for water, and market prices in the Brazos River. We also evaluate the degree to which water trading could mitigate the economic impacts of shortages, with implications for the many other river basins across the US and the world that share a similar high M&I demand and declining water availability. Lastly, we put the results from the Brazos River in context by examining patterns in water stress, M&I water demand, and water lease prices across the contiguous US (water lease prices were only available for 12 western states). This study provides new ways to characterize risk that reveals hidden water risk from previously unaccounted for economic trends in water demand and points to the limitations of markets and the importance of new solutions for water scarcity.

METHODS

To characterize future water in the Brazos River Basin, we examined nine combined climate and demand scenarios with and without cross-sector market trading. This involved four main steps. First, to estimate naturalized flows, we input downscaled climate data for the three climate scenarios into a Variable Infiltration Capacity (VIC) model [for detailed methods, see [25]]. Second, we translated naturalized flows into water availability and shortages for water right holders using the version of the Water Availability Model (WAM) used in regional planning for the Brazos River [for detailed methods, see [25]. Third, we estimated the economic losses and willingness to pay or marginal value for water from different sectors (mining, industry, municipalities, and agriculture) under each climate-demand scenario. Fourth, we simulated cross-sectoral trading of water to forecast market prices and evaluate opportunities to mitigate economic losses. We also assessed water stress, water M&I demand, and water lease prices across the contiguous US in order to assess hidden water risk from M&I demand in other watersheds and to take a first step in exploring how these spatial trends relate to spatial trends in market prices for water in 12 western states.

Scenarios

Future water availability was modeled under three demand scenarios (1999 levels, 2040 levels, and full permit) (Fig. 1) and three climate scenarios (characterized by high, medium, and low flows) for a total of nine scenarios of climate and demand over the period 1950–2098. Demand scenarios were estimated using the Texas Regional Water Planning process. The 1999 demand level was based on measured historic use. Demand for 2040 was modeled based on trends in population growth, economic development, technology change, and historic demand. Fully permitted demand is a scenario in which water users divert their entire legal water right.

Climate scenarios were based on three emission possibilities based on the Intergovernmental Panel on Climate Change's 4th assessment scenarios for future greenhouse emissions (A2, "high"; A1B, "medium"; B1,"low"). Global climate projections were taken from the World Climate Research Programme's (WCRP's) Coupled Model Intercomparison Project phase 3 (CMIP3) multi-model dataset [27]. The effect of greenhouse gas emissions on climate was estimated in 16 individual General Circulation Models (GCMs), to capture model uncertainty in the response of the climate system to a given level of GHG emissions. Each WCRP CMIP3 climate projection was bias-corrected and spatially downscaled to 1/8-degree spatial resolution to remove the bias in the simulations and to produce a higher resolution dataset useful for regional hydrologic applications.

Hydrologic Modeling of Naturalized Flow

The hydrologic simulations based on these climate projections were commissioned by the United States Bureau of Reclamation (USBR) as part of its West-Wide Climate Risk Assessments (WWCRAs) to provide risk assessment information for climate change risks to snowpack, changes in the timing of streamflow, and changes in the quantity of runoff [28]. Their analysis used GCM weather data as input to the VIC model (version 4.0.7), a spatially distributed land surface hydrology model that solves the water balance at a daily time-step at each 1/8-degree model grid cell [29–31]. The hydrological simulation methods resulted in spatial fields of runoff and baseflow that were then routed through a schematized stream network of the Brazos River Basin to produce streamflow estimates at specified flow locations (control points for the WAM model, see below) for each of the climate projections.

Water Availability Modeling

To bring together information on changes in water supply and water demand, we utilized the version of the WAM used for the most recent regional planning process for the lower Brazos River [32]. We fed the bias-corrected estimates of naturalized flow from the VIC model into the WAM. Due to the complexity of the WAM, it was not possible to use the full ensemble of GCMs in this analysis since estimating every permutation would be computationally excessive and provide little insight. Instead, three representative GCMs that spanned a range of predicted flows in the Brazos were selected to define our three climate scenarios: 1) "Medium flow": near 50% of all GCMs in terms of predicted river flow, 2) "High flow": 80% of all GCMs in terms of predicted river flow, and 3) "Low flow": 20% of all GCMs in terms of predicted river flow. Note that all three GCMs were for the "medium" (A1B) emissions scenario.

The Brazos WAM also requires a time-series of evaporation rates for 67 control points (e.g., reservoirs). To forecast future evaporation rates, we used a statistical model that describes for historical data how estimated net evaporation (potential evapotranspiration, calculated from temperature, rainfall, and other metrological data, minus rainfall) correlates with actual measurements of net evaporation. We then applied this model to forecast future evaporation rates using climate projection data.

To examine trends in water availability, we calculated annual water reliability (percent of demand met) for individual industrial water rights and for aggregate water rights by sector. We used a locally weighted regression model to estimate the temporal trends in water reliability for individual industrial water rights over the nine climate-demand scenarios. We plotted the temporal trend in water reliability for aggregate water rights by sector. We then used a Tobit regression model to estimate the effect of the nine climate-demand scenarios on the aggregate water reliability for each sector over thirty year periods from 1951 to 2098. A Tobit model was appropriate because water reliability is a limited dependent variable: reliability only takes on values from 0 to 100.

Economic Modeling

Overview

We estimated the economic losses from future water shortages over the next 30 years under each climate-demand scenario without a market that facilitates trading across sectors and with a market that facilitates trading across sectors. This allowed us to examine how water trading could reduce economic losses and to forecast water prices. This model is based on sector-specific demand functions. We estimated economic losses without market trading (the status quo) by inputting the projected shortages from the WAM into each demand function. To estimate the economic losses with market trading, we use a simple linear optimization model to find the allocation that maximizes total value. From this optimal allocation, losses were then calculated for each sector without including compensation for trades because in reality it is unclear who would capture the welfare gains from trade. It should be noted that because the demand functions are parameterized with historic lease price data they may underestimate the loss of value due to decreased production. This modeling approach makes some necessary simplifying assumptions in order to balance ease of use and information needs for strategic business planning.

Economic Losses without Market Trading

Ideally, demand functions would be estimated empirically for individual water users or sectors. However, given that this type of information is not available, we chose a functional form, quadratic, for the water demand models that is consistent with economic theory and parameterized the sector-specific demand functions with data from the literature. Water demand and supply were aggregated by use type defined in the WAM to create four users (mining, industrial, municipal, and agricultural). A limitation to this approach is that it does not account for non-extractive direct use values and indirect use values, such as recreation, amenity values, and environmental values. Aggregating by use type is somewhat akin to assuming that users within a type can trade, which is a reasonable simplifying assumption because trades within use groups require considerably less oversight than across use groups under Texas law. Note that hydropower was not included because it represented a very small part of total demand and water withdrawals are mostly returned to the river.

In order to represent diminishing marginal returns of water and to be consistent with the literature [15,33], we chose a quadratic benefit function and used the corresponding linear inverse demand function to equate marginal values. Thus, the total benefit across every user was modeled as:

$$V(Q)=\sum_{t}^{30}[(1+\delta)^{-t}\sum_{i}^{4}(\alpha_i-\beta_i q_{it})q_{it}] \tag{1}$$

where V is the present value of the water withdrawals across all four sectors over the next 30 years. This is the discounted sum of the benefits to each sector (i) in each time period (t) associated with water withdrawn (q). The sector-specific parameters for the benefit functions (,) were estimated using the historical quantity demanded and lease prices paid (see Section 2.4.4). The discount rate ( ) is set at 5%, which represents a middle value between the return on private investments [34] and a social discount rate [35]. To calculate the economic loss due to shortages, we used the benefit function to calculate the difference between benefits from the amount of water demanded and the amount of water received for each sector in each year. Given that these functions are parameterized with historic lease price data as a measure of willingness to pay for water, the results can be thought of as the theoretical cost of leasing water within each sector.

While the derived function makes sense for producers (industry, agriculture, and mining), we recognize that the benefits in municipal use may reflect consumer utility instead of profit. Thus consumer theory may dictate the use of a utility function to estimate true value. The key difference being that in the case of water, utility functions would spike towards infinite as supply approaches zero. However, we can ignore this caveat by acknowledging that municipalities themselves demand water, which they sell to consumers. Thus utility is not a concern of this model and it is assumed that residential use will be met but that the cost of meeting these demands creates the municipality's demand function.

Economic Losses with Market Trading and Market Clearing Prices

To estimate the reduction in economic losses due to market trading across sectors and forecast market prices, we implemented a static trading model. Trading occurs in aggregate across sectors ignoring geographic limitations, which means that these estimates should be thought of as describing the upper bound for mitigating losses through trading. The market equilibrium conditions were calculated annually and trade between time periods was not allowed. Although a dynamic model could be used to obtain an optimal solution over the entire period, it is unrealistic to assume such planning and collaboration would occur in the real world, especially given storage constraints. Thus, in each year, it is straight forward to equate the marginal benefit across users by solving the first order conditions of the maximization problem:

$$V_t(q_{it}) = \max_{q_{it}} \sum_{i}^{4} (\alpha_i - \beta_i q_{it}) q_{it} \tag{2}$$

$$\text{s.t.} \quad \overline{Q}_t \geq \sum_{i}^{4} q_{it} \tag{3}$$

where $\overline{Q}_t$ is the total water available in time *t* and q_{it} is the amount used by sector *i* in time *t*.

The constraint ensures that the sum of the water allocated in each period to each sector must be less than or equal to the total water available in that period. P_t^*, or the equilibrium market price, can be calculated from the first order conditions derived from the above equation, in which all marginal values of water are set equal. The total quantity in each use type is also solved from these first order conditions and is represented as q_i^*. Once P_t^* and q_{it}^* are calculated, the surplus, or the total value to the water user, can be estimated under trade scenario allocations and compared to the total value without a free market. The present value of the avoided economic losses, therefore, was calculated as:

$$\sum_{t}^{30} (1+\delta)^{-t} \left[\sum_{i}^{4} (\alpha_i - \beta_i q_{it}^*) q_{it}^* - \sum_{i}^{4} (\alpha_i - \beta_i q_{it}^s) q_{it}^s \right] \tag{3}$$

where subscript *s* denotes the amount allocated to each sector under the status quo, i.e., no trade. Given that these functions are parameterized with historic lease price data as a measure of willingness to pay for water, the results can be thought of as the avoided cost of leasing water due to facilitating a market for trading across sectors.

We ran two simulations of the model: 1) with free trade across sectors, and 2) with restricted trading to represent more realistic limitations to trading that may exist in the Brazos. In the restricted simulations, we assume that only 50% of the volume of agricultural water rights would trade water to reflect the possibility that agricultural water rights may be unreliable and that agriculture may not be willing to trade water due to non-monetary benefits associated with agricultural livelihoods. We also assumed that municipalities would not lease any water, which could be reasonable if the primary use of municipal water is for drinking water. Importantly, we were unable to account for technical or logistical limits to trading such as conveyance and storage that may further restrict trading. Note also that our model ignores the fact that currently municipalities sell effluent water in secondary markets. This is a reasonable simplification because increased water use efficiency in the future may make these markets less important.

Model Parameters and Data

To derive the benefit function and calibrate the alpha and beta parameters, initial prices, quantities, and elasticity estimates were required. The elasticity estimate and the benefit function are straight forward derivations from the linear demand function, $P_i^0 = \alpha_i - B_i Q_i^0$. The benefit function is then calculated as the integral such that the value of water in a use type is: $\int_0^Q (\alpha_i - B_i Q)dQ \rightarrow \alpha_i Q - \frac{1}{2} B_i Q^2 \rightarrow (\alpha_i - \frac{1}{2} B_i Q)Q$. Thus use-type benefit functions are straight-forward derivation once initial prices, quantities, and elasticities are available.

Initial quantities demanded were calculated as the historical allocation, (q_i^s), plus the shortfall in a given time period estimated by the WAM for each demand scenario such that $Q_i^0 = q_{it}^s + s_i$. Initial marginal values (P_i^0) and elasticity (ε_1) estimates were taken from the academic literature, data from the Water Transfer Database [36], and the Brazos River Authority pricing structure (Table 1).

Table 1. Water leasing prices and own-price elasticities from a database of water leases in the 12 western states of the US, the published literature, and the Brazos River Authority.

Use type	Prices 2012 USD/m³(2012 USD/ acre-ft)	Elasticities
Agriculture	\$0.03 (\$43[a])	−0.51b
Municipal	\$0.10 (\$125[a])	−0.51c
Industrial	\$0.24 (\$302[d])	−0.12e
Mining	\$0.29 (\$363[f])	−0.32g

a To represent prices paid during shortages, we took the 3rd quartile values from the lease price reported monthly between 1987 and 2009 in the journal *Water Strategist* and compiled in the Water Transfer Database [36]. These values are consistent with real-world prices, such as the Brazos River Authority interruptible water leasing prices—\$0.04/m³ (\$43.75/acre-ft) for agricultural use and \$0.05/m³ (\$62.50/acre-ft) for other uses.

b [16].

c [15]. Other studies are consistent with this value estimate [38].

d \$302/per acre-ft is the one time lease rate for industry from the Brazos River Authority [37]

e [17]. Estimates for the petrochemical industry were used because according to the 2011 Region H Water Planning document, "Two thirds of all U.S. petrochemical production and almost a third of the nation's petroleum industries are located in Region H" [40].

f [14].

g [18]

The marginal values for municipal, agricultural, and domestic and livestock uses were calculated using records of transactions reported in the journal the *Water Strategist*and its predecessor the *Water Intelligence Monthly* between 1987 and 2009 and compiled in the Water Transfer Database. The database includes the purchase price of permanent water rights as well as temporary leases to and from three user categories: municipal, agricultural, and environmental. Although the dataset is missing important information like priority dates, it provides key insights into the real-world prices paid by agricultural users, municipalities, and environmental organizations and agencies. Although data exists for both sales and leases, we chose to use lease transactions only, because they are more likely to represent the use value of water during a shortfall or an expected shortfall. Sale prices are likely to incorporate a substantial speculative and security value, which should not be included in the use value losses associated with shortfalls. In addition, lease prices should also be seen as a lower bound of willingness to pay since most leasers are agricultural users and lessees are municipal users. Given the distribution of power in water markets anecdotally, it is often accepted that municipalities buy water for well below their willingness to pay.

Given that the Water Transfer Database did not have price data specific to industry and that industry is extremely heterogeneous, the marginal values for industry were assumed to be equal to the local price ($0.24/m^3 or $302/acre-ft) available to industry for leasing large quantities of water under a one time, short-term contract from the Brazos River Authority's Water Supply Agreement [37].

After initial values, (P_i^0), were calculated, demand elasticity estimates were taken from the economic literature. The value of −0.51 for municipalities was chosen based on Espey et al.'s [15] meta-analysis where, after adjusting for outliers and theoretical inconsistencies, 90% of their estimates fell between −0.75 and 0 with a mean of −0.51. It is worth noting that there are many other studies examining this phenomenon, and while the results vary slightly, they are consistent with the chosen range [38]. Elasticity estimates for agriculture were taken from another meta-analysis paper in which we used values from studies conducted after 1980 [16]. Although the year 1980 is somewhat arbitrary, considerable technological and efficiency changes have occurred over the past 50 years and price elasticity estimated from the 60 s and 70 s may be less indicative of today's agricultural practices. Dropping these pre-1980 values also creates a range in line with more recent estimates [39] and coincidentally creates a mean of −0.51.

Industrial elasticity estimates were more difficult to determine, and as one would expect, largely depend on the industries involved. According to the 2011 Region H Water Planning document, "Two thirds of all U.S. petrochemical production and almost a third of the nation's petroleum industries are located in Region H" [40]. Thus, it seems reasonable to use an elasticity estimate based on petrochemical industrial use, which Renzetti estimated to be −0.12 in his 1988 paper [17]. The elasticity chosen for mining is somewhat less robust, −0.14 was chosen, but there is relatively little literature available in this area [18]. Given the well-recognized gap in the economics literature when it comes to deriving industrial demand for water [8], these were the best available estimates.

Sensitivity Analyses

We explored the sensitivity of the economic loss and price results to three assumptions. First, we examined the sensitivity of the results to the choice of functional form of the benefit function by comparing results from the quadratic benefit function with results from a linear benefit function. The linear benefit function essentially holds prices constant, however, this seems unlikely to reflect reality. Second, we calibrated and ran the economic model under two specifications for the initial starting price for industry: 1) the local price for short-term contracts ($0.24/m^3 or $302/acre-ft) and 2) a price for industry reported in the academic literature ($0.06/m^{3}or $78/acre-ft) [14]. The reported price for industry is much lower than the local price because it reflects a range of high- and low-value industry, which may make it inappropriate for this application to the high-value petrochemical industry in the Brazos Basin [40].

Water Stress, Municipal and Industrial Demand, and Prices in the US

Following methods from Averyt et al. [23], we calculated the Water Supply Stress Index (WaSSI) in all of the watersheds (USGS hydrological unit code 8) in the contiguous US using the following equation:

$$WaSSI_i = \frac{WD_i}{SW_i + GW_i} \tag{5}$$

where WD_i is the water demand based on water withdrawals, SW_i is the surface water supplies, and GW_i is the groundwater supplies in watershed *i*. The water withdrawal data is based on 2005 USGS water withdrawal estimates for the following sectors: public supply, domestic, irrigation, livestock, aquaculture, industrial, mining, and thermoelectric power. One short-coming is that these data do not account for return flows. The surface water supplies are based on an average of annual estimates for 2000–2010 from the US Forest Service Water Supply Stress

Index (WaSSI) Ecosystem Services Model, which simulates a full water balance for each watershed using watershed specific information on land cover and soils [41]. The groundwater supplies were based on 2005 USGS groundwater withdrawal estimates.

We also calculated the WaSSI due to municipal and industrial water withdrawals alone, $WaSSI_{MI}$. In this case, $WaSSI_{MI}$ only included water withdrawals from the following sectors: public supply, domestic, industrial, mining, and thermoelectric power. To illustrate the additional risk that may result from demand from municipal and industrial users that have inelastic demand and a high willingness to pay for water, we mapped an index we call the water demand stress index for watersheds with WaSSI>1. We define the demand stress index (WaDSI) as $WaSSI \times (1+WaSSI_{MI})$.

As an initial visual exploration of the potential spatial correlation between WaSSI, $WaSSI_{MI}$, and prices faced by municipalities and industry, we mapped average water lease prices to municipalities and industry in the 12 western states (Arizona, California, Colorado, Idaho, Montana, New Mexico, Nevada, Oregon, Texas, Utah, Washington, and Wyoming) from 2000 to 2010, which were obtained from the Water Transfer Database [36]. Unfortunately, meaningful statistical analyses of the relationship between water price and these indices is not possible because time series data is currently not available on water demand and this is critical for controlling for state-level fixed effects. Note that data on water lease prices were unavailable from other states and lease data for municipal and industrial users were pooled and are unavailable as individual categories. We chose to only use data on leases because lease prices are more likely to represent the value of water during a shortage or an expected shortage than permanent sales prices, which capture speculative values as well.

RESULTS

Trends in Water Demand and Water Supply Reliability

The most recent estimate of total water demand (1999) is 1760 million m^3/year (~1.426 million acre-ft/year) with the majority of demand from municipalities (56%), followed by industrial uses (29%), agriculture

(14%), and mining (1%) (Fig. 1). The projections we used for 2040 demand predict a 38% increase in demand over 1999 levels resulting in a total water demand of 2440 million m^3/year (~1.978 million acre-ft/year) (Fig. 1). The increase in total demand reflects an increase in the proportion of water used by municipalities, industry, and mining and a decrease in the proportion of water used by agriculture. If all legal water rights were used to their fully permitted levels, total water demand would increase over 1999 levels by 67% (to 2,940 million m^3) (~2.383 million acre-ft) and increases in demand from municipalities and industry would account for most of this growth in demand (84%) (Fig. 1).

Increased demand in the basin and climate change are projected to decrease the reliability of individual industrial water rights (percent of water right demand met), based on trends in reliability estimated using a locally weighted regression model (Fig. 2). Reliability of individual industrial water rights is expected to decrease under all future climate change conditions, with the impact increasing over time (Fig. 2). Decreases in reliability of individual industrial water rights are predicted to be disproportionately greater under climate scenarios characterized by low flow as opposed to those characterized by medium or high lows (Fig. 2).

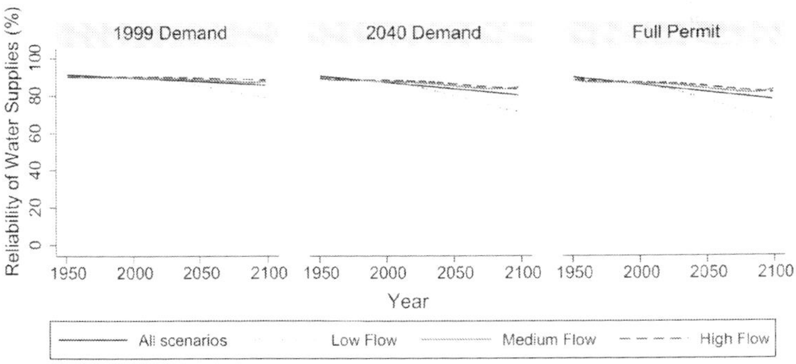

Figure 2. Water supply reliability for industrial water rights by demand scenario. Fitted values from locally weighted regressions of annual water supply reliability for each industrial water right and demand scenario are plotted for three climate conditions that are characterized by low, medium, and high flows. The fitted values from a linear model for all scenarios are also plotted.

Although there is some potential to make water rights reliable through intra and inter-sector trading, our analysis of aggregate water reliability for each sector (percent of total sector demand met) under these three demand scenarios and three climate scenarios shows an increase in the gap between water supply and water demands in the future (Fig. 3, Table S1). Even if demand stayed at 1999 levels, agricultural, industrial, and municipal sectors would experience decreased reliability under a future climate scenario characterized by low water flows; however, water reliability may be unchanged under climate scenarios characterized by medium or high flows (Fig. 3, Table S1). In contrast, under 2040 or full permit demand scenarios, agricultural, mining, and municipal sectors would be less reliable under any climate scenario and the industrial sector would be less reliable under a climate scenario characterized by low flow (Fig. 3, Table S1). Across sectors, aggregate industrial and municipal water supplies are more reliable than agricultural and mining water supplies because of the seniority of water rights and the location of water rights (Fig. 3).

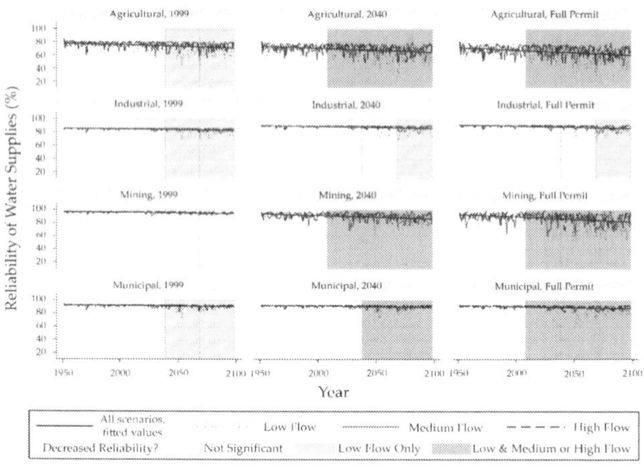

Figure 3. Water supply reliability by sector and demand scenario. Annual aggregate supply reliability for each sector and demand scenario is plotted for three climate conditions that are characterized by low, medium, and high flows. The fitted values from a linear model for all scenarios are also plotted. Regression analysis of the effect of demand conditions, climate scenario, and time period on average reliability over 30 year time periods (Table S1) shows that reliability decreases in the future for all sectors, with industrial water supplies being the least affected.

Future Economic Losses and Limited Mitigating Effects of Market Trading

Under a climate scenario characterized by medium flow, we estimate total economic losses from water shortages increasing from $680 million (30 yr NPV) under 1999 demand levels to $760 million (30 yr NPV) under 2040 demand levels and $900 million (30 yr NPV) under full permit. Although trends in absolute losses are expected to increase, percentage losses are expected to remain similar or decrease. For example, under medium flow conditions, percentages of value losses due to shortages range from 4.8% under 1999 demand levels to 3.6% under 2040 demand levels and full permit (Fig. 3). This is in part because the WAM follows the legal or design specifications for water rights, which may overestimate the water available. We focus on the percentage change in losses in Fig. 3 because the absolute value is sensitive to the specific modeling assumptions while the percentage change is more robust.

To report trends in economic losses across sectors as well as for the effect of a water market and water prices, we focus primarily on one scenario, full permit-medium flow, because our medium flow scenario represents a likely scenario for future climate conditions and, when shortages occur, reliable senior water rights are being used at their fully permitted levels.

Our projected shortages are predicted to yield very different levels of economic losses across sectors. Under the full permit-medium flow scenario, industry and agriculture may represent similar percentages of the total basin shortages (23% and 30%, respectively) over the next 30 years (2013–2042). However, industry may account for a disproportionate amount of the total economic losses (50%) compared to agriculture (8%).

Changing water governance in the Brazos River Basin to facilitate and encourage market trading may reduce, but is unlikely to eliminate, future economic losses due to shortages. Under these assumptions, we estimated that unrestricted trading across sectors could reduce losses by 2.3 percentage points (or 64 percent) under medium flow-full permit (Fig. 3). However, restricted trading, which may be a more reasonable assumption for the Brazos, would only reduce losses by about half as much (1.3 percentage points or 36 percent) (Fig. 3).

Projected Increase in Willingness to Pay for Water and Market Prices

Modeling of the marginal value of water to each sector under future climate and demand scenarios predicts that all sectors would be willing to pay more than historic observed lease prices (Table 1) or the current Brazos River Authority (BRA) lease price of $0.05/m³ ($62.50/acre-ft) (Table 2). This is indicative of the potential of future shortages, combined with high willingness to pay and inelastic demand, to put upward pressure on lease prices through market forces. We focus on reporting the ratio of modeled prices to the BRA price because the absolute value of prices is sensitive to the specific modeling assumptions while the relative change is more robust. Under medium flow and full permit conditions, the projected median willingness to pay for water from industry and mining is 9X the BRA rate, while municipalities are expect to be willing to pay 2X. In contrast, the 95th percentile willingness to pay from mining is 19X, from industry is 11X, and from municipalities is 3X (Table 2). Fig. 5 shows the projections for the willingness to pay from industry ranging from 9–13X across all climate and demand scenarios, with an increasing trend in willingness to pay due to climate change over the period from 2013–2042.

Table 2. Observed and modeled prices for the Brazos River Basin, TX and the Western US

	5th percentile	50th percentile	95th percentile
Modeled average value by sector in the Brazos (2013–2042)			
Mining	$1.37	$1.43	$1.67
Industry	$1.35	$1.36	$1.40
Municipalities	$0.21	$0.21	$0.22
Agriculture	$0.08	$0.08	$0.09
Modeled marginal value by sector in the Brazos (2013–2042)			
Mining	$0.35	$0.45	$0.94
Industry	$0.43	$0.43	$0.53
Municipalities	$0.11	$0.12	$0.13
Agriculture	$0.05	$0.06	$0.06
Modeled market price for the Brazos (2013–2042)			

No restriction	$0.11	$0.11	$0.13
Restrictions	$0.22	$0.28	$0.54
Observed market price in the Western US (2000–2010)			
To municipalities or industry	$0.01	$0.08	$0.53
To agriculture	$0.01	$0.05	$0.22
To environmental use	$0.05	$0.08	$0.10

a:All values are in 2012 USD per m^3.

b:Average value and marginal value by sector and market prices are for the full permit and medium flows scenario.

c:Note that the corresponding median value of an industrial water right in perpetuity is \$8.90/m^3 (as compared to \$0.43/m^3 for the marginal value or \$1.36/m^3 for the average value).

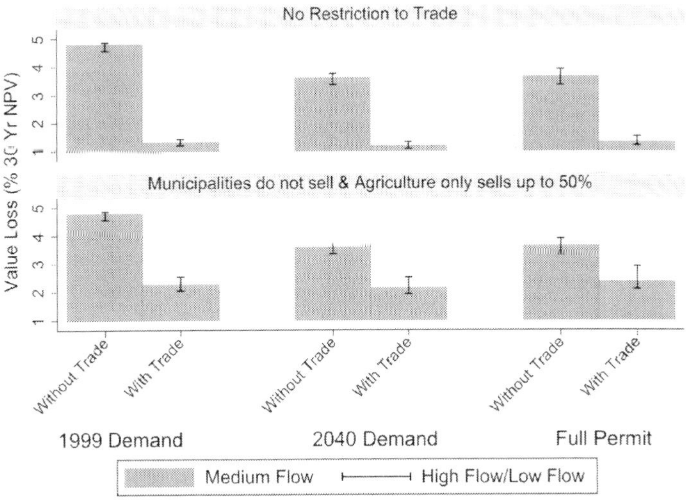

Figure 4. Projected economic losses as a percentage of the total economic loss from water shortages for the entire basin over the period 2013–2042 under scenarios with and without water trading. Economic losses are estimated for the three demand scenarios. The range for each economic loss estimate is created by the three climate scenarios (low, medium, and high flow). Losses are presented for different assumptions for market trading: (1) no restriction on trade and (2) municipalities do not lease water and agriculture only leases up to 50% of the total water volume associated with agricultural water rights. Fig. S1 shows results for additional assumptions on the functional form of the benefit function (linear or quadratic) and the starting price for industry (\$0.06/m^3 or \$0.24/m^3).

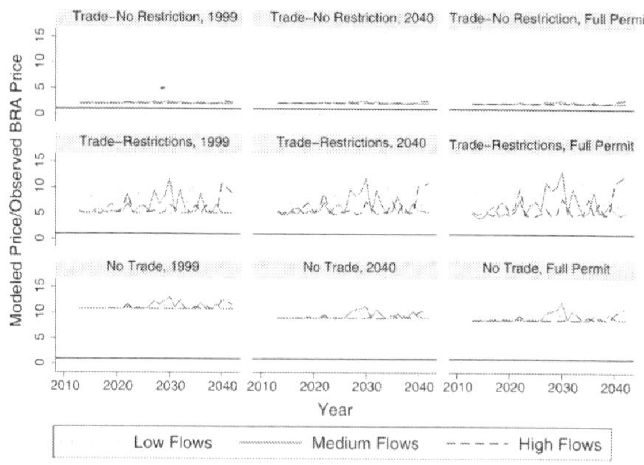

Figure 5. Ratio of modeled prices to the observed Brazos River Authority price ($0.05/m³) over the period 2013–2042 for three climate scenarios and across three demand scenarios. Three prices are modeled: willingness to pay or the marginal value of water to industry (No Trade), the market price with restrictions that municipalities will not lease water and agriculture will only lease up to 50% of the total volume of agricultural water rights (Trade-Restrictions), and the market price with no restrictions (Trade-No Restrictions).

Market trading could keep lease prices considerably below the marginal value to buyers. If water could be easily reallocated from low value users to high value users until marginal values were equivalent, the estimated market price becomes much more stable and hovers around 2–3X the BRA rate across all climate and demand scenarios (Fig. 5). However, if trade is restricted, market prices would still increase by 4–13X across all climate and demand scenarios (Fig. 5).

Limitations and Sensitivity Analyses of the Economic Model

These loss, willingness to pay, and market price estimates are based on our estimate of use-type specific benefit functions that enable us to model how much a user from a particular sector (e.g., mining, industry, municipalities, or agriculture) may be willing to pay for water based on historical water leases and economic analyses of water use. If historic market prices paid for water were far below the user's actual willingness to pay for water, future economic losses, willingness to pay, and market prices may be far greater than those we project here. Specifically, these models do not account for value loss due to lost production directly; this is only captured if it is reflected in historic lease price data. In addition, the model does not include non-extractive direct use values or indirect use values, such as recreation, amenity values, or environmental values. Economic losses also have ripple effects through the economy that will magnify total economic losses and that are not captured by these models, and while an input-output model could be used to estimate such effects, it is beyond the scope of this paper. These models also aggregate demand annually and for each sector, which would smooth and hence limit the effect of extreme shortages experienced within a year or for particular water rights. As a result, these models are not useful in characterizing extreme short-term or intermittent drought conditions and the associated extreme increases in losses or water prices that may be expected. In addition, our demand functions are held constant across each year which does not include certain trends that are likely to exacerbate economic losses, namely population growth and demand hardening. Demand hardening is a difficult to quantify concept that describes a water system's inability to conserve during times of drought as it eliminates slack and inefficiencies overtime [42]. For these reasons, results presented herein should be considered as a lower bound estimate.

We examined the sensitivity of losses, willingness to pay, and market price projections to the functional form of the benefit functions and the initial starting price for industry. If we assume linear benefit functions, total benefits are approximately a quarter as large and losses are approximately half to three-quarters as large. As expected, benefits and losses projected using a lower initial starting price for industry are lower than those projected using a higher initial starting price; however, these differences are less pronounced for results using the quadratic benefit function due to diminishing marginal benefits (Fig. S1). Modeled market prices from unrestricted trading using a linear benefit function are smaller and more variable than modeled prices using a quadratic benefit function (Fig. S2). Lower initial starting prices for industry also result in lower modeled market prices; however, this difference is less in the case of the linear benefit function (Fig. S2).

We also examined the effect of different combinations of restrictions on trading on projected market prices (Fig. S3). We found that only restricting municipalities from selling water or only restricting the volume of sale from agriculture to 50% has minimal impact on project prices (Fig. S3). However, implementing both restrictions resulted in higher and more variable market prices than in the unrestricted market simulation (Fig. S3).

Water Stress, Municipal and Industrial Water Demand, and Water Lease Prices in the Contiguous and Western US

Looking beyond the Brazos River to the 2142 watersheds of the 48 contiguous states, 6% of watersheds have water stress measured by WaSSI≥1.0, indicating that water withdrawals are greater than water supplies (Fig. 6). These watersheds are concentrated in western states with the most extreme water stress occurring in the arid southwest near major population centers. There is also evidence of water stress >1 around Lake Michigan and in the southeast. Based on these annual data, the Brazos River is not classified as water stressed, which may reflect the highly seasonal flows in the river and the fact that the WaSSI also accounts for groundwater supplies. The WaSSI calculated for municipal and industrial withdrawals only ($WaSSI_{MI}$) shows that in 2% of watersheds the $WaSSI_{MI}$ is ≥1 meaning that municipal and industrial users alone withdraw more water than is locally available. Importantly, some of the most water stressed watersheds also have high water stress due to municipal and industrial users. Looking just at watersheds with a WaSSI≥1.0 and using $WaSSI_{MI}$ as an indicator of additional hidden water risk due to the inelastic demand and high willingness to pay of these users, we calculate a new water demand stress index ($WaDSI = WaSSI \times (1 + WaSSI_{MI})$). We show that on average the WaDSI is 63% higher than the WaSSI. Importantly, patterns in water stress, water stress from municipal and industrial users, and our new index, WaDSI, appear to correspond spatially with higher water lease prices to industrial and municipal water users in the Western US over the same time period (2000–2010).

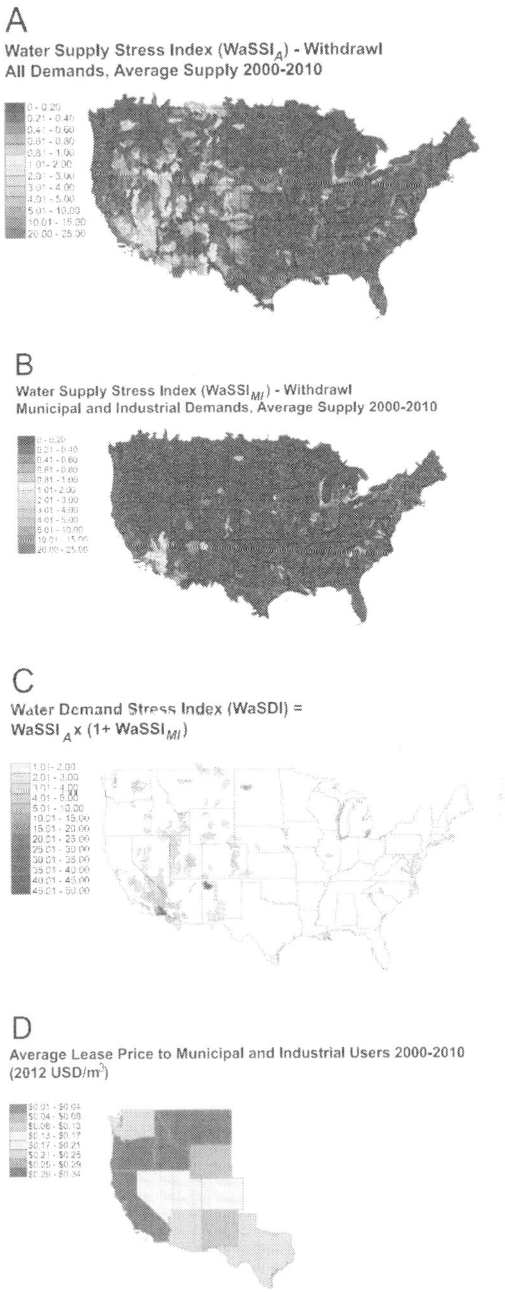

Figure 6. Characterization of water risk in 2142 watersheds in the contiguous US using (A) the water supply stress index (WaSSI), (B) the $WaSSI_{MI}$ for municipal and industrial demands, (C) a new indicator called the water demand

stress index (WaDSI) that uses $WaSSI_{MI}$ as an amplification factor for WaSSI in watersheds with WaSSI≥ 1 in order to represent the potential impact of inelastic demand and high willingness to pay from these users on economic losses and water prices, and (D) observed water leases prices paid by municipal and industrial users.

DISCUSSION

Water stress is increasingly recognized as a global problem. Its consequences for people and entire economies will depend very much on the details of demand for water, as well as future climate scenarios. Our analysis shows how businesses can use hydro-economic modeling to integrate information on economic demand in assessments of water risk at a basin-scale. Specifically, we show how this can enable a business to estimate water shortages and economic losses to multiple water users in a basin as well as forecast water prices under different future climate, demand, and market conditions. Modeling water shortages and economic losses provides important insights into potential future trends for the business and other water users across a basin. Modeling future water prices provides, for the first time, information on water that is similar to forecasts for oil, gas, and other critical inputs and that can be used for strategic planning [43]. Importantly, these water prices are based on economic information about demand from different sectors as opposed to being regulated prices [9,21]. Both the detailed hydro-economic modeling results and more coarse modeled and historic data for watersheds across the US suggest that industrialized river basins may have hidden water risk that can be accounted for with tools such as those presented here. Businesses that use these tools to integrate economic information into their water risk assessments and to look across broader spatial and temporal scales may have a strategic advantage.

Strategic planning based on water risk scenarios that incorporate hydrology and economics may provide more realistic scenarios and uncover hidden water risk. The hydro-economic model we employ here is just one possible version that could be employed by businesses. To be consistent with regional planning processes, we chose to use Brazos River Water Availability Model (WAM) to explore different scenarios of climate change and demand growth. The naturalized water data inputs for the WAM were based on using detailed down-scaled climate data to drive a Variable Infiltration Capacity model (VIC). These choices meant that we were using state-of-the-art and policy-relevant water supply and demand data that was consistent with Dow's previous water modeling work. However, the WAM still may underestimate future shortages and economics losses because it only models the legal or design specifications and not actual behavior. Our approach required substantial effort and time, which may be out of scope for different users and different applications. With this in mind, we designed the economic modeling to be simpler and flexible enough to apply using input data coming from a variety of hydrological models or data sources. For example, it would be possible to use WaSSI water model data to drive this model at a watershed scale and the online WaSSI tool enables users to explore some future scenarios based on climate change and population growth [23]. In contrast, there may be important applications that relax some of the strict assumptions employed here and develop more complex models [10]. For instance, we chose to use observed prices paid for water as an indicator of the willingness of various sectors to pay for water, but this means that we may have underestimated the willingness to pay for water during droughts due to a lack of observed data. In order to explore impacts during extreme droughts, future applications could use surveys to elicit willingness to pay for water or apply data obtained from more extreme drought conditions in other locations. In addition, the results of these modeling exercises could be linked to regional economic models and used to examine the ripple effect of droughts on the economy, which would likely increase estimates of economic losses reported here. Similarly, building on recent advances in hydro-economic modeling, future applications could include non-marketed values for water such as recreational and environmental uses [8]. Lastly, structural changes to the model could help explore finer spatial and temporal patterns, as well as incorporate dynamic feedbacks between the hydrological model and economic model over time. Currently, the model used in this paper smooths out effects of extreme short-term (sub-annual) drought conditions or extreme effects on individual water right holders by aggregating at the annual and sectoral level. Despite these limitations, our estimates of market prices are consistent with observed water lease prices and prices paid for water rights in perpetuity in the Western US.

Regardless of the exact approach, hydro-economic modeling may be a tool for scenario planning that helps businesses develop water strategies and plan for investments in technological solutions inside their fence-line and to seek out nature-based, collaborative or policy solutions that reach across a basin and to other stakeholders. Many stakeholders view water markets that enable trading of water from agriculture to M&I as the solution to future climate change and demand-driven water shortages because of the potential efficiency gains from these transfers [44, 45]. Water markets have helped mitigate the impacts of shortages in places like the southwestern US, Australia, Chile, China, and South Africa [46]. Our results also show that more efficient trading of water can mitigate impacts of shortages; however, the benefits are diminished in industrialized basins because there are few water users that are willing to supply water during droughts and many of these water rights are unreliable due in part to a lack of storage. The current global trends of industrialization, urbanization, and climate change suggest that the gap between water demand and water supply is only going to widen [47–49]. Our hydro-economic modeling of the Brazos River and data from watersheds across the US suggests that these conditions of water demand stress may be associated with increases in economic losses and prices. A future with high water prices and large economic losses could lead to conflicts between industrial, municipal, and agricultural users that feel that each other's water use is infringing on their business, livelihood, or basic right to water [50]. We show that growth in demand, especially from M&I users will only exacerbate these problems, while climate change, which has not been accounted for in many regional water plans, including the Brazos River plans, may also make them more frequent. Importantly, although the impacts of climate change were less important than the impacts of demand over the next 30 years, our modeling results out to 2098 indicate a greater impact of climate on water supplies in the second half of the century [25].

To avoid future economic loss and conflict from water shortages, our results suggest that multiple, additional strategies are urgently needed to reduce risk in demand stressed basins. Increasingly, in places like Texas, people are recognizing that comprehensive solutions that involve diverse strategies will be needed because water trading is not a panacea, despite the fact that water trading can improve market efficiency [45]. The particular geography, ecology, culture, and governance of a basin will determine which suite of strategies is best. For instance, strategies beyond water trading may be particularly important in demand stressed basins where agriculture also provides large benefits, such as food security and amenity or cultural values [51,52], that are not well-captured in the market prices for agricultural water. To build robust water plans in these demand stressed basins, strategies that fit into three major categories should be explored: 1) enhance supply (e.g., surface flows and groundwater infiltration), 2) reduce demand, and 3) increase mechanisms for transfers across space, time, and users (e.g., reservoir storage or facilitating trading). These strategies may rely on new technology such as covered reservoirs that reduce losses from evaporation or nature-based solutions such as restoring native plants that enhance infiltration and river flows. In fact, a recent investigation of nature-based or collaborative solutions for this basin suggests that these solutions may be cost-competitive with traditional solutions, although important technical and logistical hurdles remain because these are relatively new technologies [25]. Enacting many of these solutions will require basin-level collaboration to make potentially large investments to sustain river flows [53]. The good news is that the high value of water in demand stressed basins also means that returns on water investments in these basins may be high.

CONCLUSIONS

Hydro-economic modeling is a tool that businesses may use to assess water risk at a basin-scale under a variety of scenarios in order to inform strategic planning. This is an important advance from existing water risk tools that focus primarily on hydrological factors and treat water demands as fixed requirements rather than based on economic and financial decision-making [10]. Specifically, the analysis presented here helped The Dow Chemical Company at Freeport to make plans to track indicators of supply shortages such as evaporation in their water management plans, communicate the hidden risk in basins dominated by industrial and municipal demand, and seek out solutions beyond water markets.

These models could be used alone or, as in the case of The Dow Chemical Company at Freeport, as part of a suite of models or analyses that 1) estimate the current and future economic value of water to the business using replacement cost or production function approaches [25], 2) as shown in this study, assess water risk by modeling basin-scale trends in water shortages, economic benefits and losses, and water prices that may feedback to affect the value of water to the business and other stakeholders, and 3) identify and evaluate solutions to water scarcity using triple bottom line accounting (e.g., value to the business, society, and the environment) [25].

From a Brazos River Basin perspective, the analysis highlights that M&I users are at risk of supply interruption due to deficient basin storage and supply reserves and that risk increases as demands grows in the basin. Climate change and the effects on rainfall and evaporation have potential to further increase supply risk for basin users. Conservative estimates of loss based simply on historic lease prices are estimated to be $900 million (30 year NPV). However, the actual financial consequences of shortages is potentially orders of magnitude larger because the price of water is currently far less than what individual users may be willing to pay based on the value derived through the use of purchased water.

ACKNOWLEDGMENTS

This analysis was done as part of the collaboration between The Nature Conservancy and The Dow Chemical Company and was supported by funding from The Dow Chemical Company Foundation. We would also like to thank: Neil Hawkins, Mark Weick, Judy Gunderson, Jeff Opperman, Brian Richter, Kari Vigerstol, and Andrew Warner for their advising; Peter Kareiva and Matt Miller for review of earlier versions of the manuscript; Zachary Stein and Brian Perkins of HDR, Inc. and Bart Nijssen of University of Washington for their application of the WAM and VIC, respectively.

Appendix A. Supplementary Materials

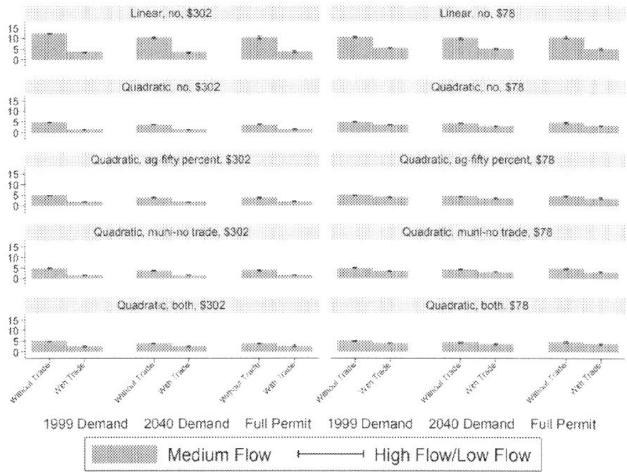

Figure S1. Projected economic losses as a percentage of the total economic loss from water shortages for the entire basin over the period 2013-2042 under scenarios with and without water trading. Economic losses are estimated for the three demand scenarios. The range for each economic loss estimate is created by the three climate scenarios (low, medium, and high flow). Losses are presented for different assumptions on the functional form of the demand function (linear or quadratic) and the starting price for industry ($0.06/m3 or $0.24/ m^3). The quadratic function is also modeled under three different restrictions on trading: municipalities will not lease water, and agriculture will only lease up to 50% of the total volume of agricultural water rights, and both restrictions.

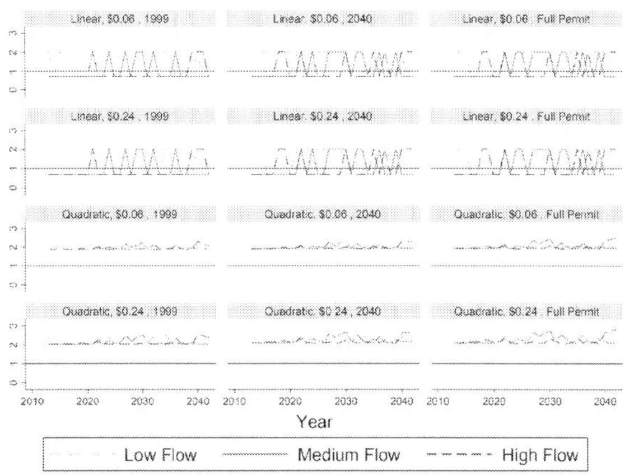

Figure S2. Ratio of modeled prices to the observed Brazos River Authority price ($0.05/$m^3$) over the period 2013-2042 for three climate scenarios and across three demand scenarios. All simulations assume unrestricted trading. Prices are modeled for different assumptions on the functional form of the demand function (linear or quadratic) and the starting price for industry ($0.06/m^3 or $0.24/m^3).

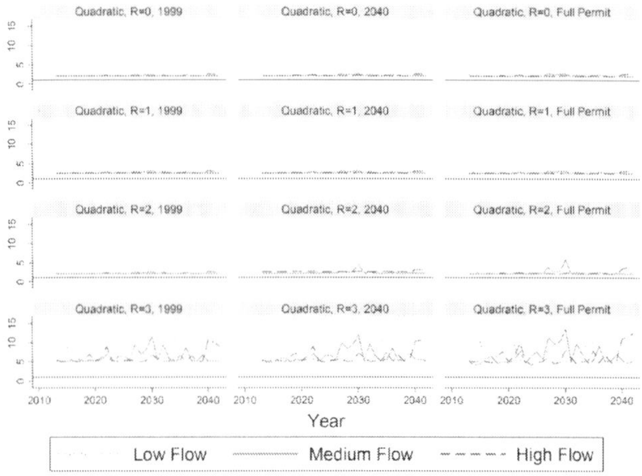

Figures S3. Ratio of modeled prices to the observed Brazos River Authority price ($0.05/m^3) over the period 2013-2042 for three climate scenarios and across three demand scenarios. The quadratic function is also modeled for under no restriction (R=0) and three different restrictions on trading: R=1) agriculture will only lease up to 50% of the total volume of agricultural water rights, R=2) municipalities will not lease water, and R=3) both restrictions.

Table S1. Estimate of aggregate annual water supply reliability for each sector between 1950 and 2098.

	Agricultural		Industrial		Mining		Municipal	
VARIABLES	model	sigma	model	sigma	model	sigma	model	sigma
30 yr Period=1965, Demand=1999, Climate=Medium Flow	0.000		0.000		0.000		0.000	
	(0.000)		(0.000)		(0.000)		(0.000)	
30 yr Period=1965, Demand=1999, Climate=Low Flow	-0.003		0.001		0.001		0.000	
	(0.019)		(0.006)		(0.014)		(0.006)	
30 yr Period=1965, Demand=1999, Climate=High Flow	0.004		0.003		0.003		0.002	
	(0.019)		(0.006)		(0.014)		(0.006)	
30 yr Period=1965, Demand=2040, Climate=Medium Flow	-0.045**		0.045***		-0.043***		-0.003	
	(0.019)		(0.006)		(0.014)		(0.006)	

30 yr Period=1965, Demand=2040, Climate=High Flow	-0.039**	0.046***	-0.027*	0.000
	(0.019)	(0.006)	(0.014)	(0.006)
30 yr Period=1965, Demand=Full Permit, Climate=Medium Flow	-0.071***	0.053***	-0.070***	-0.008
	(0.019)	(0.006)	(0.014)	(0.006)
30 yr Period=1965, Demand=Full Permit, Climate=Low Flow	-0.077***	0.053***	-0.058***	-0.006
	(0.019)	(0.006)	(0.014)	(0.006)
30 yr Period=1965, Demand=Full Permit, Climate=High Flow	-0.064***	0.053***	-0.037***	-0.004
	(0.019)	(0.006)	(0.014)	(0.006)
30 yr Period=1995, Demand=1999, Climate=Medium Flow	-0.008	0.000	-0.000	-0.001
	(0.018)	(0.006)	(0.014)	(0.006)
30 yr Period=1995, Demand=1999, Climate=Low Flow	0.006	0.003	0.002	0.002
	(0.018)	(0.006)	(0.014)	(0.006)
30 yr Period=1995, Demand=1999, Climate=High Flow	0.006	0.003	0.004	0.003
	(0.018)	(0.006)	(0.014)	(0.006)
30 yr Period=1995, Demand=2040, Climate=Medium Flow	-0.057***	0.044***	-0.045***	-0.004
	(0.018)	(0.006)	(0.014)	(0.006)
30 yr Period=1995, Demand=2040, Climate=Low Flow	-0.038**	0.047***	-0.030**	-0.000
	(0.018)	(0.006)	(0.014)	(0.006)
30 yr Period=1995, Demand=2040, Climate=High Flow	-0.037**	0.048***	-0.023	0.001
	(0.018)	(0.006)	(0.014)	(0.006)
30 yr Period=1995, Demand=Full Permit, Climate=Medium Flow	-0.089***	0.051***	-0.070***	-0.010*
	(0.018)	(0.006)	(0.014)	(0.006)
30 yr Period=1995, Demand=Full Permit, Climate=Low Flow	-0.067***	0.055***	-0.046***	-0.008
	(0.018)	(0.006)	(0.014)	(0.006)
30 yr Period=1995, Demand=Full Permit, Climate=High Flow	-0.065***	0.056***	-0.032**	-0.005
	(0.018)	(0.006)	(0.014)	(0.006)
30 yr Period=2025, Demand=1999, Climate=Medium Flow	-0.015	-0.004	-0.005	-0.007
	(0.018)	(0.006)	(0.014)	(0.006)
30 yr Period=2025, Demand=1999, Climate=Low Flow	-0.030	-0.001	-0.002	-0.005

	(0.018)	(0.006)	(0.014)	(0.006)
30 yr Period=2025, Demand=1999, Climate=High Flow	0.008	0.005	0.003	0.004
	(0.018)	(0.006)	(0.014)	(0.006)
30 yr Period=2025, Demand=2040, Climate=Medium Flow	-0.061***	0.040***	-0.057***	-0.009
	(0.018)	(0.006)	(0.014)	(0.006)
30 yr Period=2025, Demand=2040, Climate=Low Flow	-0.084***	0.038***	-0.049***	-0.007
	(0.018)	(0.006)	(0.014)	(0.006)
30 yr Period=2025, Demand=2040, Climate=High Flow	-0.034*	0.049***	-0.028**	0.002
	(0.018)	(0.006)	(0.014)	(0.006)
30 yr Period=2025, Demand=Full Permit, Climate=Medium Flow	-0.090***	0.047***	-0.084***	-0.014**
	(0.018)	(0.006)	(0.014)	(0.006)
30 yr Period=2025, Demand=Full Permit, Climate=Low Flow	-0.119***	0.043***	-0.071***	-0.014**
	(0.018)	(0.006)	(0.014)	(0.006)
30 yr Period=2025, Demand=Full Permit, Climate=High Flow	-0.059***	0.057***	-0.042***	-0.002
	(0.018)	(0.006)	(0.014)	(0.006)
30 yr Period=2055, Demand=1999, Climate=Medium Flow	-0.005	-0.001	-0.004	-0.004
	(0.018)	(0.006)	(0.014)	(0.006)
30 yr Period=2055, Demand=1999, Climate=Low Flow	-0.095***	-0.026***	-0.016	-0.027***
	(0.018)	(0.006)	(0.014)	(0.006)
30 yr Period=2055, Demand=1999, Climate=High Flow	-0.028	-0.002	-0.004	-0.009
	(0.018)	(0.006)	(0.014)	(0.006)
30 yr Period=2055, Demand=2040, Climate=Medium Flow	-0.055***	0.037***	-0.054***	-0.008
	(0.018)	(0.006)	(0.014)	(0.006)
30 yr Period=2055, Demand=2040, Climate=Low Flow	-0.173***	-0.001	-0.101***	-0.037***
	(0.018)	(0.006)	(0.014)	(0.006)

REFERENCES

1. KPMG, Sustainable Insight – Water Scarcity: A dive into global reporting trends. Switzerland: KPMG International. No. 121050. (2012). url:http://www.kpmg.com/ES/es/ActualidadyNovedades/ArticulosyPublicaciones/Documents/sustainableinsight-water-scarcity-global-reporting-trends.pdf.
2. G.P. Zhang, A.Y. Hoekstra, R.E. Mathews, Water footprint assessment (WFA) for better water governance and sustainable development, Water Resour. Ind. 1–2 (2013) 1–6.
3. WWF, Water risk filter tool. Washington, D.C.: World Wildlife Fund. ⟨http://waterriskfilter.panda.org/⟩, (2012).
4. WRI, Aqueduct: water risk Atlas. Washington, D.C.: World Resources Institute. ⟨http://www.wri.org/our-work/project/ aqueduct⟩, (2013).
5. WBCSD, Global Water Tool. New York, NY: World Business Council for Sustainable Development. ⟨http://www.wbcsd.org/work-program/sector-projects/water/global-water-tool.aspx⟩, (2011).
6. WBCSD, Water valuation: building thc business case. Geneva, Switzerland: world business council for sustainable development, (2012).
7. WBCSD, Water valuation: business case study summaries. Washington, D.C.: World Business Council for Sustainable Development, (2012).
8. J.F. Booker, R.E. Howitt, A.M. Michelsen, R.A. Young, Economics and the modeling of water resources and policies, Nat. Resour. Model. 25 (2012) 168–218.
9. R.A. Young, Determining The Economic Value Of Water: Concepts And Methods, Washington, D.C, Resources for the Future 357.
10. J.J. Harou, M. Pulido-Velazquez, D.E. Rosenberg, J. Medellín-Azuara, J.R. Lund, et al., Hydro-economic models: concepts, design, applications, and future prospects, J. Hydrol. 375 (2009) 627–643.
11. P. Gober, C.W. Kirkwood, Vulnerability assessment of climate-induced water shortage in Phoenix, Proc. Natl. Acad. Sci. 107 (2010) 21295–21299.

12. A.Y. Hoekstra, M.M. Mekonnen, A.K. Chapagain, R.E. Mathews, B.D. Richter, Global monthly water scarcity: blue water footprints versus blue water availability, Plos One 7 (2012) e32688.
13. UNESCO, Water for People, Water for Life. Paris, France: United Nations Organization for Education, Science and Culture. ⟨http://www.unesco.org/new/en/natural-sciences/environment/water/wwap/wwdr/wwdr1-2003/⟩, (2003).
14. B. Aylward, H. Seely, R. Hartwell, J. Dengel, The economic value of water for agricultural, domestic and industrial uses: a global compilation of economic studies and market prices, Ecosystem Economics LLC 31 (2010) 1–29. Prepared for UNC FAO.
15. M. Espey, J. Espey, W.D. Shaw, Price elasticity of residential demand for water: a meta-analysis, Water Resour. Res. 33 (1997) 1369–1374.
16. S.M. Scheierling, J.B. Loomis, R.A. Young, Irrigation water demand: a meta-analysis of price elasticities, Water Resour. Res. 42 (2006).
17. S. Renzetti, An econometric study of industrial water demands in British Columbia, Canada, Water Resour. Res. 24 (1988) 1569–1573.
18. S. Renzetti, Estimating the structure of industrial water demands: the case of Canadian manufacturing, Land Econ. 68 (1992) 396–404.
19. G.D. Libecap, Insitutional path depdence in climate adaptation: coman's some unsettled problems of irrigation, Am. Econ. Rev. 101 (2011) 64–80.
20. Pacific Institute, Corporate Reporting on Water: A Review of Eleven Global Industries. ⟨http://www.pacinst.org/ wp-content/uploads/sites/21/2013/02/corporate_reporting_on_water3.pdf⟩, (2007).
21. S.M. Olmstead, The economics of managing scarce water resources, Rev. Environ. Econ. Policy 4 (2010) 179–198.
22. M.L. Livingston, Designing water institutions: market failures and institutional response, Water Resour. Manag. 9 (1995) 203–220.
23. K. Averyt, J. Meldrum, P. Caldwell, G. Sun, S. McNulty, et al., Sectoral contributions to surface water stress in the coterminous

United States, Environ. Res. Lett. 8 (2013) 035046.

24. HDR, Study 1: Updated Drought of Record and Water Quality Implications for Reservoirs Upstream of Possum Kingdom Reservoir. Brazos G Planning Group,(2009).
25. S.M.W. Reddy, R.I. McDonald, A. Maas, A. Rogers, E.H. Girvetz, J. North, J. Molnar, T. Finley, G. Leathers, J. DiMuro, Finding solutions to water scarcity: incorporating ecosystem service values into business planning at The Dow Chemical Company's Freeport, TX facility, Ecosyst. Serv. 12(April), 2015, 94-107. http://dx.doi.org/10.1016/j.ecoser.2014.12.001.
26. S. Combs, The impacts of the 2011 drought and beyond. Office of the Comptroller. ⟨http://www.window.state.tx.us/ specialrpt/drought/⟩, (2012).
27. G. Meehl, C. Covey, T. Delworth, M. Latif, B. McAvaney, et al., The WCRP CMIP3 multimodel dataset: a new era in climate change research, Bull. Am. Meteorol. Soc. 88 (2007) 1383–1394.
28. Bureau of Reclamation, SECURE Water Act Section 9503(c)-Reclamation Climate Change and Water, Report to Congress, US Department of the Interior, Bureau of Reclamation, Washington, DC206.
29. X. Liang, D. Lettenmaier, E.F. Wood, S. Burges, A simply hydrologically based model of land surface water and energy fluxes for general circulation models, J. Geophys. Res. 99 (1994) 14415–14428.
30. X. Liang, E.F. Wood, D. Lettenmaier, Surface soil moisture parameterization of the VIC-2L model: evaluation and modifications, Glob. Planet. Change 13 (1996) 195–206.
31. B. Nijssen, D. Lettenmaier, X. Liang, S. Wetzel, E.F. Wood, Streamflow simulation for continental-scale river basins, Water Resour. Res. 33 (1997) 711–724.
32. Region H Planning Group, Regional Water Plan. Houston, TX: Region H Water Planning Group,(2011).
33. F. Arbués, M.Á. Garcıa-Valiñas, R. Martınez-Espiñeira, Estimation of residential water demand: a state-of-the-art review, J. Socio-Econ. 32 (2003) 81–102.

34. OMB, Guidelines and Discount Rates for Benefit-Cost Analysis of Federal Programs. US Office of Management and Budget, (1992).
35. M.L. Weitzman, Gamma discounting, Am. Econ. Rev. 91 (2001) 260–271.
36. Z. Donohew, G. Libecap, Water Transfer Database, University of California, Santa Barbara, CA, 2010.
37. BRA, Water Supply Contracts. Brazos River Authority. ⟨http://www.brazos.org/Water-Contracts.asp⟩, (2014).
38. J.M. Dalhuisen, R.J. Florax, H.L. de Groot, P. Nijkamp, Price and income elasticities of residential water demand: a metaanalysis, Land Econ. 79 (2003) 292–308.
39. K.Schoengold,D.L.Sunding,G.Moreno,Priceelasticityreconsidered: panel estimation of an agricultural water demand function, Water Resour. 42 (2006). W09411 10.1029/2005WR004096.
40. TWDB, Region H Planning Document, Texas Water Development Board, Austin, TX, 2010.
41. G. Sun, P. Caldwell, A. Noormets, S.G. McNulty, E. Cohen, et al., Upscaling key ecosystem functions across the conterminous United States by a water-centric ecosystem model, J. Geophys. Res., 116 (2011). G00J05, 10.1029/2010JG001573. pages. 1-16.
42. C.W. Howe, C. Goemans, Manager to manager–the simple analytics of demand hardening, Am. Water Works Assoc. 99 (2007) 24–25.
43. P.J. Schoemaker, Scenario planning: a tool for strategic thinking, Sloan Manag. Rev. 36 (1995) 25–40.
44. A.M. Michelsen, R.A. Young, Optioning agricultural water rights for Urban water supplies during drought, Am. J. Agric. Econ. 75 (1993) 1010–1020.
45. C.W. Howe, The increasing importance of water transfers and the need for institutional reforms, J. Contemp. Water Res. Educ. 79 (2011) 4.
46. R.Q. Grafton, G. Libecap, S. McGlennon, C. Landry, B. O'Brien, An integrated assessment of water markets: a cross-country comparison, Rev. Environ. Econ. Policy 5 (2011) 219–239.

47. R.I. McDonald, P. Green, D. Balk, B.M. Fekete, C. Revenga, et al., Urban growth, climate change, and freshwater availability, Proc. Natl. Acad. Sci. 108 (2011) 6312–6317.

48. McKinsey, Charting Our Water Future: economic frameworks to inform decision-making. 2030 Water Resources Group, (2009).

49. IPCC, The physical science basis. Contribution of working group I to the fourth assessment report of the intergovernmental panel on climate change, in: S. Solomon, S. Solomon, D. Qin, M. Manning, Z. Chen, M. Marquis, et al., (Eds.), Climate Change, Cambridge University Press, Cambridge, United Kingdom and New York, NY, USA, 2007.

50. P.H. Gleick, Water conflict chronology. The World's Water, 2008–2009: The Biennial Report on Freshwater Resources: 151– 196, (2010).

51. H.C.J. Godfray, J.R. Beddington, I.R. Crute, L. Haddad, D. Lawrence, et al., Food security: the challenge of feeding 9 billion people, Science 327 (2010) 812–818.

52. K. Daugstad, K. Rønningen, B. Skar, Agriculture as an upholder of cultural heritage? Conceptualizations and value judgements—a Norwegian perspective in international context, J. Rural Stud. 22 (2006) 67–81.

53. S. Mackenzie, P.-M. Grondin, Financing Water for All, World Water Council, Marseille, France, 2003.

CITATIONS

CHAPTER 1

Jasper Montana, Accommodating consensus and diversity in environmental knowledge production: Achieving closure through typologies in IPBES, In Environmental Science & Policy, Volume 68, 2017, Pages 20-27, ISSN 1462-9011, https://doi.org/10.1016/j.envsci.2016.11.011.

CHAPTER 2

Camino Liquete, Angel Udias, Giulio Conte, Bruna Grizzetti, Fabio Masi, Integrated valuation of a nature-based solution for water pollution control. Highlighting hidden benefits, In Ecosystem Services, Volume 22, Part B, 2016, Pages 392-401, ISSN 2212-0416, https://doi.org/10.1016/j.ecoser.2016.09.011.

CHAPTER 3

S. Vialet-Chabrand, J.S.A. Matthews, O. Brendel, M.R. Blatt, Y. Wang, A. Hills, H. Griffiths, S. Rogers, T. Lawson, Modelling water use efficiency in a dynamic environment: An example using Arabidopsis thaliana, In Plant Science, Volume 251, 2016, Pages 65-74, ISSN 0168-9452, https://doi.org/10.1016/j.plantsci.2016.06.016.

CHAPTER 4

Yajie Liu, Trond W. Rosten, Kristian Henriksen, Erik Skontorp Hognes, Steve Summerfelt, Brian Vinci, Comparative economic performance and carbon footprint of two farming models for producing Atlantic salmon (Salmo salar): Land-based closed containment system in freshwater and open net pen in seawater, In Aquacultural Engineering, Volume 71, 2016, Pages 1-12, ISSN 0144-8609, https://doi.org/10.1016/j.aquaeng.2016.01.001.

CHAPTER 5

Thomas Hahn, Constance McDermott, Claudia Ituarte-Lima, Maria Schultz, Tom Green, Magnus Tuvendal, Purposes and degrees of commodification: Economic instruments for biodiversity and ecosystem services need not rely on markets or monetary valuation, In Ecosystem Services, Volume 16, 2015, Pages 74-82, ISSN 2212-0416, https://doi.org/10.1016/j.ecoser.2015.10.012.

CHAPTER 6

Carsten Nesshöver, Timo Assmuth, Katherine N. Irvine, Graciela M. Rusch, Kerry A. Waylen, Ben Delbaere, Dagmar Haase, Lawrence Jones-Walters, Hans Keune, Eszter Kovacs, Kinga Krauze, Mart Külvik, Freddy Rey, Jiska van Dijk, Odd Inge Vistad, Mark E. Wilkinson, Heidi Wittmer, The science, policy and practice of nature-based solutions: An interdisciplinary perspective, In Science of The Total Environment, Volume 579, 2017, Pages 1215-1227, ISSN 0048-9697, https://doi.org/10.1016/j.scitotenv.2016.11.106.

CHAPTER 7

Brendan P. Walsh, Sean N. Murray, D.T.J. O'Sullivan, The water energy nexus, an ISO50001 water case study and the need for a water value system, In Water Resources and Industry, Volume 10, 2015, Pages 15-28, ISSN 2212-3717, https://doi.org/10.1016/j.wri.2015.02.001.

CHAPTER 8

Geoffrey P. Hammond, Áine O'Grady, Indicative energy technology assessment of UK shale gas extraction, In Applied Energy, Volume 185, Part 2, 2017, Pages 1907-1918, ISSN 0306-2619, https://doi.org/10.1016/j.apenergy.2016.02.024.

CHAPTER 9

Muhammad Waseem Ahmad, Monjur Mourshed, David Mundow, Mario Sisinni, Yacine Rezgui, Building energy metering and environmental monitoring – A state-of-the-art review and directions for future research, In Energy and Buildings, Volume 120, 2016, Pages 85-102, ISSN 0378-7788, https://doi.org/10.1016/j.enbuild.2016.03.059.

CHAPTER 10

Sheila M.W. Reddy, Robert I. McDonald, Alexander S. Maas, Anthony Rogers, Evan H. Girvetz, Jennifer Molnar, Tim Finley, Gená Leathers, Johnathan L. DiMuro, Industrialized watersheds have elevated risk and limited opportunities to mitigate risk through water trading, In Water Resources and Industry, Volume 11, 2015, Pages 27-45, ISSN 2212-3717, https://doi.org/10.1016/j.wri.2015.04.001.

INDEX

D

E

F

G

H

I

L

M

N